ACS SYMPOSIUM SERIES **806**

Chemicals in the Environment

Fate, Impacts, and Remediation

Robert L. Lipnick, Editor
U.S. Environmental Protection Agency

Robert P. Mason, Editor
University of Maryland

Margaret L. Phillips, Editor
University of Oklahoma Health Sciences Center

Charles U. Pittman, Jr., Editor
Mississippi State University

American Chemical Society, Washington, DC

Library of Congress Cataloging-in-Publication Data

Chemicals in the environment : fate, impacts, and remediation / Robert L. Lipnick ... [et al.], editor.

    p. cm.—(ACS symposium series ; 806)

Includes bibliographical references and index.

    ISBN 0–8412–3776–X

    1. Pollution—Environmental aspects—Congresses.

    I. Lipnick, Robert L. (Robert Louis), 1941- II. Series.

TD196.C45 C482   2001
628.5′2—dc21   2001046445

The paper used in this publication meets the minimum requirements of American National Standard for Information Sciences—Permanence of Paper for Printed Library Materials, ANSI Z39.48–1984.

The design motif for this book cover consists of some examples of molecules discussed in the chapters floating above a representation of water, land, and sky. The artwork was provided by David H. Lipnick, a student in the School of the Arts at Virginia Commonwealth University, and the three-dimensional chemical structures generated by Robert L. Lipnick.

PRINTED IN THE UNITED STATES OF AMERICA

Foreword

The ACS Symposium Series was first published in 1974 to provide a mechanism for publishing symposia quickly in book form. The purpose of the series is to publish timely, comprehensive books developed from ACS sponsored symposia based on current scientific research. Occasionally, books are developed from symposia sponsored by other organizations when the topic is of keen interest to the chemistry audience.

Before agreeing to publish a book, the proposed table of contents is reviewed for appropriate and comprehensive coverage and for interest to the audience. Some papers may be excluded to better focus the book; others may be added to provide comprehensiveness. When appropriate, overview or introductory chapters are added. Drafts of chapters are peer-reviewed prior to final acceptance or rejection, and manuscripts are prepared in camera-ready format.

As a rule, only original research papers and original review papers are included in the volumes. Verbatim reproductions of previously published papers are not accepted.

ACS Books Department

Contents

Environmental Impacts and Monitoring

Remediation

Indexes

Preface

This book is derived from a symposium sponsored by the American Chemical Society (ACS) Division of Environmental Chemistry. "Environmental Chemistry: Emphasis on EPA Research and EPA Sponsored Research," which was organized for the ACS National Meeting in Washington, D.C., August 20–24, 2000. The goal of the symposium was to highlight this U.S. Environmental Protection Agency (EPA) research and to foster the exchange of information among research groups. Papers were solicited with a call for papers, as well as direct contacts with grantees.[1]

The identification of EPA research goals is guided by Goal 8 of the 1993 Government Performance and Results Act, to apply sound science for improved understanding of environmental risk, along with state-of-the-art scientific knowledge and methods to detect, abate, and avoid environmental problems. Increased availability of environmental measurements and models for extending these data permit assessing environmental exposures and potential risks posed by contaminants, particularly to children. How these goals are expressed each year with respect to internal research and new grant solicitations reflects both long term, as well as new priorities that arise, given the demands placed upon a regulatory agency.

The symposium consisted of 32 oral presentations in 5 half-day sessions, and an evening poster session with 15 additional papers.[2] The structure of the symposium followed from the major topics identified in the original call for

---

[1] No effort was made to identify the mechanism by which individual grantees were funded, or how this work reflects EPA priorities in a more specific fashion.

[2] The complete set of extended abstracts for the symposium are part of the Preprints of Extended Abstracts for this Division (Vol. 40, No. 4, ISSN 1520–0520–0507) and are available through the Division's Business Office: Ruth A. Hathaway, 1810 Georgia Street, Cape Girardeau, MO 63701–3816 (telephone: 573–334–3827).

papers, and a need to divide these up accordingly. The oral presentation sessions were entitled: Soil/Sediment: Fate and Transport: Parts 1 and 2; Atmospheric Fate and Transport; Environmental Impacts and Monitoring; and Remediation: New Methodologies. The poster session included papers specifically requested to be posters, as well as those outside the scope of these major topics.

Following the symposium, the ACS Division of Environmental Chemistry encouraged Robert L. Lipnick to develop the presented papers as an ACS Symposium Series volume. Three of the symposium participants (Robert L. Mason, Margaret L. Phillips, and Charles U. Pittman, Jr.) agreed to participate as editors, each applying specific expertise to papers derived from symposium sessions in which their oral presentations were given. With this approach, a book proposal was reviewed and approved by the ACS Books Department. Pittman has expertise in chemical remediation treatments and organic chemistry. Mason has expertise in the fate and transport of inorganic chemicals, and especially heavy metals, in the atmosphere and in aquatic systems. Phillips' research interests focus on human exposure assessment. Lipnick has expertise in the correlation of chemical structure with physicochemical properties, quatitative structure–activity relationship (QSAR), and molecular mechanism of toxic action.

Chapters were recruited from the symposium's oral presentations and from poster presentations, as well as from specialized authorities to fill gaps. In structuring the book the chapters derived from the sessions Soil/Sediment: Fate and Transport: Parts 1 and 2 as well as Atmospheric Fate and Transport were grouped together, organizing the book into the following three sections: (1) Fate and Transport in Soil/Sediment, Waters, and Air; (2) Environmental Impacts and Monitoring; and (3) Remediation, reflecting the original session themes. Within each section of the book, the chapters were arranged first by environmental medium (soil/sediment, waters, air, and biota), with chapters addressing transport between media being placed in intermediate sequence. Secondly, chapters are ordered by breadth of focus, from more general to more specific; and thirdly, by placing chapters on inorganic species before chapters on organic species. In a few cases the chapter has been assigned to a section that doesn't correspond to the session in which the original presentation occurred, because the focus of the chapter differed from the presentation.

No single book could ever capture the breadth of EPA's research interests and technical regulatory concerns about chemicals in the environment. This is a vast, ever evolving mixture of topics. However, this volume provides a representative sampling of current topics and conveys a sense of recent progress and future directions in the field of environmental science.

Acknowledgements

The Editors gratefully acknowledge the continuing support and encouragement provided by Kelly Dennis, Associate Acquisitions Editor, and Margaret Brown, Associate Production Manager, and Stacy VanDerWall, Books Acquisitions Assistant, of the ACS Books Department in preparing this ACS Symposium Series volume. Gratitude is expressed to the 46 anonymous reviewers whose comments contributed to improved final versions of the chapters. In addition, thanks to Dr. Alan Ford, then Programming Chair of the ACS Division of Environmental Chemistry, for suggesting the topic for the symposium, and to Dr. Barbara Karn, EPA National Center for Environmental Research and Quality Assurance, for contributing to the early phases of planning of this symposium.

Robert L. Lipnick
Office of Pollution Prevention and Toxics
U.S. Environmental Protection Agency
Washington, DC 20460
lipnick.robert@epa.gov

Robert P. Mason
Chesapeake Biological Laboratory
University of Maryland
Solomons, MD 20688
mason@cbl.umces.edu

Margaret L. Phillips
Department of Occupational and Environmental Health
University of Oklahoma Health Sciences Center
Oklahoma City, OK 73104
margaret-phillips@ouhsc.edu

Charles U. Pittman, Jr.
Department of Chemistry
Mississippi State University
Mississippi State, MS 39762
cpittman@ra.msstate.edu

Chapter 1

Chemicals in the Environment: An Overview

Robert L. Lipnick[1], Robert P. Mason[2], Margaret L. Phillips[3], and Charles U. Pittman, Jr.[4]

[1]Office of Pollution Prevention and Toxics, U.S. Environmental Protection Agency, Washington DC 20460
[2]Chesapeake Biological Laboratory, University of Maryland, Solomons, MD 20688
[3]Department of Occupational and Environmental Health, University of Oklahoma Health Sciences Center, Oklahoma City, OK 73104
[4]Department of Chemistry, Mississippi State University, Mississippi State, MS 39762

Since prehistoric times, humans have been exposed to chemicals from food, medicinal agents, and occupational activity such as ancient smelting of metals. Tens of thousands of organic and inorganic chemicals are currently in commercial production. A societal interest exists in identifying chemicals that are potentially hazardous to human health or to the environment, and determining to what degree the manufacture, use, and disposal of such chemicals needs to be controlled. Many of these chemicals have undergone little investigation into their physicochemical, chemical, or toxicological properties, or their fate in the environment. This book provides examples of studies addressing the fate and transport of chemicals in soil/sediment, water, and air, and their environmental impacts. In addition, the importance of monitoring chemicals in the environment, new methods for remediation, and identification of adverse effects to humans and other organisms in the environment is a focus of this volume.

Historical Background

Humans have been exposed to chemicals since prehistoric times from the food they ate, the fires they used for warmth and cooking, and the materials they handled. Through a process of trial and error, early mankind discovered which foods were safe to eat, and which were poisonous. It was also found that certain plants, though not useful as foods, contain substances beneficial for medicinal purposes. In addition to plants, minerals were also found to be useful sources of medicines, and pigments. Man's exposure to potentially harmful chemicals was likely enhanced by the use of fire in cooking, and the subsequent production of aromatic hydrocarbons, and from the mining and smelting of ores. We list here briefly some examples of the fate and impact of chemicals, and the recognition of the impact of these activities on man and the environment.

Metals and Inorganic Compounds: Unintended Exposure

Mercury toxicity has been known since ancient times. Romans, for example, used criminals in the cinnabar mines in Spain as the life expectancy of a miner was just three years (*1*). Other examples of metals that were mined for a variety of uses include lead (Pb) and arsenic (As). Egyptians were known to use lead back to 5000 B.C., and mines in Spain date to about 2000 B.C. (*2*). It has been suggested that the use of lead in making wine and other products, and the use of lead in pipes could have contributed to widespread lead poisoning in Roman times (*3*).

Evidence for lead contamination in the environment from anthropogenic activities including mining, smelting, and combustion is historically preserved in various sites such as ice in Greenland, and lake sediments and peat in Sweden and elsewhere. Recent studies of such historical samples have shown, for example, that the $^{206}Pb/^{207}Pb$ ratios of about 1.17 in 2000 year-old corings are similar to ratios of the lead sulfide mined by the Greeks and Romans during this time period. By contrast, local northern sources such as those in Sweden that are uncontaminated by long-range atmospheric transport, exhibit a much higher ratio (about 1.53). In fact, these studies confirm that lead use and deposition greatly decreased following the decline of the Roman Empire (*4*).

Although lead has been used since ancient times for medicinal purposes, its toxic properties were also understood. Thus, lead colic was reported by Hippocrates, and about 50 A.D., poisoning of lead workers was documented by Pliny. Ramazzini observed toxicity to potters working with lead in 1700, but it was not until 1933 that Kehoe demonstrated wide exposure to lead in the environment. Lead produces adverse effects on children with respect to behavior and reduced I.Q. scores, even at very low levels (*5*).

Metals and Inorganic Compounds: Use as Medicines

Perhaps the earliest written record of the use of medicines is the Ebers papyrus, from about 1500 B.C., which describes more than 800 recipes, some of which contain substances today known to be toxic, including hemlock, aconite (ancient Chinese arrow poison) opium, and metals, including lead, copper, and antimony (6). Use of mercury in medicine in ancient Greece was described by Dioscorides, and by the Persian Ibn Sina of Avicenna (980-1036), with use against lice and scabies. He also reported observations of chronic mercury toxicity (7).

Arsenic is another metal known to the ancients with toxic as well as medicinal properties. The sulfides of arsenic, which were roasted, were described by Dioscorides in the first century A.D. as medicines as well as colors for artists. There is evidence that arsenic was used as a poison in Roman times (8). Medieval alchemists were well aware of the poisonous nature of arsenic compounds, which were used in various recipes. Paracelsus, the Swiss physician, used arsenic compounds as medicinal agents (9). Arsenic was widely used as a pesticide in the form of calcium arsenate following the turn of the 20th century.

Paracelsus understood the relationship between medicines and poisons as stated in the third of his 1538 Sieben Defensiones (10): "What is not a poison? All things are poisons and nothing is without toxicity. Only the dose permits anything not to be poisonous. For example, every food and every drink is a poison if consumed in more than the usual amount; which proves the point."

Environmental Exposure to Chemicals

Chemicals in Foods

We are now aware of just how complex is the mixture of chemicals present in food. For example, for the common potato (Solanum tuberosum), besides many substances of known nutritional value, about 120 additional substances have been detected, including solanine alkaloids, oxalic acid, arsenic, etc. An additional 228 chemicals have been identified in potatoes roasted in their skins. Naturally occurring chemicals in food have been associated not only with acute toxicity, but also with cancer and other chronic diseases (11).

Chemical Pollution in Air

In addition to potentially toxic chemicals that are produced by plants that are eaten as food, humans and other organisms gain unwanted exposure to chemicals through other means. Exposure can occur through natural sources such as emissions of volcanoes, as well as local and long-range anthropogenic sources. Such sources existed even in ancient times, including occupational exposure to lead and other heavy metals as discussed above. Technological development was accompanied by the creation of new chemical hazards. For example, in the late 17th century, a citizen of a small town in the Duchy of Modena, Italy, filed what was perhaps the earliest reported environmental lawsuit against a chemical manufacturer, claiming that "he poisoned a whole neighborhood whenever his workmen roasted vitriol in the furnace to make sublimate (12)." This may refer to the manufacture of "corrosive sublimate" (mercuric chloride) from cinnabar (mercuric sulfide) using "vitriol" (ferrous sulfate) and common salt. Increased air pollution was widely noted during the Industrial Revolution. In more recent times, substantially increased exposure resulted from disasters involving chemical factories, such as the explosion in Seveso, Italy, on July 10, 1976, in which an estimated 300 g to 130 kg of dioxin (2,3,7,8-tetrachlorodibenzo-*para*-dioxin) were released as a result of an accident during the production of 2,4,5-trichlorophenol by the alkaline hydrolysis of 1,2,3,4-tetrachlorobenzene. In December, 1984 more than 2000 people in Bhopal, India died when exposed to methyl isocyanate from an explosion at a factory using methyl isocyanate to produce the insecticide carbaryl (13,14).

Water Pollution

Modern river pollution in England is traced to about 1810 with the introduction of the modern "water-carriage system" for towns and cities to dispose of sewage directly into rivers. This problem was exacerbated during the Industrial Revolution, with factories built along rivers, and with discharge of wastes from tanneries, paper mills, chemical works, and gas works. This resulted by the early 19th century in the virtual disappearance of fish, aquatic organisms, and plants in these areas.

With reports from several British government commissions, the 1875 Public Health Act was passed. Unfortunately, its limited implementation was due to the lack of available waste treatment technology (15). Modern waste treatment plants are able to handle many of these pollutants, but persistent chemicals are resistant to such treatment and may pose a threat of contaminating aquatic bodies.

Some Industrial Revolution wastes contaminating rivers as noted in an 1874 British report were from lead, zinc, and copper industries and mining operations. For example, streams in lead mining areas were found to be "...turbid, whitened by the waste of lead mines in their courses, bringing down 'slimes' at flood time which were spread over the fields, fouling and destroying the grass." The means by which fish were killed by lead and other metals wasn't understood until the experiments of Carpenter, beginning in 1919, demonstrating that minnows and trout could survive during low flow times when lead concentrations were low, but not during flood times when 0.2-0.5 ppm might be present. The toxicity of lead and other metals was found to be due to asphyxiation from reaction with gill consititutents, leading to a mucus coating, which was found to contain lead (*16*). Later, as synthetic chemical manufacturing advanced, environmental exposure also increased with the deliberate use of pesticides to increase crop productivity or to control insect vectors of disease, such as the widespread use of DDT for malaria eradication.

Identification and Assessment of Chemical Hazards

The earliest means by which adverse effects of chemicals were found were observations on exposed humans. On the other hand, the discipline of experimental toxicology can be traced to the experiments performed on thousands of dogs by M.J. Orfila (1787-1853). Orfila also studied antidotes to poisons, as well as forensic toxicology (*17*).

Current toxicological test methods address the need to evaluate chemicals for both acute and chronic exposure with respect to a variety of effects on humans and organisms in the environment. Laboratory tests begin with acute tests. Chronic tests are performed in cases where high exposure or releases take place and/or where there is an indication from acute studies, chemical structure, or chemical and physicochemical properties indicating that further study is needed.

Octanol/Water Partition Coefficient

An especially valuable physicochemical property for chemical assessment is the octanol/water partition coefficient, usually expressed on a logarithmic scale (log K_{ow}). This property correlates with partitioning from water to soil and sediment, as well as bioaccumulation in fish and other aquatic organisms. The ability to estimate log K_{ow} from chemical structure provides a powerful tool to developers of new chemicals to model environmental partitioning and bioaccumulation prior to manufacture or even chemical synthesis (*18*). Log K_{ow}

is one of the fundamental tools for the green chemistry (*19*) approach to designing new, safer chemicals, as well as for the application of pollution prevention (*20*), whose goal is to avoid pollution at the source.

Quantitative Structure-Activity Relationships (QSAR)

At the turn of the century, Charles Ernest Overton (*21*) at the University of Zürich and Hans Horst Meyer (*22*) at the University of Marburg, independently demonstrated that anesthetic potency depends upon the partition coefficient between the site of administration and site of action, and not specific structural features as had been proposed by earlier workers. Both investigators used tadpoles for their studies, and found that measured olive oil/water partition coefficients correlate with anesthetic potential, and at high concentrations of chemicals, to lethality. Overton, in more extensive studies (*23,24*) of the pharmacology of a large number of chemicals, found that non-electrolyte organic chemicals can be more toxic than expected by this correlation if the chemical acts by a more specific mechanism. Overton hypothesized that all chemicals act by a narcosis mechanism, which is masked by more specific effects that occur at a lower concentration. For example, esters initially show a narcosis effect in tadpoles, but the duration of undisturbed narcosis or anesthesia is related to the structure of both the carboxylic acid and alcohol moieties. Overton attributed this finding to the slow hydrolysis of the ester within the organism to release the corresponding acid and alcohol. He noted that this change took place in accordance with the reported rates of chemical hydrolysis. The correlation of toxicity with partition coefficient is a simple example of a quantitative structure-activity relationship (QSAR).

QSARs for estimating the toxicity of new industrial chemicals have been employed at the U.S. Environmental Protection Agency since the early 1980s under Section 5 of the Toxic Substances Control Act, and are now used routinely by the chemical companies themselves. QSARs are also being used by EPA and other national and international regulatory bodies to assess those chemicals already in commercial production ("existing chemicals") for which few or no test data are available.

Exposure to Synthetic versus Non-Anthropogenic Chemicals

The impact of man's activities, such as mining and exploitation of petroleum products and coal for energy, may be considered in terms of the removal of chemicals from deep reservoirs to the Earth's surface and their redistribution and transformation within the biosphere and the resultant impact on human health and the environment. The contents of this volume, and indeed much environmental research overall, focus on industrial chemicals and by-products of technology such as mine tailings and combustion products. But as noted by Bruce Ames (*25*), developer of the widely-used Ames screening test for mutagenicity, "half of all chemicals, whether natural or synthetic, are positive in high-dose rodent cancer tests… Human exposure to naturally occurring rodent carcinogens is ubiquitous, and dwarfs the general public's exposure to synthetic rodent carcinogens." Ames's inference is not that synthetic chemicals are benign, but that concern over exposure to traces of synthetic chemicals for many sectors of the population is often out of proportion to their likely impact, relative to other non-anthropogenic sources.

The focus on industrial chemicals reflects public concern over the devastating effects of past as well as current high-level exposures to industrial chemicals through their manufacture, use, and disposal. Rachel Carson, in her book "Silent Spring," (*26*) brought to public attention the bioaccumulation of certain lipophilic, highly persistent chemicals such as DDT, and their adverse effects such as reduced predator bird reproduction resulting from egg shell thinning. Her book, together with tragic incidents such as the mercury contamination of Minamata Bay, Japan (*27*), the "killer fogs" of the Meuse Valley, Belgium, London, UK, and Donora, PA which resulted from combustion of sulfur-laden fuels (*28*), and other cases already noted, aroused serious concern resulting in governmental action with respect to the individual incidents, as well as more general control of the chemicals in question.

As research and regulation focus increasingly on the subtle effects of low-level contamination, further resources need to be directed to industrial and other chemicals, with respect to cancer, as well as effects such as endocrine disruption (*29,30*). Increasing interest exists, for example, in the potential effects of pharmaceuticals and personal care products when these are released into the environment. As known biologically active agents, these chemicals could potentially produce subtle, targeted effects on aquatic life which is continuously exposed through sewage discharge, albeit at very low levels (*31*).

The Process and Scope of Ecological Impairment

Studies aimed at understanding the sources, fate and effects of environmental contaminants, such as those described in this book, have provided much of the information used by managers and regulators to limit and regulate inputs to the environment as a result of human activities. These inputs include direct discharge to air, soil and water from point sources and more diffuse inputs, from so-called non-point sources, such as agricultural and urban runoff and atmospheric inputs to aquatic systems. Overall, much has been achieved in the last decades in documenting changes in pollution inputs to air, soil and water, yet much needs to be done, especially in the regulation of diffuse or non-point sources of pollutants to the nation's waters. A recent report on marine pollution in the USA, for example, states that "direct discharges of pollutants into the ocean and coastal waters... have been greatly reduced over the past 30 years as a result of the Clean Water Act and other federal statutes. As a result, diffuse sources now contribute a larger portion of many kinds of pollutants than the more thoroughly regulated direct discharges" (*32*).

Clean Water Act: Total Maximum Daily Loads (TMDLs)

Reducing these non-point source inputs is indeed the challenge, and studies of contaminant fate and transport are key to identifying the magnitude of all the potential sources to a specific environment, so that the appropriate measures can be taken to reduce the inputs to that system. For aquatic systems, this approach is formulated under the Clean Water Act as "Total Maximum Daily Loads (TMDLs)" (33). TMDLs determine the maximum amount of a pollutant an impaired waterway can receive and still meet the water quality standard. If the inputs from all the sources of that pollutant to the waterbody can be determined, load reductions can be proportionally allocated such that the new levels meet the water quality standard. In essence, a TMDL is a pollution budget for a waterbody. Understanding the fate and transport of the chemicals of concern in the environment forms the basis for such an assessment. Although the TMDL concept is simple, implementation is not straightforward, since it requires defining multiple input pathways for many pollutants. Complexity arises because many of the inputs come from non-point sources that are spatially and temporally variable and are dependent on such variables as rainfall, stream flow and other climatic factors.

The list of impaired waters in the USA, the so-called 303(d) list, includes 21,845 waterbodies with 41,318 impairments (*33*). The impaired waters make up 300,000 miles of rivers and streams and 5 million acres of lakes. The major impairments are sediment impacts (i.e. siltation, turbidity, excess sediment input) which account for 15% of all impairments; pathogens (fecal coliform, bacteria and *E. coli* exceedances) 13%; and nutrients (nitrogen, phosphorous, excessive algae) 12%. Other dominant causes of impairment are metals (specifically mercury), pesticides,and other organics, as well as low concentrations of dissolved oxygen, excessive temperature or pH, habitat alteration and an impaired biological community. Some of these impairments are addressed relatively easily in terms of identifying the sources and reasons for the impairment and allocating load reductions, while others involve complex issues such as transport through the atmosphere across state and country boundaries, and runoff from agricultural and urban environments.

Human Health Advisories for Fish: Mercury, PCBs, and other PBTs

TMDLs provide only one measure of the impact of contaminants. Listing of impaired waters in terms of human health provides a different emphasis although the same major contaminants are the primary concern. For example, many states have advisories based on health concerns associated with the consumption of contaminated fish. Currently, there are 2,838 advisories covering 325,500 miles of rivers (9.3% of the total river miles), 63,288 lakes (23% of the total lakes) and 71% of the USA coastline. Approximately 80% of these advisories are for mercury, although many water bodies have other advisories as well (34). The next most important class of contaminants is polychlorinated biphenyls (PCBs), followed by chlordane, dioxins, and DDT. Some organic contaminants are banned in the US, and their presence reflects their high degree of persistence in the environment from earlier use. The situation is similar in Canada where more than 97% of all fish advisories are for mercury, with the other contaminants being PCBs, dioxin/furans, toxaphene and mirex. Much research in fate and transport is aimed at addressing questions associated with understanding the factors controlling these elevated levels in fish and other organisms.

Fate and Transport

Quantitative measurement of contaminants in the environment is complex and costly for substances like dioxin, which consist of a number of different isomers at very low detection levels. Therefore, such empirical, detailed, site specific monitoring data cannot be gathered independently for each body of water. Models developed based upon available data provide a theoretical framework for understanding the movement of chemicals in the environment and for developing principles to assess contributions of major sources to a given waterbody.

Fate and Transport Modeling

Three chapters in this book address issues related to modeling. These include modeling the movement of chemicals through the subsurface (Chapter 2 by Bryant and Johnson) and across the sediment-water interface (Chapter 7 by Thibodeaux et al.). A recent book also details such modeling approaches (35)

Chapter 8 by Medine describes the development of speciation models for metals in the environment.

These chapters provide examples of the types of approaches that are needed to better understand contaminant fate and transport. Some examples of aquatic models currently in use include the EPA-funded Green Bay and Lake Michigan Mass Balance Studies (*36*), and Chesapeake Bay Water Quality Model (*37,38*). Models have also been developed for atmospheric transport of pollutants including the Regional Lagrangian Model of Air Pollution (RELMAP) and the Regional Acid Deposition Model (RADM) (*39,40*)

Fate and Transport of Specific Chemicals

The remaining chapters in the first section of the book cover the measurement and investigation of the fate and transport of specific chemicals in the soil, water and atmosphere and exchange among these media. Chapter 3 by West and Wilson reviews subsurface natural attenuation of contaminants. Chapters 4, 5, and 6 by Stone *et al.*, Soderstrom *et al.*, and Butler and Hayes, respectively, all relate to the movement and fate of organic chemicals in soils and sediments, while Chaper 9 by Pedersen and Suffet and Chapter 10 by Salmun and Farhan focus on runoff from the terrestrial environment to nearby waters. These last two chapters illustrate the need to understand the complexities of non-point source pollutant inputs. Recent books have also focused on similar topics for persistent and bioaccumulative toxics (PBTs) (41,42).

A growing need exists to model extremely complex systems. For example, the migration and transport of both organic and inorganic pollutants from throughout the Mississippi River basin into the Gulf of Mexico and the interaction of these pollutants with living ecosystems leads to a huge dead zone depleted of oxygen. This "dead zone," which fluctuates in size between 3,000 and 20,000 square miles, reaches a maximum after heavy spring run off from Mississippi basin farmlands, bringing nitrates, phosphates and a large quantity of organic materials into the Gulf of Mexico. Municipal and industrial wastes also contribute to this mixture. Modeling of transport and mixing needs to consider weather effects on water circulation in the Gulf, and vertical mixing complexities. Modeling becomes particularly complex because pollutant transport models must be integrated into models of the metabolic degradation of the nutrients by the ecosystem as a function of depth, temperature and life cycles of the species involved. Dissolved oxygen is depleted in response to this complex processing of nutrients. Increased understanding of these processes would permit a better focus for addressing such problems through prevention and remediation in a cost effective manner. For example, could a 30% decrease in nutrient load result in a 60% reduction of the dead zone?

Atmospheric Inputs

The atmosphere represents an important source of nutrients, metals and organics to surface waters. Three chapters in the book (Chapter 11 by Mason *et al.*, Chapter 12 by Sheu *et al.* and Chapter 13 by Church *et al.*) focus on the importance of the atmosphere as a source of metal contaminants, and on urban areas as added sources of contaminants to nearby waters. None of these three chapters addresses inputs of organic compounds from the atmosphere, which serves as an input source for these contaminants as well (*43*). Finally, the atmosphere provides a source of of nutrients, especially nitrogen, to large water bodies such as the Chesapeake Bay (*32,44*). EPA has supported many recent studies directed to better understanding urban inputs such as the Great Waters Program (*45*) and the Urban Air Toxics Program and other initiatives (*46*).

Environmental Impacts and Monitoring

Research on environmental impacts and criteria for environmental monitoring are highly interdependent. For example, when choosing or developing an analytical method for a given contaminant in air, water, or food, the analyst should seek to achieve a limit of quantitation that is below the threshold dose for the biological effect of interest. By the same token, in order to determine accurately the threshold dose for a given effect in an environmental study, the exposure must be accurately quantified. Increased sensitivity of monitoring methods has made possible the study of subtle low-dose effects. Whether such levels of exposure represent risks to human health or to the environment must await further results from toxicological studies.

The Delaney Clause

The ability to detect environmental contaminants at very low levels had important regulatory implications under the 1958 Delaney Clause of the Federal Food, Drug and Cosmetic Act (FFDCA), which prohibited the presence of carcinogenic substances as additives in processed foods (*47*). Pesticide residues were included under this rule. The Delaney Clause, based upon the theory that no risk threshold exists for chemical carcinogens, did not permit balancing benefits of food additives with potentially low carcinogenic risk when ingested in tiny amounts. The dilemma of zero tolerance was compounded by increasing analytical sensitivity with detection of progressively lower levels of pesticide residues. The Food Quality Protection Act of 1996 removed pesticide residues from coverage under the Delaney Clause. Pesticide residues in processed foods are now subject to the same risk-based standards as pesticide residues in unprocessed foods (48,49). However, other food additives are still subject to the Delaney Clause (50).

Analytical Problems with Parts-per-Trillion Detection Levels

Determination of the fate, transport, and impacts of chemicals in the environment depends upon sensitive, accurate quantitation and speciation of contaminants in soil, water, air, and biological tissues. For the most toxic contaminants, which produce adverse effects at or below the ng/L or ppt range, monitoring is costly, and requires specialized sampling protocols and analytical methods (51). So-called "clean techniques" are required for the collection and analysis of metal concentrations in natural waters. Such low level monitoring of metals requires rigorous cleaning of all equipment in strong acids, extreme care in the collection and handling of sample bottles, and the analysis of samples by trained personnel in specially prepared "clean rooms" using specialized equipment. Likewise, analysis of waters for organic contaminants such as pesticides, PCBs and polycyclic aromatic hydrocarbons (PAHs), equipment preparation and sampling, also requires adherence to stringent protocols and analysis with specialized instrumentation such as GC/MS. Innovation and more automated procedures may lead to more cost-efficient and routine analyses in the future. Chapter 15 by Young et al. describe a novel approach for the analysis of PAHs in water above contaminated sediments.

Monitoring Exposure of Humans and Other Populations

In highly dynamic or variable environments, monitoring must be not only sufficiently sensitive, but also sufficiently intensive to represent the exposure of interest over the relevant temporal and spatial scales. Chapter 17 by Phillips et al. examines temporal and spatial variability in the types of air monitoring data that have been used to study the relationship between air pollution and death and illness rates on a day-to-day basis in urban areas. Unfortunately, current regulations do not require data beyond fixed-point monitoring of a small number of air pollutants, testing of foodstuffs for pesticide residues, and periodic testing of drinking water for organic and inorganic contaminants, providing only an imprecise and incomplete measure of human exposures. Ideally, exposure assessment for a population, or for individual members of a population, should encompass all routes of exposure: inhalation, ingestion, transdermal absorption through contact with or immersion in contaminated media, and transmission from parent to offspring via placenta or yolk.

Large scale efforts at multi-route exposure assessment in human populations include the Total Exposure Assessment Methodology (TEAM) study (52,53), performed in the 1980s under EPA leadership, and the ongoing National Human Exposure Assessment Survey (NHEXAS) (54). These studies involve collecting detailed information about the activities of study participants, with simultaneous monitoring of contaminant levels in air, water, food, and surfaces. Such studies would ultimately direct control efforts towards contaminant sources with the greatest human health impact. Research focused on better characterizing children's exposure can support more stringent standards to protect this vulnerable segment of the population (55).

Environmental monitoring innovation has been directed increasingly towards the development of risk-based methods or indices that measure directly or indirectly the impact of the contamination on organisms or ecosystems. Such indices are addressed by most of the chapters in the Environmental Impacts and Monitoring section of this monograph. In Chapter 14, Ozretich et al. discuss the use of EPA's equilibrium partitioning sediment guidelines to assess the risk to benthic organisms of mixed PAH contamination in sediments. Mishra et al. report in Chapter 16 on an analytical method for quantifying organophosphorus insecticides based on their anticholinesterase activity. In Chapter 18, Adair et al. review the use of birds as sentinel species for assessing the environmental impact of metal contamination. Three chapters discuss the use of molecular biology methods in the development of ecological indicators. Chapter 20 by Rogstad et al. describes the use of dandelion genetic markers as a biomonitor in contaminated environments. In Chapter 19, Krane uses intraspecies genetic diversity as a measure of the ecological health of a habitat, and in Chapter 21, Newburn and Krane describe genetic differences between different species of midges used as indicators of ecosystem integrity.

Biomarkers

Research into biomarkers for environmental contaminant exposure, an important area in impact-based monitoring, is unfortunately not covered in any detail in this volume. Biomarkers are applicable both to toxicological studies conducted in the laboratory and to field studies of exposed populations, including human populations (56). Potential biomarkers include tissue burdens of contaminants, molecular changes, e.g. creation of DNA adducts by genotoxic substances (57) that can be detected in vitro in the tissues of exposed organisms, and physiological changes that may be detected *in vivo*, e.g. , steroid hormone levels in animals exposed to endocrine disruptors (*58*). Biomarkers are valuable because they (1) provide evidence of individuals' past exposure to contaminants, which might otherwise be only hypothetical; (2) reflect both uptake and intrinsic bioavailability of a contaminant, and (3) may be very sensitive early indicators of an adverse effect on the organism. A search of EPA STAR grant awards between 1995 and 2000 identified 77 supported projects explicitly involving biomarkers, representing many different terrestrial and aquatic species.

Remediation

Background

A National Research Council (NRC) report estimates that $234 billion to $389 billion will be spent to clean up contaminated sites on land belonging to U.S. Departments of Defense, Energy, Interior, Agriculture and NASA. Total costs of cleaning all contaminated sites publicly and privately owned are estimated from $500 billion to $1 trillion (*59*). In 1996 alone, such costs were about $9 billion. Nevertheless, remediation technologies are currently inadequate, with high costs, and limited innovative developments. In a 1994 report, NRC evaluated the performance of pump-and-treat systems for ground-water remediation at 77 sites (*60*), and found that only 10% met regulatory standards. Most companies which were founded on marketing new technologies have fared rather poorly, and venture capital disbursements to environmental technology companies have dropped by more than 50% since 1991.

Categories of Remediation Technologies

The major categories of remediation technologies can be divided into the following major classes: (*61*) bioremediation, chemical treatment (*62*), liquid extraction (soil washing, soil flushing and solvent chemical), stabilization and solidification, thermal desorption and vapor extraction/air sparging. Among the recent advances in these remediation categories, few cost-effective options are available. Use of the newer developments is limited, especially for ground-water cleanups at large complex sites. In fact, innovative groundwater restoration technologies have been selected for only 6% of sites regulated under the superfund program (*63*). A General Accounting Office audit (*64*) of contaminated federal facilities confirmed that "few new technologies have found their way into cleanups." Inadequate cost containment has delayed remediation in the public-sector market and decreased incentive for choosing new technology. Contractors are frequently insufficiently supervised after the award of a cost-reimbursable cleanup contract (*64*). Without an incentive to contain costs, quick action and cost-effectiveness in selecting the technology to be used go unrewarded. NRC has performed an analysis of these issues (*65*).

PCB Dehalogenation Methods

The development of PCB dechlorination/detoxification methods has been the subject of substantial research, but little practical application of the research has occurred. The situation is generally representative of hazardous substance remediation technology as a whole. Over the past several decades, dehalogenations of PCBs and related chlorinated organic compounds have been carried out by many methods such as incineration (*66,67*), wet air oxidation (*68*), catalytic dehydrochlorination (*69*), sodium-based reduction (*70,71*), reaction with superoxide (*72*), photolysis in the presence of hydrogen donors (*73*), sodium borohydride or alkoxyborohydride reductions promoted by Ni(0) (*74,75*), electrolytic reduction (*76*), nickel-catalyzed hydrogenolysis (*77*), silylhydride dechlorination (*78*), amine-promoted titanium complex-catalyzed borohydride dechlorination (*79*), iron-promoted dechlorination (*80*), thermolysis over solid bases like $CaO/Ca(OH)_2$ (*81*), KOH in polyethyleneglycol (KPEG), (*82-85*), KOH in tetraethyleneglycol (KTEG) (*86*), and by adding chlorinated organics directly into cement kilns.

Commercial Limitations of PCB Remediation Technologies

Despite the research progress mentioned above, none of the above techniques has been widely adopted commercially for environmental remediation applications due to one or more limitations of each method. For example, incineration of concentrated chlorinated organic compounds requires special treatments to remove the HCl generated. HCl corrodes the equipment if not removed. In addition, incineration of PCBs and other chlorinated organics often produces more toxic compounds (e.g. dioxins) if it is not carefully controlled. Thus, Erickson *et al.* (*87*) reported that combustion of PCBs leads to the formation of small amounts of the most highly toxic polychlorinated dibenzofurans (PCDFs) and polychlorinated dibenzodioxins (PCDDs). Moreover, disposal of chlorinated solvents, neat PCBs, and related chemicals by incineration in cement plants (which cannot be used on soils and sludges), can elicit an adverse community response. General Electric tried to make the KPEG method practical; unfortunately, it required inconveniently high temperatures for high conversions to polyalkoxyarenes from highly chlorinated aromatics (*82-85*). KTEG was sometimes unable to destroy olefinic or aliphatic carbon-chlorine bonds (*86*). Many of the methods noted above seem impractical for commercial PCB remediation or the remediation of other chlorinated aromatic contaminated materials (such as soils). Thus, other effective methods for dechlorination of PCBs and related chlorinated organic molecules are needed. Finally, a large number of bioremediation schemes have been tested, but highly chlorinated PCBs, pesticides etc. appear resistant to practical approaches, and *in situ* methods are complicated by large variability in microbial populations within the strata.

Recently, Commodor Solutions Technology Inc. (*88,89*) and Pittman and co-workers (*90-93*), have pioneered the ambient temperature, solvated-electron reduction of PCBs, both neat and in wet soils, using Ca/NH_3 and Na/NH_3. Remediation of PCB- and chlorinated aliphatic hydrocarbon-contaminated soils, wet and dry by solvated electron reductions at ambient temperatures, are addressed by Pittman *et al.* in chapter 25. Dechlorination rates using Na/NH_3 are huge (diffusion controlled) and the chlorine is mineralized as NaCl. High dechlorination efficiencies are available at reasonable consumptions of sodium metal. In this technique, wet soils were slurried in liquid NH_3 and then either calcium or sodium metal was dissolved. The solvated electrons dechlorinated PCBs to biphenyl at far faster rates than the solvated electrons were consumed by water. While this method is promising for environmental remediation, it requires a suitable reaction vessel able to sustain the moderate pressures produced by ammonia at temperatures between 0 and 50 °C. Ammonia boils at –33 °C, but its use as a liquid in the 0 to 50 °C range is standard practice in industry.

The third section of this monograph also covers some other representative recent research to develop new remediation options. When an aquifer or soil is contaminated, careful site characterization must be integrated with the remediation design to clean the site. This is the topic of Chapter 22 by Reeves and coworkers. Chemical reductions are the topic of Chapter 23 by Farrell *et al.* and Chapter 25 by Pittman *et al.* Passing water from a contaminated plume over a bed of finely divided iron to remediate contamination from chlorinated organics has recently attracted significant attention. A consideration of the mechanisms controlling process is provided by Farrell.

In-situ electrochemical processes have been applied and investigated with increasing frequency in the past decade. Chapter 26 by Chen and coworkers describes the electrochemical dechlorination of trichloroethylene using granular-graphite electrodes which can provide high surface areas. Chlorinated aromatics can sometimes be destroyed by bioremediation. One key problem concerning bioremediation of chlorinated organics is the bioavailability of these compounds within soils or strata. Soil washing or flushing has often been used. In Chapter 27, Pennell *et al.* combine bioremediation with a nonionic surfactant mobilization approach to enhance the bioavailability of chlorinated benzenes to microbial reductive dechlorination. A basic study of how heavy metals can be removed by bacteria which display synthetic phytochelatins is discussed by Mulchandani and members of his research group in Chapter 24. Taken together these chapters provide a glimpse of the type of R & D that might lead to improved technologies for future application.

Outlook

Control of chemicals in response to new scientific information and public concern over environmental degradation have led to an improvement in the quality of the air and waters of the United States. The public, government, and industry continue to rely on multi-disciplinary environmental scientists to develop effective, practical, and cost-effective solutions. Increased knowledge in correlating both desired industrial properties and undesired environmental properties (toxicity, bioaccumulation, and persistence) with mechanism and chemical structure, have led to the development of safer chemicals, as well as ones with better industrial performance characteristics. Although pesticides are specifically designed for their toxic action, advances have been made as well in their environmental safety due to increased selectivity and lower persistence in the environment. Just as chemistry has contributed to environmental problems, it is also leading the way towards solutions. Heightened public awareness has led to the recognition that safer, greener products are also more profitable. Moreover, the avoidance or reduction of costly by-products, i.e., pollution, is also good business practice. These methods represent a part of the evolving industry of environmental technology. Societal control of chemicals that are already in commercial production is more complex involving risk assessment and risk management. The latter includes both scientific (risk assessment) as well as public policy considerations. The latter which affect risk management decisions include economic factors, public values, political factors, technological factors, social factors, and legal factors.

Smaller businesses may lack technical expertise to make such changes, and government programs such as EPA's Design for the Environment Program (DfE) have provided needed support. In other instances, regulation is needed. Much of the reduction in air pollution has come from greater control of release of valuable substances or industrial by-products. Changing industrial practices have more efficient synthetic process with less hazardous waste produced (green chemistry). Remediation processes are discussed in six chapters in part 3 of the book. While it is important to clean up existing hazardous waste sites, to become a more sustainable society, we need to reduce or virtually eliminate the release of such chemicals into the environment. The ultimate goal for society is not only to avoid release of chemicals into the environment, but also that of any disposal of hazardous wastes. This process has come to be known under the paradigm of pollution prevention (P2). This process will lead to less need for remediation technologies such as the ones discussed earlier in this chapter.

The ability to monitor chemicals in the environment at increasingly low levels using methods such as gas chromatography/mass spectrometry have led to the earlier identification of problems. Chemicals found through such monitoring include not only the desired industrial products, but also by-products that may be unknown to the manufacturer. For example, octachlorostyrene was discovered in the Great Lakes, and determined to be a by-product of the electrolytic production of chlorine using a graphite electrode. Octachlorostyrene is not produced commercially. TCDD and TCDF and other dioxins and furans, are also industrial by-products, as well as combustion products of chlorine and carbon. Reduction of these emissions from industrial reactions has been easier than more diffuse combustion sources, such as backyard burning.

Much of the fundamental studies underlying the fate and effects of chemicals in the environment has come from government sources including the U.S. Environmental Protection Agency (EPA). To a lesser extent than the National Science Foundation, EPA supports broad areas of fundamental environmental research under its Exploratory Research Program. To support its mandate, however, as a regulatory agency, EPA funds much of its external research programs under the EPA STAR (Science to Achieve Results) Program to meet changing needs. This is illustrated by the recent focus of grant solicitations on air pollution in 1998/99 (e.g. urban air toxics, particulate matter, atmospheric chemistry); mercury in 2000 and 2001; and the proposed 2001 emphasis on nutrient fate and transport. Despite the need to focus on current environmental concerns, the short timescale of many funded research projects (1-3 years) and the changing emphasis may not encourage needed long term studies of fate, transport and impact; monitoring of emerging chemicals; and development of remediation technologies.

Although non-persistent chemicals released into the environment at low concentrations can undergo chemical transformation and biodegradation, they can produce acute effects if highly toxic, e.g., to fish and other aquatic organisms. Moreover, such chemicals can even produce chronic effects if release into the environment is continuous. Persistent chemicals may present a direct hazard through direct or indirect hazard to humans and the environment. Moreover, highly lipophilic persistent chemicals tend to undergo bioconcentration and food chain transfer whereby chronic exposure can result in adverse effects to both fish and predators (e.g., eagles) as well as human dietary exposure. Control of chemicals in the environment has benefitted by recognition of hazards not previously known such as bioaccumulation of DDT in aquatic organisms and food chain transfer leading to lowered reproduction and other effects. Except for pesticides, chemicals are not designed to be toxic but to satisfy other industrial properties such as surfactant activity.

It is the mandate of scientists to support the sound measurement and use of high quality data to identify, characterize, and solve environmental problems. The chapters in this book provide an overview of the progress made in understanding the fate, transport, impacts, and remediation of chemicals in the environment, and the breadth and extent of these studies. These chapters provide examples of rigorous science supporting the control of chemicals in the environment.

Disclaimer

The contents of this manuscript do not necessarily reflect the official views of the U.S. Environmental Protection Agency, and no official endorsement should be inferred.

References

1. Mercury: A fact sheet for health professionals. www.orcbs.msu.edu/AWARE/pamphlets/hazwaste/mercuryfacts.html
2. Evers, B; Hawkins, S; and Schulz, G. *Ullmann's Encyclopedia of Industrial Chemistry*, Vol A15, VCH, Weinheim, pp. 194-233, 1990 [article on Pb]
3. Nriagu, J. Lead and *Lead Poisoning in Antiquity*, Wiley and Sons, NY 1983.
4. Brännvall, M.-L.; Bindler, R.; Renberg, I.; Emteryd, O.; Bartnicki, J.; Billström, K. *Environ. Sci. Technol.* **1999**, 33, 4391-4395.
5. Evers, B; Hawkins, S; and Schulz, G. *Ullmann's Encyclopedia of Industrial Chemistry*, Vol A15, VCH, Weinheim, pp. 194-233, 1990 [article on Pb].

6. Friberg, L.; Nordberg, G.F. Introduction. In Friberg, L; Nordberg; and Vouk, V.B., Eds., *Handbook on the Toxicology of Metals*, 2nd edition, Vol. I: General Aspects, Elsevier, Amsterdam, 1986.

7. Holmstedt, B. and Liljestrand, G. *Readings in Pharmacology*, Raven Press, New York, 1981, pp. 22-24.

8. Hanusch, K; Grossmann, H.; Herbst, K.-A.; Rose, G. *Arsenic and Arsenic Compounds.* In *Ullmann's Encyclopedia of Industrial Chemistry*, 5th Edition, Vol. A3, pp. 113-115, Weinheim, 1985.

9. Holmstedt, B. and Liljestrand, G. *Readings in Pharmacology*, Raven Press, New York, 1981, pp. 27-30.

10. Albert A., Xenobiosis: *Food, Drugs, and Poisons in the Human Body*, Chapman and Hall, London, New York, 1987, p. 1.

11. Albert A., Xenobiosis: *Food, Drugs, and Poisons in the Human Body*, Chapman and Hall, London, New York, 1987, p. 58.

12. Ramazzini, B. *Diseases of Workers*, Wright, W. C., Trans.; Hafner Publishing : New York, 1964.

13. Costa, D. L.; Amdur, M. O. In *Casarett & Doull's Toxicology: The Basic Science of Poisons*; Klaassen, C. D., Ed.; 5th Edition; McGraw-Hill: New York, 1996; p. 879.

14. Ramondetta, M.; Repossi, A. (Eds.), Seseo: *20 Years After: From dioxin to the Oak Wood*, Fondazione Lombardia per l'Ambiente, Milan, 1998.

15. Klein, L. River Pollution: II. Causes and Effects, Butterworth, London, 1962, pp. 3-9.

16. Jones, J.R.E. Fish and River Pollution, Butterworths, London, 1964, 53-56.

17. Holmstedt, B. and Liljestrand, G. *Readings in Pharmacology*, Raven Press, New York, 1981, pp. 355-359.

18. Lipnick, R.L. Structure-Activity Relationships, in Rand, G.M., Ed. Fundamentals of Aquatic Toxicology, 2nd edition, Taylor and Francis, Washington, DC, pp. 609-655.

19. EPA's Green Chemistry Program, Office of Pollution Prevention and Toxics, http://www.epa.gov/greenchemistry/

20. Pollution Prevention, U.S. Environmental Protection Agency, Office of Pollution Prevention and Toxics, http://www.epa.gov/p2/

21. Lipnick, R.L. *Trends Pharmacol. Sci.* **1986**, 7, 161-164.

22. Lipnick, R.L., *Trends Pharmacol. Sci.* **1989**, 10, 265-269; Erratum: **1990**, 11 p. 44.

23. Lipnick, R.L. (ed.) *Charles Ernest Overton: Studies of Narcosis and a Contribution to General Pharmacology*, Chapman and Hall, London, and Wood Library-Museum of Anesthesiology, 1991.

24. Lipnick, R.L. A QSAR study of Overton's Data on the Narcosis and Toxicity of Organic Compounds to the Tadpole, Rana temporaria. In: *Aquatic Toxicology and Hazard Assessment*: 11th Symposium, G.W. Suter, II and M. Lewis, eds., American Society for Testing and Materials, STP 1007 Philadelphia, PA, 1989, pp. 468-489.

25. Ames, B. N.; Gold, L. S. *Mutat. Res.* **2000**, 447, 3-13.

26. Carson, R. *Silent Spring*, Houghton Mifflin, Boston, 1962.

27. Goyer, R. A. In *Casarett & Doull's Toxicology: The Basic Science of Poisons*; Klaassen, C. D., Ed.; 5th Edition; McGraw-Hill: New York, 1996; p. 712.

28. Costa, D. L.; Amdur, M. O. In *Casarett & Doull's Toxicology: The Basic Science of Poisons*; Klaassen, C. D., Ed.; 5th Edition; McGraw-Hill: New York, 1996; p. 863.

29. Coburn, T., Dumanoski, D. and Myers, J.P. Our Stolen Future, 1996, Plume: New York.

30. Special Report on Environmental Endocrine Disruption: An Effects Assessment and Analysis, Risk Assessment Forum, U.S. Environmental Protection Agency, EPA 630/R-96/012, February, 1997: http://www.epa.gov/ORD/WebPubs/endocrine/

31. Daughton, C. G.; Ternes, T. A. *Environ. Health Perspect.* **1999**, 107 (suppl. 6), 907-938.

32. Boesch, D. F.; Burroughs, R. H.; Baker, J. E.; Mason, R. P.; Rowe, C.L.; Siefert, R. L. *Marine Pollution in the United States: Significant Accomplishments, Future Challenges*; Pew Oceans Commission: Washington, DC, 2001.

33. EPA. Guidance for Water Quality-Based Decisions: The TMDL Process, EPA 440/4-91-001, 1991 and Total Maximum Daily Load (TMDL) Program, EPA-84-F-00-004, 2000, Office of Water, Washington, DC. EPA Office of Water Website, 2001, http://www.epa.gov/owow/tmdl/status.html.

34. EPA. National Listing of Fish and Wildlife Advisories, EPA 823-F-01-010, 2001, Washington, DC. EPA Office of Science and Technology Website, 2001 http://www/epa.gov/ost/fish/listing.html.

35. DiToro, D.M. *Sediment Flux Modeling*. John Wiley and Sons, NY, 2001.

36. EPA Great Lakes National Program Office, 2001. Green Bay Website: www.epa.gov/glnpo/gbdata; Lake Michigan: www.epa.gov/glnpo/lmmb.

37. Cerco, C.F.; Cole, T.M. Technical Report EL-94-4, US Army Corp of Engineers, Vicksburg, MS.

38. Cerco, C.F. *J. Environ. Engin.* **1995**, 121, 549-557.

39. Eder, B.K.; Coventry, D.H.; Clark, T.L.; Bollinger, C.E. RELMAP Users Guide, EPA/600/8-86/013, Research Triangle Park, NC.

40. Dennis, R.L.; Barchet, T.L.; Clark, T.L.; Seilkop, S.K. NAPAP SOS/T Report 5. In *Acidic Deposition: State of Science and Technology*, National Acid Precipitation Assessment, Washington, DC.

41. Lipnick, R.L.; Hermens, J.L.M.; Jones, K.C.; Muir, D.C.G (Eds.) *Persistent, Bioaccumulative, and Toxic Chemicals I: Fate and Exposure*, ACS Symposium Series 772, ACS, Washington, DC, 2000.

42. Lipnick, R.L.; Jansson, B.; Mackay, D; and Petreas, M. (Eds.) *Persistent, Bioaccumulative, and Toxic Chemicals II: Assessment and New Chemicals*, ACS Symposium Series 773, ACS, Washington, DC, 2000.

43. Baker, J.E. (Ed.) *Atmospheric Deposition of Contaminants to the GreatLlakes and Coastal Waters*. SETAC Press, 1997.

44. Pearl, H.W.; Aguilar, C.; Fogel, M.L. In *Atmospheric Deposition of Contaminants to the GreatLlakes and Coastal Waters,* Baker, J.E., Ed.;. SETAC Press, 1997, pp 415-429.

45. EPA. Deposition of Air Pollutants to the Great Waters: Third Report to Congress. EPA-453/R-00-005, 2000, Office of Air Quality, Research Triangle Park, NC.

46. EPA National Center for Environmental Research Website, 2001. http://es.epa.gov/ncerqa/grants/

47. Vogt, D. U. *The Delaney Clause: The Dilemma of Regulating Health Risk for Pesticide Residues*; CRS Report 92-800 SPR; Congressional Research Service: Washington, DC, 1992. http://www.cnie.org/nle/pest-3.html

48. Schierow, L.-J. *Pesticide Policy Issues*; CRS Report 95016; Congressional Research Service: Washington, DC, 1992. http://www.cnie.org/nle/pest-2.html

49. *United States Code*, Title 21, Section 346a, 1999.

50. *United States Code*, Title 21, Section 348(c)(3)(A), 1999.

51. EPA. Methods and Guidance for the Analysis of Water. CD ROM, Product Code #4093, Government Institutes: Washington, DC.

52. Wallace, L. A.; Pellizzari, E. D.; Hartwell, T. D.; Sparacino, C.; Whitmore, R.; Sheldon, L.; Zelon, H.; Perritt, R. *Environ. Res.* **1987**, 43, 290-307.

53. Wallace, L., Nelson, W.; Ziegenfus, R.; Pellizzari, E.; Michael, L.; Whitmore, R.; Zelon, H.; Hartwell, T.; Perritt, R.; Westerdahl, D. *J. Expos. Anal. Environ. Epidem.* **1991**, 1, 157-192.

54. Lioy, P. J.; Pellizzari, E. *J. Expos. Anal. Environ. Epidem.* **1995**, 5, 425-444.

55. Cohen Hubal, E. A.; Sheldon, L. S.; Burke, J. M.; McCurdy, T. R.; Berry, M. R.; Rigas, M. L.; Zartarian, V. G.; Freeman, N. C. G. *Environ. Health Perspect.* **2000**, 108, 475-486.

56. Schulte, P. A. In *Biomarkers and Occupational Health: Progress and Perspectives*; Mendelsohn, M. L.; Peeters, J. P.; Normandy, M. J., Eds.; Joseph Henry Press: Washington, DC, 1995; pp. 1-6.

57. Perera, F. P.; Weinstein, I. B. *Carcinogenesis* **2000**, 21, 517-524.

58. Hutchinson, T. H.; Brown, R.; Brugger, K. E.; Campbell, P. M.; Holt, M.; Länge, R.; McCahon, P.; Tattersfield, L. J.; van Egmond, R. *Environ. Health Perspect.* 2000, 108, 1007-1014.

59. Committee on Innovative Remediation Technologies, National Research Council, *Chemtech* **1998**, April, 46-52.

60. National Research Council, *Alternatives for Ground-Water Cleanup*, National Academy Press: Washington, DC, 1994.

61. *Innovative Site Remediation Technology: Design and Application* Volumes 1-7 1997 (Wastech Monograph Series, Phase 2).

62. Weitzman, L.; Gary, K; Kawahara, F. K.; Peters, R. W.; and Verbicky, J. *Innovative Site Remediation Technology-Chemical Treatments*, American Academy of Environmental Engineers, 1994.

63. U.S. EPA, *Innovative Treatment Technologies: Annual Status Report* (8th ed.), Report No. EPA-542-R-96-010, EPA, Office of Solid Waste and Emergency Response: Washington, DC 1996.

64. Guerrero, P. F. *Federal Hazardous Waste Sites: Opportunities for More Cost-Effective Cleanups*; GAO/T-RCED-95-188. U.S. General Accounting Office: Washington, DC 1995.

65. National Research Council's Committee on Innovative Remediation Technologies, *Innovations in Ground Water and Soil Cleanup: From Concept to Commercialization*, National Academy Press: Washington, DC 1997; ISBN 0-309-06358-2 (see http://www.nap.edu).

66. Wentz, C.A. In: Chemical Engineer Series, Clark, B.J.; Moriss, J.M. (Eds.), McGraw Hill, New York, 1989.

67. Exner, J.H. Detoxification of Hazardous Waste, Ann Arbor Science Books, Ann, Arbor, MI, 1982, p. 185.

68. Baillod, R.C., Lampartes, R.A., Laddy, D.G. In Proceedings of the Purdue Industrial Waste Conference, West Lafayette, IN, 1978.

69. Biros, F.J., Walker, A.C., Medbery, A. Bull. Environ. Contam. Toxicol. **1970**, 5, 317-323.

70. Oku, A., Ysufuku, K.K., Dataoka, H. Chem. Ind. **1978**, 84, 1.

71. Davies, W.A., Prince, R.G.H., Process Saf. Environ. Prot. **1994**, 72, 113.

72. Sugimoto, H. ; Shigenobu, M.; Sawyer, D. T. *Environ. Sci. Technol.* **1988**, 22, 1182-1186.

24

73. Epling, G. A.; Florio, E. M.; Bourque, A. J. *Environ. Sci. Technol.* **1988**, 22, 952-956.
74. Tabaei, S.M.H., Pittman, C.U., Jr., *Tetrahedron Lett.* **1993**, 34, 3263-3266.
75. Tabaei, S.M.H., Pittman, C.U., Jr., and Mead, K. T. *J. Org. Chem.* **1992**, 57, 6669-6671.
76. Zhang , S. and Rusling , J. F., *Environ. Sci. Technol.* **1993**, 27, 1375-1380.
77. Roth, J. A., Dakoji, S. R., Hughes R. C., and Carmody, R. E. *Environ. Sci. Technol.* **1994**, 28, 80-87.
78. Romanova, V. S., Parnes, Z. N., Dulova, V. G., and Volpin, M. E., Russ. RU 2,030,377, Izobreteniya 7 (1996) 135.
79. Liu, Y., Schwartz, J., and Cavallaro, C. L. *Environ. Sci. Technol.* **1995**, 29, 836-840.
80. Chuang, F.-W., Larson, R. A., and Wessman, M. S., *Environ. Sci. Technol.* **1995**, 29, 2460-2463.
81. Yang, C.-M., and Pittman, C.U., Jr., Hazardous Waste and Hazardous Materials **1996**, 13/4, 445-464.
82. Brunelle, D. J. and Singleton, D. A., *Chemosphere* **1983**, 12, 183-207.
83. Brunelle, D. J. and Singleton, D. A., *Chemosphere* **1985**, 14, 173-181.
84. Brunelle, D. J., *Chemosphere*, **1983**, 12, 167-173.
85. Brunelle D. J., Merdirotta, A. K. and Singleton, D. A., *Environ. Sci. Technol.* **1985**, 19, 740-746.
86. Rogers, C.J. and Kornel, A., *U. S. Patent* 4,675,464, June 23, 1987.
87. Erickson, M. D., Swanson, S. E., Flora, J. D., Jr., and Hinshaw, G. D., *Environ. Sci. Technol.* **1989**, 23, 462-470.
88. Weinberg, N., Mazer, D. J., and Able, A. E., U. S. Patent No. 4,853,040, 1 August 1989.
89. Weinberg, N., Mazer, D. J., and Able, A. E., U. S. Patent No. 5,110,364, 5 May 1992.
90. Pittman, C. U., Jr., and Tabaei, S. M. H., Emerging Technologies in Hazardous Waste Management, in: D. W. Tedder (Ed.), Vol. 11, Atlanta, GA, 27-29 September 1993, pp. 557-560.
91. Pittman, C. U., Jr., and Mohammed, M. K., Extended Abstracts 1, in: Proceedings of the EC Special Symposium on Emerging Technologies in Hazardous Waste Management, Birmingham, AL, 9-11 September 1996, pp. 720-723.
92. Sun, G.-R., He, J.-B., and Pittman, C. U., Jr., *Chemospere* **2000**, 41, 907-916.
93. Pittman, C. U., Jr., Technical Report, Project No.1434-HQ-96-GR-02679-21, Department of the Interior, April 2000, pp. 1-13.

Fate and Transport in Soil/ Sediment, Water, and Air

Chapter 2

Theoretical Evaluation of the Interfacial Area between Two Fluids in a Model Soil

Steven L. Bryant[1] and Anna S. Johnson[1,2]

[1]Center for Subsurface Modeling, The University of Texas at Austin, Austin, TX 78712
[2]Department of Chemical Engineering, The University of Texas at Austin, Austin, TX 78712–1062

We present *a priori* estimates of the area of the fluid/fluid interface during drainage of a simple model soil. The geometric configuration of the fluid phases depends on the pressure difference between the phases, the history of the fluid displacement, and the geometry of the pore space. The latter is a critical feature, and to determine it we use a random, dense packing of equal spheres as a model soil. Drainage in this model soil (nonwetting fluid displacing wetting fluid) is simulated by invasion percolation in a network model of the pore space. The total interfacial area includes contributions from isolated (trapped) volumes of wetting phase and from the bulk (connected) wetting phase. The relative contributions depend strongly on the connectivity of the wetting phase during drainage. Predictions of irreducible wetting phase saturation are consistent with experimental data. Predicted interfacial areas are consistent with measurements when the contribution of thin wetting films on grain surfaces in drained pores is included.

Introduction

The interface between immiscible phases in porous media controls many mass transfer processes of scientific and economic interest, including the contamination and remediation of groundwater by non-aqueous phase liquids (NAPLs). In the latter application the overall rate of mass transfer is a critical parameter in assessing risk from a given contaminant source, in designing remediation strategies, and in interpreting results from allied technologies such as inter-well partitioning tracer tests used to assess the volume of NAPL in place. The rate of mass transfer depends upon the thermodynamic driving force and the area of the interface between the phases. This is commonly expressed as

$$N_i = ka(C^* - C_i) \qquad (1)$$

where N_i [=] $M/L^2/t$ is the flux of chemical species i across the interface, k [=] L^2/t is the intrinsic mass transfer coefficient, a [=] L^{-1} is the specific area of the interface per unit volume, C_i [=] M/L^3 is the concentration of species i in one of the phases, and C^* is a reference or equilibrium concentration in that phase.

Though the importance of interfacial area is clear, it is rarely handled as a distinct parameter in theoretical or empirical studies of subsurface environmental problems. One reason is that measuring the interfacial area between fluids in a porous medium is difficult. Another reason is that estimating the interfacial area from first principles is hampered by the inability to specify the appropriate boundary conditions, particularly the geometry of the pore space confining the fluids.

The practical response to these difficulties has been to lump the interfacial area with the intrinsic rate constant for the mass transfer process. The lumped parameter is known as the effective mass transfer coefficient. This approach permits useful analysis of data such as effluent concentration histories from laboratory columns or wells in the field, and it permits the construction of tractable mathematical models of macroscopic transport (1-5). However, an effective mass transfer coefficient determined in this way for one set of conditions may not apply in a different soil, for a different volume fraction of NAPL, at a different flow rate, etc.

The difficulty of quantifying the pore space geometry has been a principal obstacle in obtaining quantitative predictions of interfacial area. This paper presents a way to overcome this obstacle by means of a geometrically determinate porous medium that is physically representative of simple soils and sandstones. We demonstrate the approach for drainage, the displacement of a wetting fluid by a nonwetting fluid under the control of capillary forces.

Technical Aproach

The approach combines developments from three areas: characterization of the model porous medium; simulation of the drainage process within the model medium; and computation of interfacial area and phase volumes for a given phase configuration.

A Physically Representative Model Porous Medium

In naturally occurring porous media, the grains are randomly positioned, and consequently the pore space is an extremely irregular collection of intersecting conduits of converging/diverging cross-section. Important insights have come from research on networks and lattices that capture qualitative features of pore space such as connectivity and variable pore sizes. Because of the near impossibility of quantifying the geometry of pore space, however, most previous attempts at pore-scale modeling have been prevented from making quantitative *a priori* predictions.

The introduction of physically representative model porous media overcame many of these limitations (6,7). The idea is to create models whose geometry is completely determined and which capture essential physical attributes of naturally occurring materials. Mason and Mellor (7) pioneered this method with the Finney pack, a dense (void fraction ~ 0.36) random packing of equal spheres. Finney measured the spatial coordinates of several thousand spheres within his packing (8). These coordinates completely determine the geometry of the grain space and hence of the pore space within the packing. Properties and processes that depend on microstructure can therefore be computed directly for this packing.

Monodisperse, smooth spheres are undoubtedly an oversimplification of the polydisperse, angular grains that comprise soils, sands and sedimentary rocks. Nevertheless, by capturing the essential feature of random grain locations, the Finney pack has proven to be a remarkably powerful tool for translating microscopic geometry into macroscopic properties of simple sandstones (9-13). In this work we adopt the Finney pack as a simple model soil.

Network Model Construction

The foundation for the computations presented here is a network model of the pore space of the model soil. In this section we summarize the process for obtaining this model; more detailed descriptions may be found in (6,7,9,10,14,15). A network model (Figure 1) is a collection of sites and bonds, in which the sites correspond to pore bodies and the bonds to throats connecting the sites. Sites and bonds may be assigned geometric attributes such as volume, cross-sectional area, diameter etc. In creating a network model of the Finney pack, these geometric attributes are calculated directly from the known geometry of the pore space in the pack. This is an important distinction from the usual construction of network models, in which the geometric attributes are assigned from frequency distributions. Frequency distributions are difficult to measure directly and cannot be inferred uniquely from standard indirect measurements. Moreover, pore scale geometric attributes in dense, granular media are known to be spatially correlated, rather than randomly distributed (14, 15). The network models used here capture these pore space features directly, and this is critical for obtaining physically representative, *a priori* predictions.

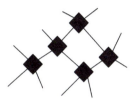

Figure 1. Schematic of a network model. The diamonds represent sites, which are connected by bonds. In this network each site has four neighbors. Geometrical attributes such as volume can be associated with the sites, while the bonds may be assigned features such as critical curvature for drainage.

In order to extract a network model from the Finney pack, a method is required for identifying pore bodies and pore throats. The mathematical device of Delaunay tessellation is ideally suited for this purpose. A Delaunay tessellation subdivides a set of points by grouping nearest neighbors together. Applied to the centers of the grains in the Finney pack, it defines a set of tetrahedral cells, each cell having vertices corresponding to the centers of four nearest neighbor grains (Figure 2).

The Delaunay cells conveniently and unambiguously define pore bodies and throats within the pore space. The cross-sectional area of the pore space leading into a Delaunay cell reaches a local minimum at the cell face. Each face of a Delaunay cell thus defines the locally narrowest constriction in the pore space, and this corresponds naturally to a pore throat, controlling access to the interior of the cell. The terms 'pore' and 'cell' will be used interchangeably, and 'pore throat' will refer either to the void area within a cell face or to a small void volume extending on either side of a cell face. A network model of the pore space of the Finney pack is readily obtained by associating sites in the network with Delaunay cells (pore bodies) and by associating bonds with the cell faces (pore throats). Because every cell is a tetrahedron, the topology of this network is fixed: each site has four bonds. Because the dimensions of each cell are determined by the known spatial coordinates of the cell vertices, the dimensions of the pore throats (void areas in cell faces) and pore bodies (void volumes within cells) are also known. Thus the network model is completely defined and provides a faithful representation of the microstructure of a real porous medium. In this work we use the central 3367 spheres of the Finney pack to obtain a network model of about 15000 pores and 30000 throats.

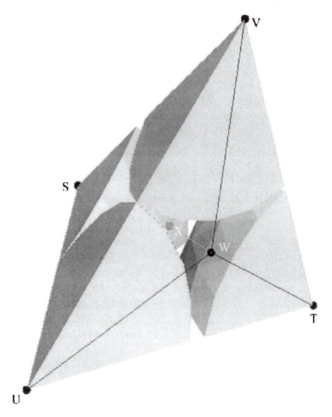

Figure 2. A Delaunay cell within the Finney pack is a tetrahedron that groups together four nearest neighbor spheres. Each apex of the tetrahedron (labeled S, T, U, V) is the center of a sphere. Only the segments of the spheres contained within the cell are shown; the rest of each sphere is contained in adjacent cells. The void area in each face of the cell is a local minimum in cross-section and thus controls access of the meniscus between nonwetting and wetting phases into the cell volume. The center of the throat in face TUV is labeled W, and the center of the pore body is labeled X.

Simulation of Drainage

The starting point for computing the configuration of fluid phases in the model soil is the Young-Laplace equation, which relates capillary pressure to the mean curvature of the interface between the fluids:

$$P_c = \gamma C \tag{2}$$

where the capillary pressure P_c is the pressure difference $P_{nw} - P_w$ between the non-wetting and wetting fluids; γ is the interfacial tension between the fluids and C is the mean curvature of the interface. For convenience we assume the wetting phase wets the grains perfectly, i.e. the contact angle is zero.

The Young-Laplace equation describes a static configuration, but it can also be applied to the displacement of a fluid from a porous medium by a second, immiscible fluid, if the displacement occurs at a rate sufficiently low for capillary forces to dominate viscous forces. This is the basis for simulating drainage and imbibition processes by means of invasion percolation (16). As the capillary pressure increases during drainage, the curvature of the interface increases and the non-wetting phase is able to enter smaller and smaller pore throats. The critical curvature required for non-wetting phase to pass through a pore throat is a function of the throat geometry (17-19). Here we use the simplest estimate of critical curvature, due to Haines (20), which presumes that the interface is spherical as it passes through the throat.

The simulation is implemented as described by Bryant et al. (14) and Mason and Mellor (7). The network model is initially full of wetting phase W. A small increment is added to the capillary pressure, and the resulting curvature of the W/NW interface is computed. Each cell containing NW phase is checked to see whether it will drain into a neighbor containing W phase. Newly drained cells are added to the list of NW-containing cells, and this list is re-checked to see whether additional cells will drain. When no more cells can be drained at the current capillary pressure, the fluids are at capillary equilibrium, and the capillary pressure is incremented again. This algorithm works well for small increments of capillary pressure, providing good *a priori* predictions of mercury injection experiments (14).

Trapping Of W Phase

A common assumption in drainage simulations, including the scheme outlined above, is that the W phase cannot be trapped, because film flow along the wetted grain surfaces always provides a connected path along which displaced fluid can escape. The rate of film flow is necessarily small, however, and the time scales for some drainage events in the environment may be too short for film flow to be appreciable. In order to address these cases, we extend the drainage simulation to allow trapping of the wetting phase.

We consider two morphologies for the trapped W phase: pendular rings left at grain contacts after surrounding pores have been drained (Figure 3) and lenses left in pore throats when the pores connected by a throat have been drained via other throats (Figure 4). We consider three levels of connectivity of the W phase. The baseline case assumes that all W phase in the pore space is connected. In this case pendular rings and lenses will continue to shrink as the applied capillary pressure increases. Lenses will disappear when the applied capillary pressure is sufficient to bring the opposing menisci into contact within the pore throat. Rings held at a gap between two spheres cannot exist beyond a certain maximum curvature that depends on the size of the gap. These rings therefore disappear as capillary pressure increases, but rings held at grain contacts with no gap exist for all capillary pressures. Thus the irreducible W phase saturation is zero in the baseline case, but this saturation is only reached asymptotically.

The second level of connectivity, denoted "intermediate" in this work, allows for the isolation of pendular rings but assumes that lenses of W phase remain connected to bulk W. A ring is assumed to become isolated when all the pores associated with the grain contact have drained and no lenses remain in the pore throats connecting these drained pores.

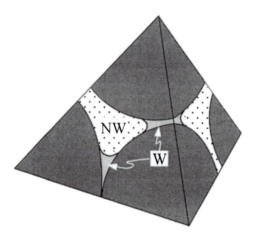

Figure 3. Schematic of pendular rings at grain contacts. In this work the term "grain contact" refers to any pair of nearest neighbor spheres, which need not actually touch. If a gap exists between the spheres, a pendular ring can exist only at curvatures below a threshold value that depends on the size of the gap.

The third level of connectivity, denoted "poor" in this work, allows for the isolation of both pendular rings and lenses. Rings are assumed to become isolated when all the pores associated with the grain contact have drained, regardless of the status of the throats connecting those pores. Lenses are assumed to be trapped at the instant they are formed.

Under these criteria, trapping of W is a local phenomenon, in the sense that W is deemed trapped depending only on the fluid configuration in the immediate vicinity of a grain contact (for rings) or pore throat (for lenses). In future publications we will discuss a third morphology for trapping, "islands" of W phase in several adjacent undrained pores. We will also examine the influence of global connectivity assumptions on the interfacial area.

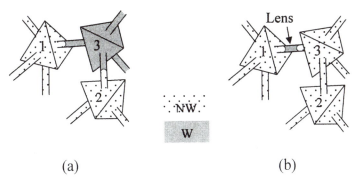

(a)   (b)

Figure 4. Schematic of lens formation in a pore throat. (a) NW has drained pores 1 and 2, but the meniscus curvature was not large enough to penetrate the throats connecting pores 1 and 2 to pore 3. (b) Suppose the critical curvature for the throat connecting pores 2 and 3 is smaller than that for pores 1 and 3. Then as the applied capillary pressure increases, pore 3 will be drained via the throat from pore 2. When this occurs the meniscus will enter but not penetrate the throat connecting pores 1 and 3, trapping a lens of W.

Computation of Area and Phase Volumes

Once the equilibrium configuration of phases is determined at a given capillary pressure, the area of the meniscus and the volume of each phase are computed on a cell-by-cell basis. Rings are assumed to be volumes of revolution, and the NW/W meniscus is assumed spherical in unpenetrated pore throats. The geometry of these morphologies is readily determined by classical methods (21-23). Our definition of interfacial area does not include the surface of grains in contact with NW in drained cells. Though this surface will in general be coated with a thin film of W, mass transfer to/from this film is negligible for many contamination/remediation processes. Nevertheless for the purposes of comparing predictions with measurements, we can readily estimate the "footprint" of rings and lenses on grain surfaces. Since the total grain area in each cell is easily computed, we can estimate the area of the film by difference.

Results

Simulations of drainage in the Finney pack were conducted with the algorithms described above for the different levels of W connectivity. The simulations began with the pore space occupied entirely by W phase and were carried out to irreducible W phase saturation; that is, the capillary pressure was increased in increments until no further movement of the meniscus was possible. When W is assumed to be completely connected, the meniscus moves indefinitely at grain contacts with no gap between the spheres, but the area and volume of the pendular rings at these contacts quickly become negligible.

The "intermediate" level of connectivity yields an endpoint W phase saturation of 0.024, existing exclusively as pendular rings at grain contacts. At the endpoint about half the nearest neighbor grains in the packing supported pendular rings. The "poor" level of connectivity yields an endpoint W saturation of 0.115, nearly five times larger than the endpoint under "intermediate" connectivity (Figure 5). The increase is due to trapped lenses of W phase.

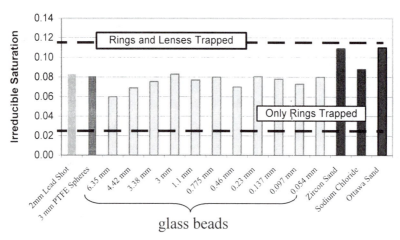

Figure 5. Irreducible wetting-phase saturations for packings of spheres of various sizes and materials fall between the predicted endpoints for the "intermediate" (W phase trapped as pendular rings at grain contacts) and "poor" levels of connectivity (W phase trapped as pendular rings and lenses in pore throats). Packings of non-spherical grains exhibit somewhat larger irreducible saturations. Data taken from Morrow (24).

Morrow (24) conducted drainage experiments on several different types of porous media to determine the effect of different particle properties on irreducible saturation. The "intermediate" level of connectivity, under which pendular rings are the only morphology for trapped W, underestimates all the data (Figure 5). Pendular rings alone are evidently not sufficient to account for the trapped W phase in these experiments. Predictions assuming the "poor" level of connectivity, under which lenses are trapped as soon as they form, overestimates the data, indicating that in these experiments many lenses remain connected to the bulk W.

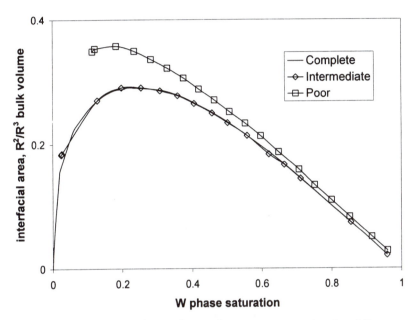

Figure 6. Interfacial area during drainage computed under different assumptions regarding the connectivity of W phase. A maximum occurs because the contribution from connected W phase vanishes at small W saturations. When W phase can become isolated, the drainage process yields an irreducible W phase saturation. The values of area in this plot include only the contributions of 3D morphologies: pendular rings at grain contacts, lenses trapped in pore throats, and the interface between the bulk connected volumes of W and NW phase. The film of W on grain surfaces in drained cells is not included.

During the drainage process the specific interfacial area exhibits a maximum at a W phase saturation of about 0.2 (Figure 6). If the W phase is completely connected, the endpoint saturation is zero, and the interfacial area declines rapidly from the maximum. For "intermediate" connectivity (W phase trapped as rings) the specific interfacial area at the endpoint (irreducible W saturation) is $0.18\ R^{-1}$, where R is the radius of the model soil grains. For comparison, the specific surface area of the grains in the packing is $1.9\ R^{-1}$, an order of magnitude greater. For "poor" connectivity (W phase trapped as rings and lenses) the specific interfacial area at the drainage endpoint is 0.35 R^{-1}. The increase in area is proportionately much smaller than the increase in irreducible W saturation. This is because the surface-to-volume ratio (S/V) of lenses is lower than that of pendular rings. For W phase saturations greater than about 0.3 the total interfacial area is not strongly sensitive to the level of connectivity, differing by less than 10% in this range.

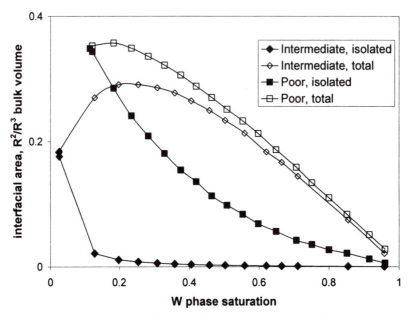

Figure 7. The relative contribution of isolated W to the interfacial area depends strongly on W connectivity. Under intermediate connectivity, significant quantities of W become isolated as pendular rings only small W saturation. Under poor connectivity, rings and lenses become isolated much earlier in the drainage process.

The evolution of the specific interfacial area during drainage reveals an important distinction between area associated with trapped (isolated) W phase and area associated with the bulk (connected) W phase. Under the assumption of intermediate connectivity, when only rings can be trapped, the contribution of the trapped W phase to the interfacial area is very small until the W phase saturation drops below 0.1 (Figure 7). This is because few pendular rings become isolated until most of the pores have drained. In contrast, under the assumption of poor connectivity, the isolated W phase contributes significantly to the interfacial area almost as soon as drainage begins, and this contribution dominates when the W phase saturation is less than 0.35 (Figure 7).

These plots show that the existence of a maximum in the total interfacial area is due to the contribution of the bulk connected W phase, which necessarily vanishes as the drainage process reaches the endpoint, and to the surface/volume ratios of the trapped phase morphologies. Pendular rings exhibit an average S/V around 19 R^{-1}; lenses, about 5 R^{-1}; bulk W phase, of order 1 R^{-1}. Thus the contributions from the isolated and bulk W phases are in very different proportions for the different assumptions of connectivity.

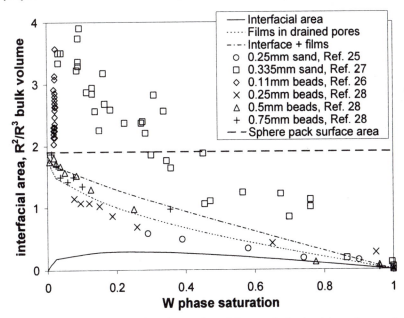

Figure 8. Predicted interfacial area from 3D morphologies (rings, lenses, bulk connected phases) is smaller than measurements reported in the literature by as much as an order of magnitude. The area of W film in drained pores increases during drainage, and when this area is added to the interfacial area, the predictions show the same trend as the data. The measurements have been normalized by the average radius of the grains or beads in each experiment.

Measurements of interfacial area reported in the literature generally exceed the predictions (Figure 8), except for very high W phase saturations. When the contribution of thin films of W in contact with NW in drained cells is included, the predictions exhibit the same trend as the data. We conclude that the experimental techniques are sensitive to the thin films as well as the interfaces between macroscopic volumes of W and NW, making a direct comparison of the predictions for 3D morphologies (pendular rings at grain contacts, lenses in pore throats, interface between bulk connected phases) impossible. The predictions show that even at quite small saturations of W the area of films in drained pores is significantly less than the grain surface area; the area of W films in drained pores does not reach 95% of the total grain area until W phase fraction is much less than 0.01. This is because pendular rings at grain contacts have large "footprints" on the grain surfaces compared to their interfacial area. Figure 8 shows that this behavior at low W saturation is consistent with experiments in which the area was measured by titrating sections of equilibrated bead packs (28). Over the entire range of saturations, the predicted sum of interfacial area and film area falls within a factor of 2 to 3 of the measurements. Part of this discrepancy reflects errors from representing polydisperse media as a monodisperse packing with the same average grain size and from idealizing the geometry of the phase morphologies. The discrepancy may also reflect a subtle aspect of the role of thin films in interfacial tracer experiments: not all the film may be thick enough permit the sorption of the tracer molecules (27). This would account for the bead pack experiments of (28) falling below the prediction and the sandpack experiments of (27) lying above the prediction for smooth spherical grains but below the corresponding prediction for rough sand grains, which have an order of magnitude more surface area (27).

Conclusions

A simple but physically representative model soil permits quantitative predictions of the interfacial area between W and NW phases during drainage. Depending on phase saturation, the specific interfacial area on a unit bulk volume basis ranges from 0.03 to 0.3 R^{-1}, where R is the radius of the grains of soil. These values include the contributions of three-dimensional phase morphologies (pendular rings at grain contacts, lenses in pore throats, the interface between bulk connected W and NW phases), but not the contribution of grain surfaces in pores that have been drained. The specific interfacial area exhibits a maximum at a W phase saturation of about 0.2, reflecting the tradeoff between the decreasing volume of bulk connected W phase and the increasing volume of isolated W phase.

The simulations show the importance of the modes of W phase trapping. Under an assumption of "intermediate" W connectivity, in which W can be isolated only as pendular rings at grain contacts, the bulk W phase contribution dominates the interfacial area until the W phase saturation is below 0.10. Under an assumption of "poor" W connectivity, in which W can be isolated at grain contacts and as lenses in pore throats, the isolated W phase makes a significantly larger contribution to the interfacial area. The total interfacial area is not sensitive to the trapping criteria in this work. Depending on the mass transfer process of interest, however, the relative contributions to this total from isolated and bulk W phase may be very important.

Predicted irreducible W phase saturations are consistent with experimental data, indicating that lenses are an important morphology of trapped W phase in many situations, though in others only pendular rings are likely. Predictions using the simple "intermediate" and "poor" categories of W connectivity bracket these measurements, indicating that global rather than local assessments of lens and ring trapping are likely to be more representative. Predictions of interfacial area underestimate measurements reported in the literature, primarily because the measurements include the contribution of W films in drained pores. Estimates of the latter contribution in the model soil show that it is as much as an order of magnitude greater than the area between macroscopic volumes of the phases. The trend of the combined contributions is the same as the trend of the reported data, which fall within a factor of 2 to 3 of the prediction.

Acknowledgements

We are grateful to EPA (R827116/01/0) and to NSF (DMS 9873326) for partial support of this research. This work has not been subjected to any EPA review and therefore does not necessarily reflect the views of the Agency, and no official endorsement should be inferred. The Geomview package from the Geometry Center at the University of Minnesota (www.geom.umn.edu) was used to investigate pore space as illustrated in Fig. 2.

References

1. Abriola, L.M., T.J. Dekker, and K.D. Pennell, "Surfactant enhanced solubilization of residual dodecane in soil columns 2. Mathematical modeling," *Env. Sci. Tech.* **27**, 2341 (1993).

2. Imhoff, P.T., P.R. Jaffé, and G.F. Pinder, "An experimental study of complete dissolution of a nonaqueous phase liquid in saturated porous media," *Water Resources Research*, **30**, 307-320 (1994).

3. Mayer, A. and C. T. Miller, "The influence of mass transfer characteristics and porous media heterogeneity on nonaqueous phase dissolution," *Water Resources Research* **32**, 1551-1567 (1996).

4. Powers, S., L. Abriola, J. Dunkin and W. Weber, Jr., "Phenomenological models for transient NAPL-water mass transfer processes," *J. Contam. Hydro.* **16** 1-33 (1994).

5. Reeves, H. and L. Abriola, "An iterative-compositional model for subsurface multiphase flow," *J. Contam. Hydro.* **15**, 249-276 (1994).

6. Bryant, S., D. Mellor and C. Cade, "Physically representative network models of transport in porous media," *AIChE J.* **39**, 387-396 (1993).

7. Mason, G. and D. Mellor, "Analysis of the percolation properties of a real porous material," in Characterization of Porous Solids II, (F. Rodriguez-Reinoso et al., eds.), Elsevier Science Publishers, Amsterdam, 41 (1991).

8. Finney, J., "Random packings and the structure of simple liquids: I. The geometry of random close packing," *Proc. Roy. Soc. Lond.* **319A**, 479 (1970).

9. Bryant, S., C. Cade and D. Mellor, "Permeability prediction from geologic models," *AAPG Bull.* **77**, 1338-1350 (1993).

10. Bryant, S. and M. Blunt, "Prediction of relative permeability in simple porous media," *Phys. Rev. A* **46**, 2004-2011 (1992).

11. Bryant, S. and N. Pallatt, "Predicting formation factor and resistivity index in simple sandstones," *J. Petrol. Sci. Eng.* **15**, 169-179 (1996).

12. Bryant, S. and S. Raikes, "Prediction of elastic wave velocities in sandstones using structural models," *Geophysics* **60**, 437-446 (1995).

13. Bryant, S., C. Cade and J. Evans, "Analysis of permeability controls in sandstones: A new approach," *Clay Minerals* **29**, 491-501 (1994).

14. Bryant, S., G. Mason and D. Mellor, "Quantification of spatial correlation in porous media and its effect on mercury porosimetry," *J. Coll. Interface Sci.* **177**, 88-100 (1996).

15. Bryant, S., P. King and D. Mellor, "Network model evaluation of permeability and spatial correlation in a real random sphere pack," *Transport in Porous Media* **11**, 53-70 (1993).

16. Wilkinson, D. and J. Willemsen, "Invasion percolation: a new form of percolation theory," *J. Phys. A: Math. Gen.* **16**, 3365-3376 (1983).

17. Lenormand, R., C. Zarcone and A. Sarr, "Mechanisms of the displacement of one fluid by another in a network of capillary ducts," *J. Fluid Mech.* **195**, 337-353 (1983).

18. Mason, G. and N Morrow, "Meniscus curvatures in capillaries of uniform cross-section," *J. Chem. Soc., Faraday Trans. 1*, **80**, 2375-2393 (1984).

19. Mason, G. and N. Morrow, "Meniscus configurations and curvatures in non-axisymmetric pores of open and closed uniform cross section," *Proc. Roy. Soc. Lond. A* **414**, 111-133 (1987).

20. Haines, W. *J. Agr. Sci.* **17**, 264 (1927).

21. Falls, A., J. Musters and J. Ratulowski, "The apparent viscosity of foams in homogeneous bead packs," *SPERE* **4**, 155-164 (1989).

22. Gvirtzman, H. and P. Roberts, "Pore scale spatial analysis of two immiscible fluids in porous media," *Water Resources Research* **27**, 1165-1176 (1991).

23. Johnson, A., "Pore level modeling of interfacial area in porous media." MS Thesis. The University of Texas at Austin. (2001).

24. Morrow, N., "Irreducible wetting-phase saturations in porous media," *Chemical Engineering Science* **25**, 1799-1818 (1970).

25. Kim, H., P. Rao and M. Annable, "Determination of effective air-water interfacial area in partially saturated porous media using surfactant adsorption," *Water Resources Research* **33**(12) 2705-2711 (1997).

26. Kim, H., P. Suresh C. Rao and Michael D. Annable, "Consistency of the interfacial tracer technique: experimental evaluation," *J. Contam Hydrol.* **40**, 79-94 (1999).

27. Schaefer, C., D. DiCarlo and M. Blunt, "Experimental measurement of air-water interfacial area during gravity drainage and secondary imbibition in porous media," *Water Resources Research* **36**, 885-890 (2000).

28. Faisal Anwar, A., M. Bettahar, and U. Matsubayashi, "A method for determining air-water interfacial area in variably saturated porous media," *J. Contam. Hydrol.* **43** 129-146 (2000).

Chapter 3

Environmental Impacts and Monitoring: A Historical Perspective on the Use of Natural Attenuation for Subsurface Remediation

Candida C. West[1] and John T. Wilson[2]

[1]University of Oklahoma Health Sciences Center, Department of Occupational and Environmental Health, Oklahoma City, OK 73190
[2]NRMRL, SPRD, U.S. Environmental Protection Agency, Ada, OK 74820

The collective processes that constitute the broadly used term "natural attenuation," as it relates to subsurface remediation of contaminants, refer to the physical, chemical, and biological interactions that, without human intervention, reduce or contain contaminants in the subsurface environment. Knowledge of these processes has been used successfully to control water quality in surface water. However, the understanding of the impact of the same processes in the subsurface environment has improved fundamentally and dramatically over the course of the last three decades. It is now known that chemicals are capable of moving deep into the subsoil and sediments and are subject to a wide variety of complex chemical reactions, water/solid interactions, and microbial degradation. This chapter discusses the evolution in our understanding of these processes, how they relate to current philosophies concerning subsurface contaminant clean-up standards, and the issues pertaining to public acceptance of natural attenuation as an important component of remediation strategy.

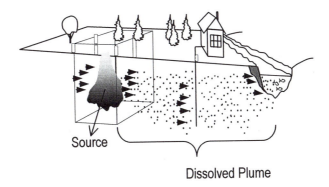

Figure 3: Plume of dissolved contaminant impacting downgradient receptors.

Background

Agronomists and soil scientists were early investigators of chemical movement in soil. Chemical sorption, particularly of pesticides such as DDT, onto organic phases of the soil structure was recognized as an important process in agriculture. The belief that the upper soils had a seemingly endless capacity to sorb chemicals also led to widespread use of land application of as a method of water treatment. However, studies conducted in the 1970s and 80s led to the knowledge that anthropogenic chemicals were indeed capable of moving deeply into the subsurface, often reaching aquifers. The contamination could continue to move downgradient as a "plume" resulting from the dissolution of the material at the "source", threatening to contaminate water supplies and pose a public health risk. These observations led to continuing investigations of instances of water contamination, which ultimately lead to the development of nationwide research programs on basic mechanisms of fate and transport of contaminants in subsurface environments.

The ability of a natural body of water, surface or subsurface, to receive anthropogenic or toxic materials without harmful effects or damage to indigenous organisms refers to the system's assimilative capacity. The processes responsible for the capacity to assimilate are both abiotic and biotic in nature and include a variety of physical, chemical, and biological processes that may reduce the mass, toxicity, mobility, volume, or concentration of potential toxicants. Natural attenuation refers to subsurface *in-situ* (in place) assimilative processes including biodegradation, dispersion, dilution, sorption, volatilization, chemical/biological stabilization, and transformations that are operative to a degree dependent on the system conditions, without the aid of human intervention.

A Matter of Definition

"The safest words are always those that bring us most directly to the facts"
-Charles H. Parkhurst-

The phrase "natural attenuation" as understood within the context of subsurface remediation of contaminated soil and water evolved over a period of several years. It has largely been accepted as the politically and scientifically appropriate term, however it is still frequently used synonymously with "intrinsic remediation", "natural recovery", and "natural assimilation". The phrases "intrinsic bioremediation" and "passive bioremediation" refer strictly to biologically mediated attenuation processes, which are obviously a less broad definition, not including all the physical and abiotically mediated processes that can immobilize or destroy contaminants. Given this, it is true that if the natural attenuation capacity of a particular contaminated subsurface system does not include significant biodegradation and biological hydrolysis, then the contaminant concentration reduction most likely results from nothing more than dilution. Ironically, up until around 1970, it was thought that the subsurface environment was essentially sterile. As stated, it is now known that there is an active microbial population functional at some level at virtually every point in the subsurface environment.

Contaminants can be found in the subsurface dissolved in water, sorbed to soil, as vapors in the unsaturated pore spaces or existing as discrete volumes of free material in its' original form (referred to as "free phase"). Generally, the zone that contains residually bound or free-phase contaminant as described earlier serves as a source (reservoir) for further migration of contaminants into the ground water. That portion of contaminant dissolved in water creates a plume that migrates with the flow of water and is subject to natural attenuation processes.

Scientific Basis for Natural Attenuation

An extensive body of literature exists documenting the validity of natural attenuation processes. Under some hydrogeologic conditions, processes such as sorption or precipitation might be a major source of contaminant removal from a system. These processes primarily contain the contaminant, which may, or may not, later be released back into the system as the geochemistry of the system changes. On the other hand, the possibility for complete contaminant destruction exists through microbial action and there is little question that biodegradation is a dominant process in natural attenuation. The presence of anthropogenic carbon (contaminant) can stimulate a pre-existing, actively degrading microbial population. The rate and extent of biodegradation is dependent on the availability of the contaminant to the microbe, the adequacy of electron acceptors and nutrients, and the degree to which the microbes need to adapt to a new carbon (or additional carbon) source. Fuel hydrocarbons, particularly benzene, toluene, ethylbenzene, and xylene (BTEX) compounds are readily utilized as a primary substrate. In essence, the microbiological degradation of these hydrocarbons can be simplified as:

Microbes

carbon source + electron acceptors + nutrients
(anthropogenic or indigenous e.g. O_2, NO_3, SO_4^{-2}, Fe^{+3})

Byproducts + CO_2 + H_2O + energy

Until recently, chlorinated hydrocarbons, comprising the second most common class of groundwater pollutant in the US, were believed to not readily biodegrade. It is now been determined that there are three primary biological pathways that can transform these compounds, albeit with somewhat more restrictions than the readily degraded fuel hydrocarbons. The predominant pathway for highly chlorinated hydrocarbons (e.g. tetrachloroethylene, PCE) is reductive dechlorination to lower chlorinated hydrocarbons (dichloroethylene, DCE and vinyl chloride, VC). PCE serves as the electron acceptor, therefore requiring an electron donor. In a situation of co-mixing of the PCE with the BTEX compounds, there is a ready supply of electron donor.

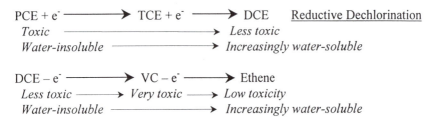

Lower chlorinated hydrocarbons may also be used as electron donors.

DCE + electron acceptor $\longrightarrow$ CO_2
VC   + electron acceptor $\longrightarrow$ CO_2

During reductive dechlorination a chlorine atom in the molecule is replaced with a hydrogen atom. The chlorine atom is released to the environment as a chloride ion. Trichloroethylene is reduced to dichloroethylene (primarily cis-dichloroethylene), which can be further reduced to vinyl chloride, then to ethylene and ethane (*1-2*). In an analysis of data from 61 sites having plumes of chlorinated ethylenes McNab et al. (*3*) found no evidence of reductive dechlorination at 23 sites, dechlorination to dichloroethene at 18 sites, and dechlorination to vinyl chloride at 20 sites.

For many years, the only established mechanism for biodegradation of vinyl chloride was reductive dechlorination to ethylene and ethane. The rate of reductive dechlorination of vinyl chloride is slower than that of dichloroethylene; therefore vinyl chloride should be expected to accumulate in ground water plumes (*2*). Contrary to this expectation, trichloroethylene and dichloroethylene disappear in many plumes without the accumulation of vinyl chloride, ethylene, or ethane. In recent years, it has been recognized that bacteria in aquifers can use a variety of electron acceptors to oxidize cis-dichloroethylene or vinyl chloride to carbon dioxide. Microbial Degradation of cis-dichloroethylene has been demonstrated using iron (III) (*4*), manganese (IV) (*5*), or oxygen (*6-7*). Microbes can also degrade vinyl chloride using iron (III) (*8-9*), or oxygen (*10-11*).

The discussion presented here is meant as the briefest description of the types of microbial processes operative in the subsurface environment. For further details concerning the mechanisms and conditions of biodegradation of contaminants, the reader is referred to the literature. Several excellent recent articles include those by Azadpour-Keeley et al. (*12*), Anderson and Lovley (*13*), Nales et al. (*14*), Weiner et al. (*15*), and Chapelle and Bradley (*16*).

The second pathway for the transformation products, cis-DCE and vinyl chloride (VC) is an anaerobic oxidation to carbon dioxide. This may occur with a variety of electron acceptors including oxygen, iron (III), and manganese (IV), or native organic matter.

The third most common biodegradation pathway for chlorinated hydrocarbons is cometabolism via enzymatic reactions occurring fortuitously with oxidation of compounds such as methane or toluene. These three pathways are, of course, complicated by the fact that; 1) each can only occur under specific chemical conditions; 2) the dominant pathway changes as the source of electron donor and/or acceptor (both anthropogenic and indigenous) is reduced; and 3) There are several different isomers, and other compounds of chlorinated hydrocarbons, that can enter these pathways from biotic or abiotic processes. In other words, it is a complicated process.

Determination of Appropriateness

Although the geochemical and hydrological conditions that favor natural attenuation have been studied for many years, there is still considerable debate concerning the contaminants for which it may be used and the manner in which the decision to use natural attenuation as part of a remedial solution is made. The efficiency of natural attenuation is highly site specific. Evaluation of the significance of natural attenuation for a particular site is dependent on obtaining a high level of understanding of the biogeochemistry and hydrology of the system.

Adequate monitoring is critical to the initial evaluation process and continues with constant monitoring throughout the lifetime of the system until a decision is made that there is no longer a threat to public health. Input of high quality monitoring data into a contaminant fate and transport model is particularly important in assessing the relative importance of biodegradation as an attenuation mechanism for a particular site. Initial efforts for site characterization require the use of appropriate ranges of decay rates and mechanisms in ground water. Suarez and Rifai (*17*) have published an extensive review of the rate constants for biodegradation of fuel and chlorinated hydrocarbons to be used in screening and modeling tools.

Two recently published texts (18-19) provide extensive coverage of the theoretical and practical aspects of understanding, evaluating, and quantifying natural attenuation as a remedial option. Methods and implications for assessing the potential for bioremediation are still a subject of investigation (20-27).

Observations of Natural Attenuation

Naturally attenuating contaminant plumes follow a variety of patterns with respect to the concentration in the dissolved contaminant mass and downgradient movement as a function of time (Figure 4). The trend of spatial variation of contaminant concentration from one point in time to another is essential to an accurate evaluation of the extent of natural attenuation. Plumes lengths exist that ranges from tens to thousands of feet downgradient from the source of contamination. The plume may be expanding (net gain in contaminant mass in the dissolved phase; increased plume length), stable (no net gain or loss; stable plume length), or shrinking (net loss in contaminant mass; decreasing plume length) (18). Based on an extensive database on plume migration, Rice et al. (28) empirically derived the term "exhausted" for fuel hydrocarbon plumes having a length of less than 70 ft and an average benzene concentration of less than 10 parts per billion.

A new class of studies, referred to as "plume-a-thon" studies, provide an opportunity to examine plume behavior at a variety of sites (29). Rice et al. (28) reported that plume lengths at 271 fuel hydrocarbon sites in California have stabilized at relatively short distances, less than 250 feet, from the release sites and that of these 59% were stable, 33% were shrinking and only 8% were growing. Likewise, Mace et al. (30) studied 217 sites in Texas and determined that 75% of benzene plumes are less than 250 feet long and are stable or decreasing in length regardless of whether active remediation is implemented or not.

Natural attenuation has been studied most extensively in groundwater contaminated by leaking underground storage tanks (USTs) containing petroleum hydrocarbons and in fact is now the leading remedy in the UST program (31-33). It is known that petroleum hydrocarbons undergo extensive aerobic biodegradation and for this reason natural attenuation has been generally accepted as a viable option for remedial option. The ASTM (34) and EPA (35) have provided guidance for the use of natural attenuation as a remedial tool. The ASTM guidance covers only petroleum hydrocarbon release sites whereas the EPA guidance applies to all contaminants at Superfund, RCRA and UST sites. It is estimated that in 1997 natural attenuation was used at more than 15,000

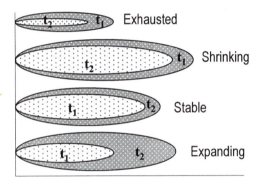

Time, t ($t_2 > t_1$)

Figure 4: Reproduced with permission from reference 7. (Copyright 1999 John Wiley & Sons

underground storage tanks leaking gasoline and that in the EPA's Superfund program, the use of natural attenuation grew from 6% in 1990 to more than 23% in 1997 (*36*).

Extensive discussions of the scientific basis for natural attenuation are widespread in the literature. The National Research Council (NRC) published a report in 2000 (*37*) which was undertaken to evaluate the current knowledge, limitations, risks, data needs, and protocols concerning the use of natural attenuation as a remedial tool.

Protocols for documentation of natural attenuation have abounded as natural attenuation has been embraced as a viable option for remediation. The purpose of a protocol is to make a public statement on how a decision to use natural attenuation is made and how implementation should be conducted through the use of a strategy and methodology. A wide range of protocols currently used to document natural attenuation is reviewed in the NRC report (*37*). Also included in the report is an up-to-date review of the scientific basis and approaches for evaluating natural attenuation and summarizes federal agency and several state environmental agency policies concerning its' use. While several protocols for evaluating natural attenuation have been published, there is not a consensus on the density of data necessary for decision-making or on the specific water quality parameters and contaminant concentration data that should be collected. The goal is to collect data that provide the least uncertainty as to the success of natural attenuation at a specific remediation site. A consensus approach to achieve the goal is still an evolving process (*38*).

Potential Advantages and Disadvantages of Natural Attenuation

There are significant advantages associated with natural attenuation as a remedial tool.

Potential advantages include:
- As with any *in situ* process, there is less remediation waste produced, the risk of cross contamination from one media to another (i.e. soil to air) is virtually eliminated, and disturbance to ecological receptors is reduced.
- Has the potential to destroy the contaminant, particularly in the case of complete mineralization.
- The remediation site may be more aesthetically acceptable to the local community than other forms of remediation requiring extensive engineering infrastructure.
- Can be used in conjunction with, or as a follow-up to, other active remedial measures and can be used on all or part of a site.

- Can potentially significantly reduce the costs of remediation to cleanup goals.
- Is not subject to maintenance or operational limitations imposed by mechanical devices

Potential disadvantages include:
- May require longer time frame to achieve remediation objectives using natural attenuation relative to more aggressive remedial options, especially when a source zone exists.
- More stringent requirements for monitoring can be complex and costly.
- Transformation products may be more toxic or mobile than the parent compounds.
- Over time subsurface conditions could change either re-mobilizing contaminants or simply reducing the efficiency of natural attenuation.
- Public acceptance to natural attenuation is slow and requires more extensive education and outreach efforts.

Regulatory Acceptance of Natural Attenuation

The goal of remediation is to achieve cleanup to an acceptable level of risk to human and environmental receptors thereby returning groundwater and soil back to their beneficial use within a timeframe that is reasonable at a cost that is reasonable. Until recently, cleanup goals were based on reducing contaminant concentrations in soil and water to the lowest concentration possible using the best available technology, generally to federal or state mandated concentrations such as maximum concentration levels (MCLs). Essentially it seemed as if we naively expected that regulatory requirements could drive the rate and extent to which contamination could be removed. With increasing experience however, it was realized that soil and water remediation is a very difficult process.

Currently there are reportedly 358,269 confirmed releases or leaking underground storage tanks in the United States and the number of releases is increasing at a higher rate than that of cleanup (*39*). Combined with RCRA and Superfund programs the sheer number of sites in need of remediation spurred the reality that there is simply not enough money and resources available to clean up all sites to MCLs. As a result, many programs are reexamining the regulatory cleanup framework.

Concentration-based cleanup goals are being replaced with performance-based goals, providing an opportunity to prioritize sites for maximum risk reduction to the public and the environment. Risk-based corrective action (RBCA pronounced "rebecca") is defined by the EPA as an approach to corrective action that integrates components of traditional corrective action with alternative risk and exposure assessment practices. This means that clean-up needs are based on human and environmental risk assessment using exposure models and economic considerations, instead of basing them on "one-size-fits-all" concentration-based regulatory mandates.

The evolution in our understanding of the circumstances under which natural attenuation processes work to clean up contaminated sites has helped provide the means to implement this paradigm shift. The recognition that natural attenuation can be, in and of itself, a remedial option, has provided a tool that can be considered for use at a substantial portion of the hundreds of thousands of contaminated sites across the nation

In April of 1999 the USEPA issued a directive (35) clarifying the EPA's policy regarding the use of natural attenuation for remediation of soil and groundwater at sites administered by EPA's Office of Solid Waste and Emergency Response. EPA clearly states that it does not consider the use of natural attenuation as a remedial action to be a "no action" or "walk-away" approach nor does it consider it to be a presumptive or default remedy. The EPA coined the term "monitored" natural attenuation (MNA) to indicate that this option is heavily dependent on long-term performance monitoring of the system

An important component in the implementation of MNA is the specification of a contingency remedy. This is a cleanup approach specified in the site remedy decision document that is triggered by performance criteria that indicate the MNA is not operating as anticipated. Contingency plans are not unique to MNA, but are included where there is uncertainty regarding performance of a selected technology.

There has been criticism that the use of natural attenuation has outpaced scientific support (36). However, the remediation community has clearly embraced natural attenuation nationwide. As performance data continue to be collected and the ability to evaluate the appropriate use of natural attenuation improves, remediators will have an additional tool to design a robust program for cleanup of contaminated sites.

Public Acceptance of Natural Attenuation

Community involvement in decision-making concerning the cleanup of contaminated sites is critical, particularly when suggesting the use of natural attenuation as a part of the remediation plan. When contaminated sites exist in residential areas there is generally a very high level of concern. In fact, it has been found that residents ranked contaminated sites in their neighborhoods as having a risk of 4.7 on a scale of 1 to 5, with 5 representing the highest risk level, as compared to health officials who ranked the same sites as having a much lower risk (40). The discovery of contaminant exposure in a community creates physiological stress due to health concerns, possible reduction in property values. An individual who lives near a hazardous waste site may even experience rejection and alienation from friends who are unaffected by the contamination.

Affected communities want contamination to be controlled quickly and natural attenuation is unlikely to be perceived as being part of the solution. There are many factors that may influence the perception of the acceptability of natural attenuation as a remediation strategy for a contaminated site. The NRC (36) has outlined many of the problems that confound community confidence in the use of MNA as determined by a panel of six community leaders representing four states affected by large industrial waste sites. Among their concerns:

- Visibility – Communities are comforted by obvious activity and infrastructure that is unlikely to be provided by the mere presence of monitoring wells.
- Uncertain time line and outcome – Unlike natural disasters which have a definite endpoint, natural attenuation is a time intensive process and there in uncertainty as to the extent to which the contamination will be removed
- Low level of engineered technology – There may be the belief that the lack of visible engineered technology is synonymous with no action and the appearance that the contaminants are simply being left in place
- Concern that byproducts will be a health threat – It is true that depending on the extent of degradation, daughter products may be more toxic than the original contaminants. The production of vinyl chloride as a byproduct of chlorinated hydrocarbon reduction is an example.
- Institutional controls are inadequate – Due to the longer time line associated with remediation using MNA there is concern of the ability to restrict site access to provide protection for the children and animals in the community.
- Lack of funds for contingency plans – There is concern that by accepting MNA funds for contingency plans may not be available at a later time if they are needed.

It is critical that a community is involved in decisions regarding the use of MNA. Ashford and Rest (40) reviewed case studies of public participation. They recommended that community participation be designed to fit the needs of a particular community. It is important to educate the community members so that they better understand the technical aspects of how decisions are being made and facilitate their participation in the decision making process.

The NRC report (37) concludes with several recommendations for public participation that are common to any decision making process in waste cleanup. They specifically include that when MNA is being considered as part of the remedial action it is imperative that the environmental agencies and responsible parties involved provide the community with clear evidence indicating which natural attenuation processes are responsible for the loss of contaminants in an easy-to-understand format.

Notice
The information in this document has been subjected to administrative review by the United States Environmental Protection Agency. It does not necessarily reflect the views of the Agency and no official endorsement should be inferred.

References:

1. Vogel, T.M.; P.L. McCarty. *Appl. Environ. Microbiol.* **1985**, 49, 1080-1083.
2. Vogel, T.M.; C.S. Criddle; P.L. McCarty. *Environ. Sci. Technol.* **1987**, 21, 722-736.
3. McNab, W.W.; D.W. Rice; C. Tuckfield. *Bioremediation Journal* **2000**, 4, 311-335.
4. Bradley, P.M.; F.H. Chapelle. *Environ. Sci. Techno.* **1997**, 31, 2692-2696.
5. Bradley, P.M.; J.E. Landmeyer; R.S. Dinicola. *Appl. Environ. Microbio.* **1998**, 64, 1560-1562.
6. Klier, N.J.; West, R.J; P.A. Donberg. *Chemosphere* **1999**, 38, 1175-1188.
7. Bradley, P.M.; F.H. Chapelle. *Environ. Sci. Technol.* **2000**, 34, 221-223.
8. Bradley, P.M.; F.H. Chapelle. *Environ. Sci. Technol.* **1996**, 30, 2084-2092.
9. Bradley, P.M.; F.H. Chapelle; J.T. Wilson. *J. Contam. Hydrol.* **1998**, 31, 111-127.
10. Davis, J.W; C.L. Carpenter. *Appl. Environ. Microbiol.* **1990**, 56, 3878-3880.
11. Hartmans, S.; J.A.M. de Bont. *Appl. Environ. Microbiol.* **1992**, 58, 1220-1226.
12. Azadpour-Keeley, A; Russell, H.H.; Sewell, G.W. USEPA, OSWER, EPA/540/S-99-001, 1999.
13. Anderson R.T.; Lovley, D.R. *Bioremediation Journal* **1999**, 2, 121-135.
14. Nales, M.; Butler, B.J.; Edwards, E.A. *Bioremediation Journal* **1998**, 2, 125-144.
15. Weiner, J.M.; Lauck, T.S.; Lovley, D.R. *Bioremediation Journal* **1998**, 2, 159-173.
16. Chapelle, F.H.; Bradley, P.M. *Bioremediation Journal* **1998**, 2, 227-238.
17. Suarez, M.P.; Rifai, H. *Bioremediation Journal* **1999**, 3, 337-362.
18. Wiedemeier, T.H.; Rafai, H.S.; Newell, C.J.; Wilson, J.T., *Natural Attenuation of Fuels and Chlorinated Solvents in the Subsurface*, John Wiley & Sons, 1999, 617 pp.

19. Testa, S.M.; Winegardner, D.L. *Restoration of Contaminated Aquifers: Petroleum Hydrocarbons and Organic Compounds*, Lewis Publishers, 2000, 446 pp.

20. Kennedy, L.G.; Everett, J.W.; Ware, K.J.; Parsons, R.; Green, V. *Bioremediation Journal* **1998**, *2*, 259-276.

21. Ravi, V.; Chen, J.; Wilson, J.T.; Johnson, J.A.; Gierke, W.; Murdie, L. *Bioremediation Journal* **1998**, *2*, 239-258.

22. Bollinger, C.; Schonholzer, F.; Schroth, M.H.; Hahn, D.; Bernasconi, S.M.; Zeyer, J. *Bioremediation Journal* **2000**, *4*, 359-371.

23. Schreiber, M.E.; Bahr, J.M. *Bioremediation Journal*, **1999**, *3*, 363-378.

24. Carey, G.R.; Wiedemeier, T.H.; Van Geel, P.J.; McBean, E.A.; Murphy, J.R.; Rovers, F.A. *Bioremediation Journal* **1999**, *3*, 379-393.

25. Holder, A.W.; Bedient, P.B.; Hughes, J.B. *Bioremediation Journal* **1999**, *3*, 137-149.

26. Sorenson, Jr., K.S.; Peterson, L.N.; Hinchee, R.E.; Ely, R.L. *Bioremediation Journal* **2000**, *4*, 337-357.

27. Holder, A.W.; Bedient, P.B.; Hughes, J.B. *Bioremediation Journal* **1999**, *3*, 137-149.

28. Rice, D.W.; Dooher, B.P.; Cullen, S.J., Everett; L.G., Kastenberg; W.E., Grose; R.D.; Marino, M.A. Recommendation to Improve the Cleanup Process for California's Leaking Underground Fuel Tanks (LUFTs), report submitted to the California State Water Resources Control Board and the Senate Bill 1764 Leaking Underground Fuel Tank Advisory Committee, California Environmental Protection Department, Sacramento, CA, 1995, 20 pp.

29. Rifai, H.S. *Bioremediation Journal* **1998**, *2*, 217-219.

30. Mace, R.E.; Fisher, R.S.; Welch, D.M.; Parra, S.P. Extent, Mass, and Duration of Hydrocarbon Plumes from Leaking Petroleum Storage Tank Sites in Texas, Bureau of Economic Geology Geological Circular 97-1, 1997, 52 pp.

31. USEPA, Office of Solid Waste and Emergency Response (OSWER), Use of Risk-Based Decision-Making in UST Corrective Action Programs, Directive 9610.17, 1996.

32. ASTM, Standard Guide for Risk-Based Corrective Action Applied at Petroleum Release Sites, ASTM Designation E 1739-95, American Society for Testing and Materials, 1995.

33. USEPA, Office of Solid Waste and Emergency Response (OSWER), Cleaning up the Nation's Waste Sites: Market and Technology Trends, EPA-542-R-96-005.

34. ASTM, Standard Guide for Remediation of Ground Water by Natural Attenuation at Petroleum Release Sites, ASTM Designation E 1943-98, American Society for Testing and Materials, 1998.

35. USEPA, Office of Solid Waste and Emergency Response (OSWER), Directive Number 9200.4-17P, Use of Monitored Natural Attenuation at Superfund, RCRA Corrective Action and Underground Storage Tank Sites, April 21, 1999, 32 pp.

36. Renner, R., *Envir. Sci. & Tech.* **2000**, 34, pp. 203A-204A.

37. National Research Council, Natural Attenuation for Groundwater Remediation, Committee on Intrinsic Remediation, Water Science and Technology Board and Board on Radioactive Waste Management, Commission on Geosciences, Environment, and Resources, National Academy Press, Washington, DC, 2000, 274 pp.

38. Nyer, E.K.; Boettcher, G. *Ground Water Monitoring and Remediation* Winter **2001**, *21*, 42-47.

39. Small, M.C., *Bioremediation Journal* **1998**, 2, 221-225.

40. Ashford, N.A.; Rest, K.M. Public Participation in Contaminated Communities, Cambridge, MA: Center for Technology, Policy, and Industrial Development, Massachusetts Institute of Technology, 1999.

Chapter 4

Speciation and Chemical Reactions of Phosphonate Chelating Agents in Aqueous Media

A. T. Stone[1], M. A. Knight[2], and B. Nowack[3]

[1]Department of Geography and Environmental Engineering,
313 Ames Hall, The Johns Hopkins University, Baltimore, MD 21218
[2]Environmental Engineering and Science Program, 138–78 Keck
Laboratories, The California Institute of Technology, Pasadena, CA 91125
[3]Institute of Terrestrial Ecology, ETH Zürich, Grabenstrasse 11a,
CH–8952 Schlieren, Switzerland

ABSTRACT

Organic chemicals possessing phosphonate functional groups (RPO_3^{2-}) are being used in a growing number of applications. Through a series of examples, the speciation and chemical reactions of phosphonates in aqueous solutions and (hydr)oxide mineral suspensions are systematically explored. Concepts are introduced that are useful for assessing the consequences of intentional or inadvertent release into environmental media. Such concepts should aid the development of more environmentally benign synthetic organic chemicals.

INTRODUCTION

Chemists mix and match various functional groups and structural moieties in an attempt to develop manufactured chemicals with desirable properties. In recent decades, phosphonate functional groups (RPO_3^{2-}) have appeared in a rapidly growing number of manufactured chemicals. Phosphonate-based chelating agents are used to prevent the formation of undesirable precipitates, and to protect others from dissolution. Such chelating agents are used to depress free metal ion activity and to increase total dissolved metal ion concentrations. Others can convert unreactive metal ions into reactive ones (and vice versa). In many of these applications, molecular charge, protonation level, and ability to bind metal ions are critical; phosphonate-based chelating agents yield properties and reactivities that are more desirable than those obtained with alternative Lewis Base functional groups.

Other classes of phosphonate-containing synthetic organic compounds (or their metal ion complexes) possess pronounced biological properties. Phosphonates are used as pesticides, as growth regulators, as pharmaceuticals, and as sterilants. Biologically-active phosphonates possess some chemical traits that are similar to a naturally-occurring biochemical, and some traits that are different. The proper balance between similar and dissimilar traits is important for achieving efficacy.

Desirable (or undesirable) attributes of manufactured chemicals must always be viewed within the context of a particular system and it contents. Phases (and interfacial regions) set the stage for partitioning. Other chemicals present in the system affect speciation, and serve as reactants, catalysts, or inhibitors. Organisms exhibit enormous variations in their susceptibilities towards different chemical species.

This chapter begins with a comparison of the Lewis Base properties of carboxylate, phosphonate, and amine functional groups. The formation of metal ion-chelating agent complexes in solution and the adsorption of phosphonates onto (hydr)oxide mineral surfaces are then discussed. In the remaining portions of this chapter, interconnections between the coordination chemistry and chemical reactivity of phosphonates are explored. The underlying message is that the environmental chemistry of metal ions (whether dissolved or particulate-bound) and phosphonates are closely linked. The more we know about coordination chemistry, the better we can predict the effects of phosphonate-containing synthetic chemicals in the environment.

Molecular Charge, Basicity, and Equilibrium Speciation in Solution

EDTA (ethylenediaminetetraacetic acid), first patented in Germany in 1935 and in the U.S. in 1946, has long dominated the market for synthetic chelating agents (1). Four carboxylate (RCOO⁻) and two amine (R₂N) Lewis Base groups are favorably placed within the EDTA structure for occupying up to six coordinative positions of a central metal ion. Suppose that a carboxylate group is replaced by a phosphonate group. How does the nature of metal ion-chelating agent binding change, and how is the equilibrium speciation of the metal ion of interest altered? These and other questions pertaining to chelating agent design will be addressed in this section.

Acetic acid, methylenephosphonic acid, and methylamine are useful archetypes for more complex molecules (Figure 1). Oxygen atoms serve as sites of proton and metal ion binding to carboxylate and phosphonate groups. Bonds are both ionic and covalent in nature. As far as ionic bonding is concerned, bond strength increases as the charge-to-radius ratio of the Lewis Base group increases. The dianionic nature of the phosphonate group boosts coordination relative to the carboxylate group, but this effect is partially offset by its' larger size (2). It is also worth noting that the carboxylate group is planar, while the phosphonate group is tetrahedral (2). Although the amine group bears no net charge, there is significant electron density directed towards the free p-orbital (3). The amine group is classified as a "hard" Lewis Base, although the mix of covalent versus ionic contribution to bonding is somewhat higher than observed with carboxylate and phosphonate groups (4).

When comparing one ligand to another, basicity is of paramount importance. As the ability of a Lewis Base to coordinate protons is increased, its' ability to coordinate metal ions also increases. Hence, with increasing pK_a, there is a corresponding increase in logK values for metal ion coordination. Of the three ligands shown in Figure 1, methylamine is by far the most basic, and yields logK values (e.g. for complexes with +II metal ions, not shown) that are higher than for the other two ligands. It is important to keep in mind, however, that metal ions must compete with protons for available ligands. As solution pH is decreased, a point is reached where protons out-compete metal ions for coordinating available ligands. Similarly, ligands must compete with hydroxide ions (OH^-) at high pH for available metal ions. As solution pH is increased, a point is reached where either hydroxo complexes in solution (e.g. $Fe^{II}OH^+$) or the precipitation of (hydr)oxide solids (e.g. $Fe^{II}(OH)_2(s)$) dominate metal ion speciation.

Replacing a carboxylate group with a phosphonate group has a number of important consequences. Near neutral pH, the phosphonate dianion (RPO_3^{2-}) is more basic than the carboxylate group, and, all other factors being equal, better able to coordinate metal ions. The methylenephosphonate dianion, for example, exhibits a pK_a of 7.82, while the corresponding pK_a for acetate monoanion is 4.67 (Figure 1).

Even under acidic conditions, phosphonates are considerably more effective chelating agents than the corresponding carboxylates (5). For example, while methylenephosphonic acid is in a monoanionic form above pH 2.3, acetic acid is in a protonated, unavailable form until a pH of 4.6 is reached

62

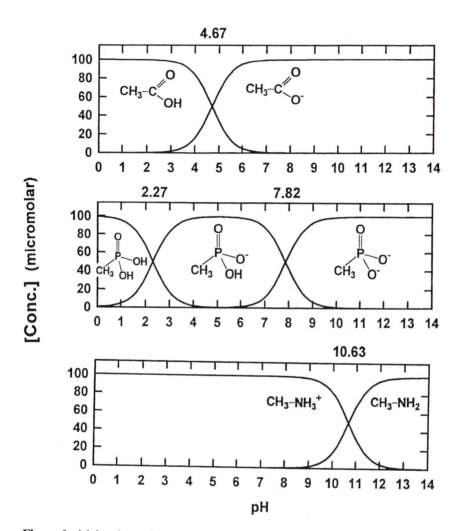

Figure 1. <u>Molecules with one Lewis Base Group</u>: *Speciation of 100 μM acetic acid, methylenephosphonic acid, and methylamine as a function of pH in 10 mM ionic strength medium.*

(see Figure 1). Greater electron density resides on the oxygen atom in P=O than in C=O (6). Thus, phosphonate diesters $(RP(O)(OR)_2)$ coordinate metal ions more strongly than carboxylate esters $(RC(O)OR)$. Coordination to the oxonate oxygen may also be important in species such as $RP(O)(OH)O^-$ and $RP(O)(OH)_2$.

Two (or more) Lewis Base groups can simultaneously coordinate a central Lewis Acid, provided that the resulting ringed structure is not unduly strained. Protons assume this arrangement via relatively weak hydrogen bonds. Metal ions much more readily exhibit coordination numbers of two or more. The corresponding structure, termed a "chelate", is often an effective means of ensuring that metal ions successfully out-compete protons for available Lewis Base Groups.

In Figure 2, four new ligands are assembled using amine, phosphonate, and carboxylate Lewis Bases and methylene ($-CH_2-$) linkages. For the most part, the methylene linkages electronically isolate each Lewis Base, minimizing inductive and resonance effects on basicity. (A modest electronic effect causes the pK_a for the amine group in glycine and iminodiacetic acid (Figure 2) to be one pH unit lower than the pK_a for methylamine.) These structures enable formation of five-membered hydrogen-bonded and chelate rings, as illustrated in Scheme 1. The dramatic lowering of pK_a values for phosphonate dianions and carboxylate monoanions in all four compounds can be attributed to hydrogen-bonding.

IDA (iminodiacetic acid) and glyphosate (N-(phosphonomethyl)glycine), used for the calculations shown in Figures 3-6 are high-volume synthetic chelating agents. (Equilibrium constants were derived from (7)). Despite having three Lewis Bases suitably placed for chelate ring formation, 1.0×10^{-4} M IDA fails to capture Fe^{II} (1.0×10^{-5} M) below pH 6.0, as shown in Figure 3. Metal coordination is driven by bond formation with the two free carboxylate groups. The amine group is protonated below pH 9.6, and requires metal ion-induced deprotonation in order for $Fe^{II}L^\circ$ to form.

Glyphosate captures Fe^{II} at slightly lower pHs than observed with IDA, a result attributed to the monoprotonated species $Fe^{II}HL^\circ$. The placement of protons within free (8) and metal ion-complexed phosphonate species is currently being investigated in a number of laboratories, frequently with NMR (9-14). The stoichiometry $Fe^{II}HL^\circ$ can either be explained by (i) Fe^{II} coordination via carboxylate and phosphonate groups, without participation by $R_2NH_2^+$ or (ii) a proton shift to one of two anionic oxygen atoms of the phosphonate group, allowing both the amine and the phosphonate to participate in Fe^{II} coordination (as evoked by Sawada et al. (10) for the coordination of Fe^{II} by NTMP).

Figure 3 illustrates one additional distinction between the dicarboxylate and the carboxylate/phosphonate ligands. With IDA, an increase in pH causes

64

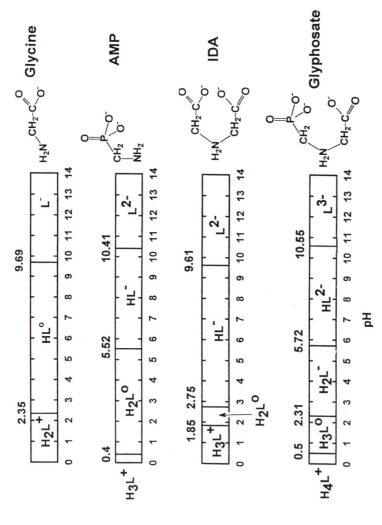

Figure 2. *Molecules with combinations of Lewis Base Groups: pK$_a$s of glycine, aminomethylphosphonic acid (AMP), iminodiacetic acid (IDA), and N-(phosphonomethyl)glycine (glyphosate) in 10 mM ionic strength medium.*

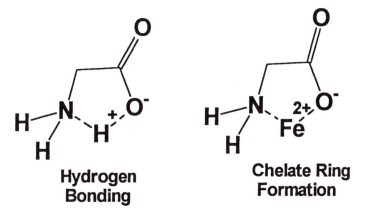

Scheme 1: *Hydrogen bonding and chelating ring formation*

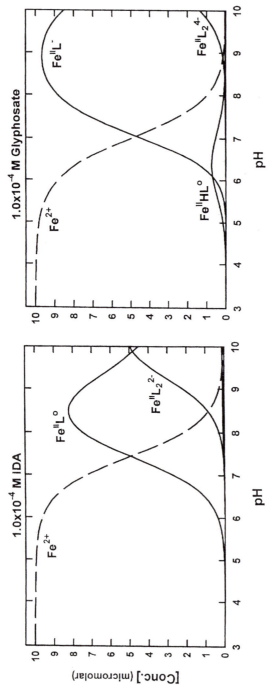

Figure 3. *Speciation of 1.0x10⁻⁵ M Feᴵᴵ in the presence of 1.0x10⁻⁴ M iminodiacetic acid and 1.0x10⁻⁴ M glyphosate (10 mM NaNO₃ constant ionic strength medium).*

a shift away from the 1:1 complex ($Fe^{II}L^{\circ}$) towards the 1:2 complex ($Fe^{II}L_2{}^{2-}$). With glyphosate, the 1:2 complex is a much less important species. This is a common observation with phosphonate chelating agents (2), and it arises from unfavorable steric and coulombic interactions between two large, negatively-charged functional groups. This phenomenon can also be seen in speciation calculations where the pH is fixed and the total ligand concentration increased (Figure 4). The 1:1 complex with glyphosate ($Fe^{II}L^{-}$) forms at lower total ligand concentrations than the corresponding complex with IDA, owing to the higher basicity of the phosphonate Lewis Base group, and persists to much higher total ligand concentrations, owing to the relative instability of the 1:2 complex, $Fe^{II}L_2{}^{4-}$.

Calculations performed with Fe^{III} are illustrated in Figure 5. The +III charge and smaller radius of this metal ion yield substantially greater logK values for the formation of hydroxo species and (hydr)oxide solids. $Fe(OH)_3$(amorphous), used in the calculations as the solubility-limiting phase, forms at pH 3.9 in the presence of 1.0×10^{-4} M IDA, and at pH 5.1 in the presence of 1.0×10^{-4} M glyphosate. The higher basicity and negative charge of the phosphonate-containing ligand clearly improves its performance as a chelating agent. It is interesting to note that the ternary complex with hydroxide ion, $Fe^{III}(OH)L^z$, is more important in the presence of IDA than in the presence of glyphosate. We can speculate that more hydroxide-ligand coulombic repulsion occurs in the glyphosate complex than in the IDA complex.

With IDA and glyphosate, logK values are available for the complexation of both Fe^{II} and Fe^{III}. Based upon this information, we can explore the effects of chelating agent concentration on the reduction potential for the Fe^{3+}(aq) + e^{-} = Fe^{2+}(aq) half-reaction. Suppose a system is comprised of a fixed total amount of reduced and oxidized forms of a metallic element. In our case, 1.0×10^{-5} M Fe^{II} and 1.0×10^{-5} M Fe^{III} have been added to solution. Each oxidation state will be distributed over a variety of species with different numbers of coordinated ligands and different protonation levels. With Fe^{II} and glyphosate, for example, $Fe^{II}{}_T$ (total dissolved Fe^{II}) will be comprised of Fe^{2+}(aq), $Fe^{II}HL^{\circ}$, $Fe^{II}L^{-}$, and $Fe^{II}L_2{}^{4-}$, plus other, less abundant species. $Fe^{III}{}_T$ (total dissolved Fe^{III}) will be comprised of $Fe^{III}L^{\circ}$ and $Fe^{III}(OH)L^{-}$, plus other less abundant species; precipitated solids (e.g. $Fe(OH)_3$(am.)) must, of course, be appropriately substracted from the mass balance equation. Once the free metal ion concentration is calculated for each oxidation state, the reduction potential is found using the Nernst Equation:

$$E = E^{\circ} + \frac{RT}{F} \ln\left[\frac{\gamma_{3+}[Fe^{3+}(aq)]}{\gamma_{2+}[Fe^{2+}(aq)]} \right]$$

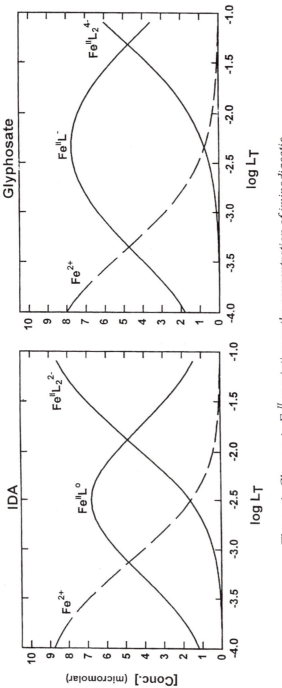

Figure 4. *Changes in Fe^{II} speciation as the concentration of iminodiacetic acid and glyphosate increases from 1.0×10^{-4} M to 0.10 M. (pH 7.0, 0.5 M NaNO$_3$ constant ionic strength medium).*

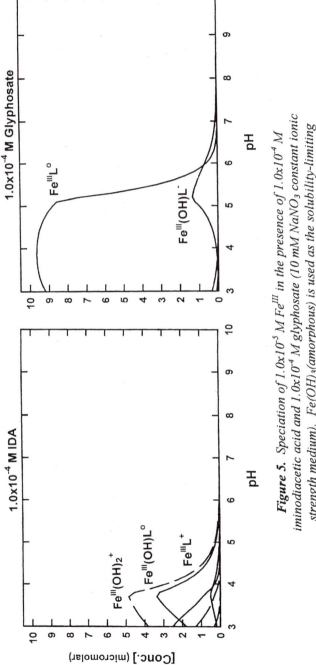

Figure 5. Speciation of 1.0×10^{-5} M Fe^{III} in the presence of 1.0×10^{-4} M iminodiacetic acid and 1.0×10^{-4} M glyphosate (10 mM $NaNO_3$ constant ionic strength medium). $Fe(OH)_3$(amorphous) is used as the solubility-limiting phase.

(In this case, $E° = +0.770$ volts.) As shown in Figure 6, increasing amounts of either IDA or glyphosate cause a <u>decrease</u> in reduction potential under acidic pH conditions, and an <u>increase</u> in reduction potential under alkaline conditions. Figures 3 and 5, along with a consideration of Fe(II) and Fe(III) solubility in the presence and absence of chelating agent, readily explain this surprising finding. Under strongly acidic conditions, $Fe^{II}(OH)_2(s)$ does not form, and addition of sufficient amounts of chelating agent brings about the complete dissolution of $Fe^{III}(OH)_3(am)$. Fe^{II}_T and Fe^{III}_T (total dissolved concentrations of Fe^{II} and Fe^{III}) are therefore constant. Complex formation constants (logKs) for oxygen-donor and most nitrogen-donor ligands are higher for Fe^{III} than for Fe^{II}. Hence, as more chelating agent is added, $[Fe^{3+}(aq)]$ decreases more than $[Fe^{2+}(aq)]$, causing the reduction potential to decrease.

Under alkaline conditions, $Fe^{II}(OH)_2(s)$ requires pHs greater than 9.6 to form, while $Fe^{III}(OH)_3(am)$ forms over a much wider pH range. As long as any $Fe^{III}(OH)_3(am)$ is present in the system, $[Fe^{3+}(aq)]$ at constant pH is fixed by the dissolution reaction $Fe^{III}(OH)_3(am) + 3H^+ = Fe^{3+}(aq) + 3H_2O$. Under these conditions, increasing total ligand concentrations can still depress $[Fe^{2+}(aq)]$, but leave $[Fe^{3+}(aq)]$ unchanged. The net result is an increase in the reduction potential.

NTA (nitrilotriacetic acid), which contains one more carboxylate-containing arm than IDA, is a much more effective chelating agent (Figure 7). Under the conditions employed (10 μM Fe^{III} and 100 μM chelating agent), IDA prevents $Fe(OH)_3(am)$ formation up to a pH of 3.7, while NTA prevents formation up to pH 7.0. With both IDA and NTA, ternary complexes with hydroxide ions (OH⁻) are important.

NTMP (nitrilotri(methylenephosphonic acid)) is the phosphonate analog to NTA. As shown in the right panel of Figure 7, NTMP performs substantially better than IDA and glyphosate in solubilizing Fe^{III}, but is slightly less effective than NTA under neutral and alkaline conditions. The advantages of a phosphonate group *vis a vis* a carboxylate group become less important as the denticity of the chelating agent increases and as the number of phosphonate groups within the molecule increases. There are several reasons for this (2). Unfavorable electrostatic interactions between arms of a multidentate ligand are more important for dianionic functional groups than for monoanionic functional groups. With multidentate ligands, the greater steric requirements of the tetrahedral phosphonate group relative to the planar carboxylate group becomes important. The driving force for protonation becomes quite large, and each monoprotonated phosphonate group is less basic (and hence capable of forming bonds to metal ions) than a deprotonated carboxylate group.

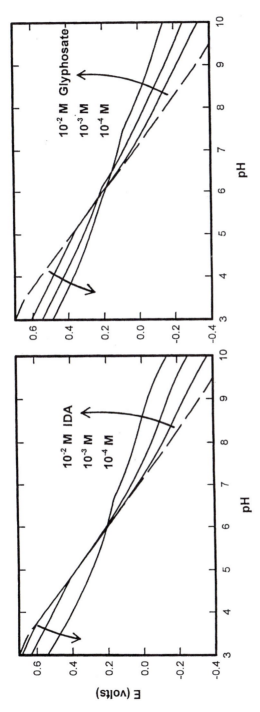

Figure 6. *Effect of increasing iminodiacetic acid and glyphosate concentrations on the reduction potential for the $Fe^{3+}(aq) + e^- \rightarrow Fe^{2+}(aq)$ half-reaction. Reaction conditions: 1.0×10^{-5} M Fe^{II}, 1.0×10^{-5} M Fe^{III}, 10 mM $NaNO_3$ constant ionic strength medium. $Fe(OH)_3$(amorphous) serves as the solubility-limiting phase for Fe^{III}. The dashed line corresponds to the reduction potential in the absence of added ligand.*

72

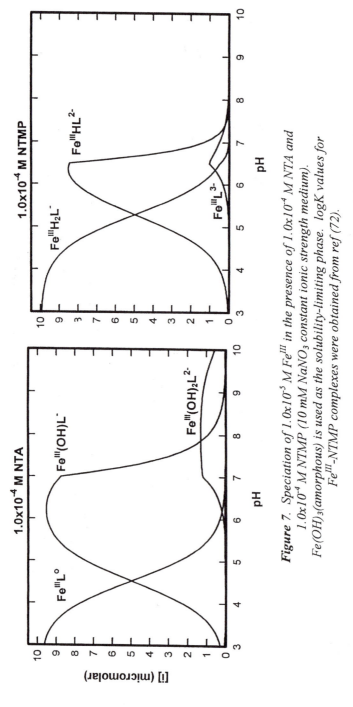

Figure 7. Speciation of 1.0×10^{-5} M Fe^{III} in the presence of 1.0×10^{-4} M NTA and 1.0×10^{-4} M NTMP (10 mM $NaNO_3$ constant ionic strength medium). $Fe(OH)_3$(amorphous) is used as the solubility-limiting phase. logK values for Fe^{III}-NTMP complexes were obtained from ref (72).

Rates of Metal Ion and Ligand Exchange

Surface waters, soils, and sediments are often divided into "parcels" that are modeled as open systems (16). Inputs and outputs of water, water-borne solutes, and water-borne particles change over time. As a consequence, chemical conditions within the parcel change. If metal ion exchange and ligand exchange reactions take place quickly, equilibrium descriptions of chemical speciation are appropriate. If these reactions take place slowly, however, then an appraisal of speciation requires knowledge of reaction kinetics (17).

In dissociative exchange, a coordinated ligand leaves the inner coordination shell before the replacement ligand enters. In associative exchange, the incoming ligand enters before the original ligand exists, yielding an intermediate with a higher coordination number. Interchange mechanisms (I_d and I_a) fill out a continuum between the two mechanisms (18).

The incorporation of isotopically-labelled water into the inner coordination sphere of aquated metal ions is one of the simplest exchange reactions observed. Rate constants for this reaction are used by chemists to make generalizations about exchange rates for different metal ions. A few illustrative rate constants for this reaction are presented below (from (19) and (20)):

$$Me(H_2O)_n^{z+} + H_2O^* \rightarrow Me(H_2O)_{n-1}(H_2O^*)^{z+} + H_2O$$

	k (seconds^{-1})		k (seconds^{-1})
$Mn^{II}(H_2O)_n^{2+}$	2.1×10^7	$Al^{III}(H_2O)_n^{3+}$	1.3×10^0
$Fe^{II}(H_2O)_n^{2+}$	4.4×10^6	$Cr^{III}(H_2O)_n^{3+}$	2.4×10^{-6}
$Co^{II}(H_2O)_n^{2+}$	3.2×10^6	$Fe^{III}(H_2O)_n^{3+}$	1.6×10^2
$Ni^{II}(H_2O)_n^{2+}$	3.2×10^4		
$Cu^{II}(H_2O)_n^{2+}$	4.4×10^9		

Differences in speciation have an enormous effect on rates of exchange. $Cr^{III}(OH)(H_2O)_{n-1}^{2+}$ exhibits water exchange rates that are 75-times higher than for $Cr^{III}(H_2O)_n^{3+}$ (18).

Under any realistic set of chemical conditions, a chelating agent of interest is likely to exist in two or more protonation levels, and is likely to be coordinated to common metal cations such as Ca^{2+} and Mg^{2+}. A metal ion of interest is likely to be coordinated by major anions (e.g. Cl^-, SO_4^{2-}, and CO_3^{2-}) or naturally-occurring organic ligands. As reflected in rate constants for water exchange (above), Cu^{II} undergoes ligand exchange far more rapidly than the other +II metal ions listed above. Despite this fact, the achievement of equilibrium between Cu^{II} and strong ligands like EDTA under seawater conditions can require timescales of minutes to hours (21-23). Similar results have been obtained with Fe^{III}. Fe^{III} speciation during infiltration into groundwater appears to be under kinetic control (24). Twenty days are sometimes required for equilibrium between Fe^{III} and strong ligands like EDTA to be achieved in streamwaters (25). $Ni^{II}EDTA^-$, formed during wastewater treatment, has been shown to persist for significant periods of time following release into estuarine waters (26).

Cr^{III} exhibits rates of water exchange that are orders-of-magnitude lower than for the other metal ions listed above. It is tempting to assume that all Cr^{III} complexes are exchange-inert, and hence immutable. Several points must, however, be kept in mind. If rates of dissociative exchange are low, exchange via associative mechanisms might become important. (When associative mechanisms are predominant, exchange rates become a function of the identity of the incoming ligand.) With multidentate ligands, the linkage between two Lewis Base groups may introduce steric strain that causes exchange rates to increase (27). Other solute species can catalyze exchange reactions. Bicarbonate ion, for example, facilitates entry of EDTA into the inner coordination sphere of Cr^{III}, and hence catalyzes $Cr^{III}EDTA^-$ formation (28).

A preliminary experiment has been conducted which compares the effects of carboxylate and phosphonate functional groups on rates of Cr^{III} ligand exchange. Laboratory practices outlined in previous publications (29-30) were followed. All chelating agents were purchased from Aldrich Chemical Co. and Fluka Chemical Co. and used without additional purification. The potassium salt of $u\text{-}fac\text{-}Cr^{III}(IDA)_2^-$ was synthesized according to procedures outlined by Weyh and Hamm (31). This is one of three possible isomers of the 1:2 complex with iminodiacetic acid shown in Scheme 2. As described in a previous publication, the $u\text{-}fac$ and $s\text{-}fac$ isomers can be readily distinguished from one another using capillary electrophoresis (29). The $mer\text{-}$ isomer is probably not important under the conditions employed in our experiments.

Track-etched polycarbonate filters and hydrophilic cellulose membrane filters (both 0.2 micron pore diameter, Whatman Filter Co.) were used to recover supernatant solutions. Total chelating agent and metal ion-chelating agent complex concentrations were determined using a Quanta 4000E capillary electrophoresis instrument (Waters Corp.) and bare fused-silica capillaries (75 microns wide, 60 cm long, Polymicro Technol.). The capillary electrolyte consisted of 25 mM phosphate buffer (pH 7.0) and 0.5 mM tetradecyl trimethylammonium bromide (TTAB) electroosmotic flow modifier (see (29)).

In the absence of other chelating agents, there is a gradual decline in the concentration of $u\text{-}fac\text{-}Cr^{III}(IDA)_2^-$, attributed to interconversion to the $s\text{-}fac$ form of the complex. Our exchange experiments employed 100 µM $u\text{-}fac\text{-}Cr^{III}(IDA)_2^-$, 5.0 mM acetate buffer, and 1.0 mM concentrations of the

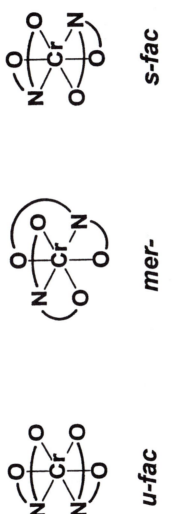

Scheme 2: *s-fac, mer, and u-fac* Co^{III}(IDA)$_2^-$

following chelating agents: NTA, EDTA, NTMP, BPMG (bi(phosphono-methyl)glycine), PMIDA (phosphonomethyliminodiacetic acid), and EDTMP (ethylenediaminetetra(methylenephosphonic acid)). NTA (with no phosphonate groups), PMIDA (with one phosphonate group), and BPMG (with two phosphonate groups) share the same three-arm structure. As shown in Figure 8, the two chelating agents possessing phosphonate groups capture Cr^{III} via ligand exchange more rapidly than NTA. Similarly, EDTMP (with four phosphonate groups) captures Cr^{III} considerably more rapidly than EDTA (with no phosphonate groups). We can conclude that phosphonate groups facilitate the kinetics of Cr^{III} capture at this pH. A second set of experiments has been performed in solutions buffered to pH 7.2 using 5.0 mM MOPS (morpholinepropanesulfonic acid). During 21 days of reaction, no decline in u-fac-$Cr^{III}(IDA)_2^-$ concentration was observed, regardless of whether strong chelating agents were added.

These experiments demonstrate that rates of exchange reactions involving Cr^{III} and phosphonate chelating agents are sensitive to pH and medium composition, and slow enough to control speciation in a number of important aqueous systems. A tempting response is to develop computer-based models for predicting toxic metal ion speciation under kinetically-controlled conditions. It should be kept in mind, however, that the data requirements of kinetic models (e.g. forward and reverse rate constants, rate constants for catalytic processes, pH and major ion concentrations as a function of time) are much greater than those of equilibrium models. Quantitative models serve a didactic purpose, and encourage us to think realistically about factors affecting speciation. On a practical level, however, it is time for attention to be focused on analytical methods for directly determining speciation in environmental samples. Direct speciation measurements provide a strong basis for making environmental management decisions, and provide a way of testing computer-based speciation models. An integration of speciation measurements with modeling efforts provides the best prospect for predicting speciation.

Adsorption and Precipitation

(Hydr)oxide and aluminosilicate products of rock weathering comprise much of the available surface area in soils and sediments. Such inorganic solids can be either amorphous or (micro)crystalline. In the presence of vapor or liquid water, surfaces are hydrated. In adsorption experiments, it is common to compare adsorption onto two inorganic solids that are chemically distinct yet possesses the same physical characteristics (e.g., surface area and surface charge). Similarly, the adsorption behavior of two chemically distinct solutes

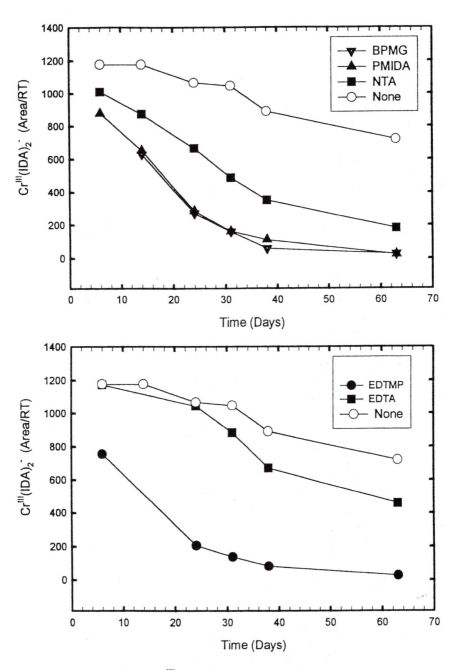

Figure 8. *Loss of* u-fac-$Cr^{III}(IDA)_2^-$ *as a function of time due to conversion to the s-fac isomer (open symbols) and due to capture by the strong chelating agents BPMG, PMIDA, NTA, EDTMP, and EDTA (filled symbols). Reaction conditions: 100 μM* u-fac-$Cr^{III}(IDA)_2^-$, *1.0 mM chelating agent, 5.0 mM acetate buffer (4.4 < pH < 5.6).*

with the same physical characteristics (e.g., molecular charge) can be compared. If these substitutions can be made without changing the extent of adsorption, the process is termed "non-specific". Long-range electrostatic interactions are primarily responsible for non-specific adsorption. The equilibrium distribution of ions near a charged surface is determined by the balance of electrostatic interactions that accumulate ions and osmotic forces that disperse them. Once the surface charge density, ionic medium composition, and solvent dielectric constant are known, this equilibrium distribution is readily calculated (32). The extent of purely non-specific adsorption can be calculated by integrating ion concentrations across the "electrical double layer" within which electrostatic accumulation of solutes takes place (33).

When the chemical identity of adsorbent and adsorbate significantly affect the extent of adsorption, the process is termed "specific". Some degree of chemical bonding is implied, but the nature of this bonding is often difficult to discern. "Inner-sphere" bonding, in which adsorbate molecules directly coordinate surface-bound atoms, and "outer-sphere" bonding, in which adsorbate and adsorbent are separated by one or more layers of water molecules, depend upon the identity of the participating species and hence constitute specific adsorption. Hydrogen-bonding and difficult-to-characterize hydration forces also contribute to inner-sphere adsorption. Surface protonation level can be measured with great accuracy, and has been used in a study of phenylphosphonic acid adsorption onto aluminum (hydr)oxide surfaces (34-35). Sophisticated IR and Raman techniques allow light absorption from surface complexes to be distinguished from light absorption by the electrolyte medium and underlying solid. As far as phosphonates are concerned, these techniques have been used to study the adsorption of phenylphosphonic acid (36) and methylphosphonic acid (37).

Specific adsorption of a charged species onto charged surfaces necessarily involves non-specific effects as well. Computer-based models which quantify all the factors contributing to adsorption are now widely employed. Early models that treated all surface sites as chemically equivalent and treated all surface charges as integer numbers (32) have been modified to allow for fractional charge and surface site heterogeneity (e.g. (38-39)). The recently developed Charge Distribution Model (40) is particularly useful for addressing the adsorption behavior of multidentate chelating agents. It recognizes that a portion of the Lewis Base groups of a molecule may engage in complex formation with surface-bound metal atoms, while others may "dangle" at some distance away from the surface, where electrostatic forces are less pronounced.

A tridentate chelating agent and one or two surface-bound Fe^{III} atoms are used in Figure 9 to illustrate modes of inner-sphere surface complex formation. Each square represents a Lewis Base within the structure of the solid. Each

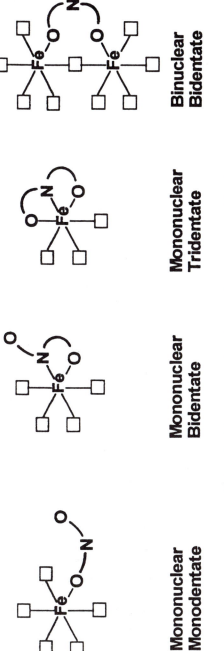

Figure 9. Modes of inner-sphere surface complex formation.

Binuclear Bidentate

Mononuclear Tridentate

Mononuclear Bidentate

Mononuclear Monodentate

"N" represents a nitrogen-donor atom (e.g., from an amine group) while each "O" represents an oxygen-donor atom (e.g., from either a carboxylate or phosphonate group). Depending upon the accessibility of one, two, or three coordinative positions around a central Fe^{III} ("mononuclear") atom, monodentate, bidentate, and tridentate surface complexes are possible. If the "arms" of the chelating agent connecting the Lewis Base groups are long enough, bridging between two surface-bound Fe^{III} atoms is possible (e.g. the binuclear bidentate complex). As mentioned above, Lewis Base groups not engaged in bond formation "dangle" at some distance away from the surface.

Details regarding the adsorption of monophosphonates (34-37) and polyphosphonates (41-42) are provided in the references just cited. Here, we focus upon the adsorption properties of polyphosphonates, which are particularly intriguing.

A comprehensive study of the adsorption of eight phosphonate-containing compounds onto FeOOH(goethite) has recently been published (41). At pH 10, the extent of adsorption of 40 μM NTMP onto 0.42 g/L FeOOH(goethite) is approximately 30 percent. A thousand-fold increase in the background electrolyte concentration (from 1.0 millimolar to 1.0 molar $NaNO_3$) has no effect on the extent of adsorption. This lack of an ionic strength effect is surprising, given that FeOOH(goethite) bears a net negative charge at this pH, and that the predominant solution phase species for NTMP is HL^{5-}.

Long-range electrostatic interactions which repulse like-charged ions away from the surface are much stronger at low ionic strength than at high ionic strength (33). Three different phenomena may be responsible for the lack of an ionic strength effect. (i) Each Fe^{III}-phosphonate bond lowers the charge of the adsorbed complex by one unit. If all three phosphonate groups of NTMP are bonded in this way, the surface complex would have a stoichiometry and charge of $(>Fe^{III})_3 HL^{2-}$. (ii) Phosphonate-containing "arms" that are not bonded to the surface may position themselves away from the surface, where electrostatic interactions with the charged surface are diminished. (iii) Phosphonate-containing "arms" that are not bonded to the surface may protonate or form ion pairs with electrolyte cations (e.g. Na^+) to a greater extent than phosphonate groups in bulk solution.

When added phosphonate concentrations are increased at fixed surface loading and pH, the extent of adsorption eventually levels out. Maximum extents of adsorption provide an estimate of the "footprint" of each molecule on the surface. Footprint size may reflect the numbers of surface-bound Fe^{III} atoms engaged in bond formation with the adsorbed molecule, or may indicate crowding by neighboring molecules. Electrostatic repulsion between like-charged adsorbate molecules on the surface may also be important.

Maximum extents of adsorption at pH 7.2 for eight phosphonate-containing compounds are shown in Figure 10. Going from left to right, the first three compounds, HMP (hydroxymethylphosphonic acid), MP (methylphosphonic acid), and HEDP (1-hydroxyethane-1,1-diphosphonic acid) do not contain amine groups. Maximum surface coverage by the diphosphonate HEDP is half that of the monophosphonate HMP and 40 % less than the monophosphonate MP. The next three compounds possess a primary, a secondary, and a tertiary amine group, and one, two, and three phosphonate groups, respectively. The effect of amine groups on footprint size does not result in clear trends; maximum surface coverage by AMP is 18 % less than by MP, but maxium surface coverage by IDMP is 24 % greater than by HEDP. (Hydrogen-bonding, chelating ring formation, and differences in adsorbed species charge may all be contributing factors.)

Maximum surface coverage by the triphosphonate NTMP is 35 % lower than the coverage than by the diphosphonate IDMP. EDTMP, with four phosphonate groups, and DTPMP (diethylenetrinitrilopentakis-(methylenephosphonic acid), with five phosphonate groups, adsorb to approximately the same extent as NTMP. The fact that the fourth and fifth phosphonate groups do not increase the molecular footprint indicates that the additional phosphonate groups are not associated with the surface. Steric considerations may prevent them from coordinating surface-bound Fe^{III} atoms, and electrostatic repulsion may force them away from the anchoring phosphonate groups (41).

Engineered systems and environmental media contain other solutes that can influence the extent of phosphonate adsorption. Other ligands (e.g. carbonate, phosphate, natural organic matter) can compete with phosphonate molecules for available surface sites, thereby lowering the extent of phosphonate adsorption. Metal ions can either raise or lower the extent of phosphonate adsorption, depending upon the balance of a number of conflicting phenomena. As discussed in previous sections, the formation of dissolved metal ion-phosphonate complexes occurs to the greatest extent in the mid-pH range. By providing another "compartment" for phosphonates in solution, complex formation in solution can lower adsorption. Metal ions that adsorb as separate entities cause the surface charge to shift in the positive direction. At high pH, this shift in surface charge lessens the extent of long-range electrostatic repulsion between negatively-charged surfaces and phosphonate polyanions, thereby raising the extent of adsorption. Multidentate ligands can bridge between surface-bound metal atoms and dissolved metal ions, forming "ligand-like" ternary complexes. Alternatively, metal ions can bridge between surface-bound oxo or hydroxo groups and dissolved ligands, forming "metal ion-like" ternary complexes. Cooperative adsorption of this kind, illustrated in Figure 11, is the subject of a comprehensive review (43).

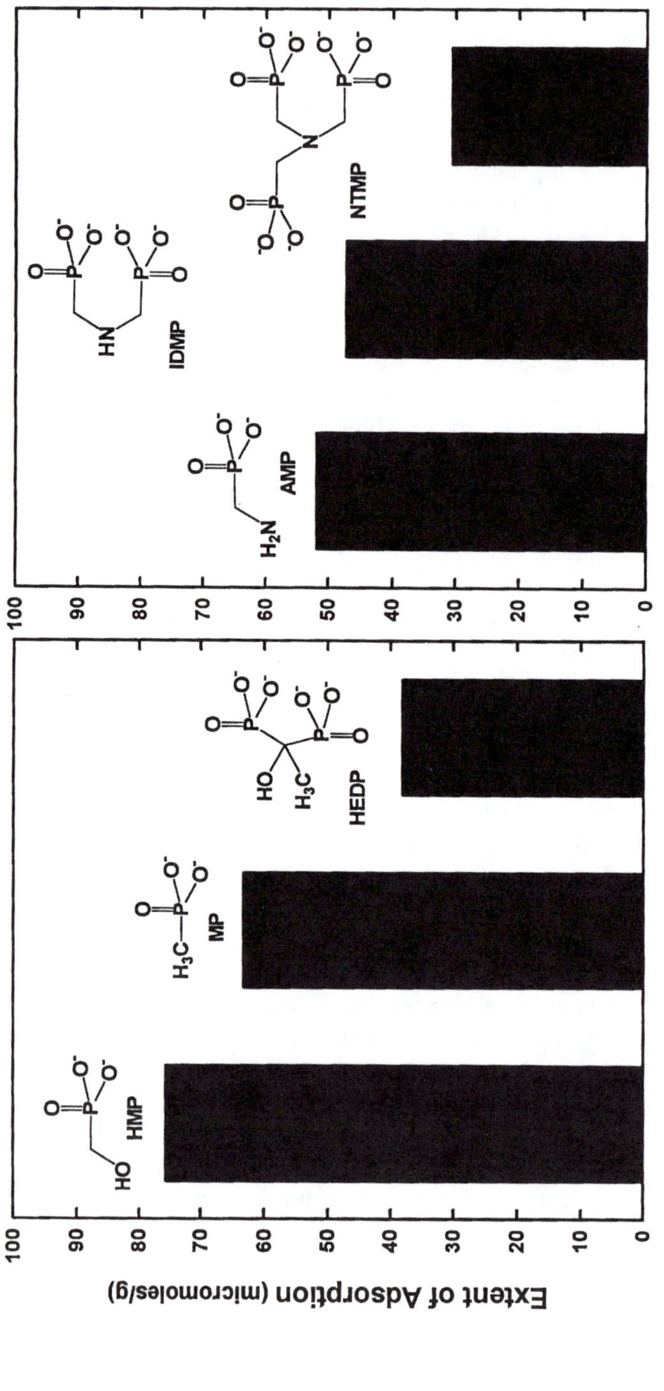
Extent of Adsorption (micromoles/g)
0
10
20
30
40
50
60
70
80
90
100
HMP
MP
HEDP
AMP
IDMP
NTMP

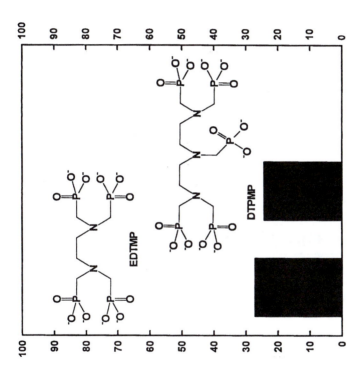

Figure 10. *Maximum extents of adsorption for eight phosphonate-containing compounds onto 0.42 g/L FeOOH(goethite). Reaction conditions: 1 mM MOPS buffer (pH 7.2), 10.0 mM NaNO₃ (data from ref. (41).*

84

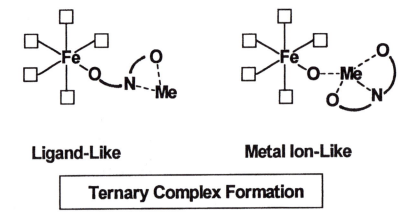

Ligand-Like   **Metal Ion-Like**

Ternary Complex Formation

Figure 11. *Ligand-like and metal ion-like ternary complex formation.*

We have completed an adsorption study in systems containing FeOOH(goethite), representative phosphonate ligands, and metal ions likely to be encountered in wastewater effluents (42). Transition metal ions are typically encountered at concentrations comparable to those of phosphonate ligands. 10 μM Fe^{III} exerts a very slight diminishing effect on the adsorption of 10 μM EDTMP and DTPMP, but no discernable effect on the adsorption of AMP, IDMP, HEDP, and NTMP. Cu^{II} and Zn^{II} (also at 10 μM) exerted no discernable effect. Ca^{II} is a major ion in most waters; concentrations from 50 μM to 1.0 mM significantly increased NTMP adsorption (42).

Examining how phosphonates affect the adsorption of metal ions is, of course, important too. At pH 4.0, 10 μM NTMP, EDTMP, and DTPMP substantially raise the extent of 10 μM Cu^{II} adsorption. "Ligand-like" ternary complex formation and the lowering of surface charge are believed responsible (42). At pH values greater than 7, 10 μM NTMP, EDTMP, and DTPMP substantially lower the extent of 10 μM Cu^{II} adsorption. Here, the formation of Cu^{II}-phosphonate complexes in solution draws Cu^{II} ion away from the FeOOH(goethite) surface. Weaker chelating agents such as AMP, HEDP, and IDMP exhibited only slight effects on Cu^{II} adsorption (42).

Poor environmental practices in past decades have left a legacy of toxic metal-contaminated sediments in many rivers, lakes, and estuaries. It would be desirable to determine whether inputs of synthetic chelating agents can solubilize toxic metal ions from such sediments. Working with sediments downstream from a copper electrorefinery plant, Bordas and Bourg (44) determined that greater than 100 μM concentrations of EDTA and NTMP were able to solubilize 10 % or more of particle-bound Cd, Pb, and Cu within a 30-hour time period. Wastewater effluents typically contain chelating agent concentrations that are tens- to hundreds-of-times lower (45).

Chemical Transformations

Preceding sections have pointed out ways in which the identity, number, and arrangement of Lewis Base functional groups within chelating agent molecules affect the complex formation and adsorption reactions of phosphonates. Here, we provide a few illustrative examples of how these properties affect the chemical reactivity of phosphonates.

Fe^{III}-EDTA and other Fe^{III} complexes with (amino)carboxylate chelating agents readily photolyze, whereas free (amino)carboxylates and their complexes with common +II metal ions do not (46). Indeed, the kinetics of Fe^{III} capture and release control ambient concentrations of Fe^{III}-(amino)carboxylate complexes, and hence control degradation rates in surface waters (25).

(Amino)phosphonate chelating agents behave in the same way. Fe^{III}-EDTMP, for example, readily photolyzes, while free EDTMP does not (47). In the laboratory, photolysis of Fe^{III}-EDTMP yields orthophosphate ion and N-methyl HMP. Intermediates that chelate Fe^{III} (e.g. N-methyl IDMP) are themselves rapidly photolyzed (47). By analogy, NTMP and most of the phosphonate-containing chelating agents discussed in this chapter should also be subject to photolysis.

It is also important to identify chemical breakdown processes in environments where photolysis cannot take place, e.g. deep portions of lakes and rivers where sunlight does not penetrate, as well as within soils and aquifers. In carefully prepared transition metal ion-free solutions, chemical breakdown of a wide range of (amino)phosphonates (e.g. by hydrolysis) was found to be very slow, requiring timescales of 100 days or more (48). Using hydrogen peroxide concentrations in the 0.1 mM range and comparable concentrations of Fe^{II}, appreciable degradation was observed (48). The exceedingly low hydrogen peroxide concentrations found in actual groundwater samples (e.g. 20 nM reported by (49)) are probably not enough to induce degradation.

The complete degradation of NTMP into IDMP, HMP, and AMP within approximately 48 hours in non-illuminated surface and ground waters has been reported (50). Comparable degradation was observed in a laboratory test medium that contained a number of inorganic anions (NO_3^-, Cl^-, SO_4^{2-}, $B(OH)_4^-$, MoO_4^{2-}, and EDTA) and metal ions (Na^I, Mg^{II}, Ca^{II}, Mn^{II}, Co^{II}, Cu^{II}, Zn^{II}, and Fe^{III}).

NTMP undergoes degradation in non-illuminated Baltimore City tap water. A 70 % loss was observed over a 7-day period, corresponding to a half-life of approximately 4 days (51). A series of single metal ion experiments indicated that ambient concentrations of Mg^{II} and Ca^{II} and 10 μM concentrations of Cu^{II}, Zn^{II}, and Fe^{III} yielded no discernable degradation of NTMP. Rapid abiotic degradation was observed in O_2-saturated solutions containing 10 μM Mn^{II}; no degradation was observed in O_2-free solutions. It can be concluded that Mn^{II}-catalyzed autooxidation is responsible for NTMP degradation. The reaction scheme is presented in Figure 12. Interested readers should refer to Nowack and Stone (51) for details regarding reaction product identification, the effects of pH and medium composition on reaction rates, and comparisons with other phosphonate-containing ligands.

The mechanism of Mn^{II}-catalyzed NTMP autooxidation is unique in several respects. Although the oxidation of Mn^{2+}(aq) by O_2 is thermodynamically favorable at pH values greater than 4.7 (52), reaction

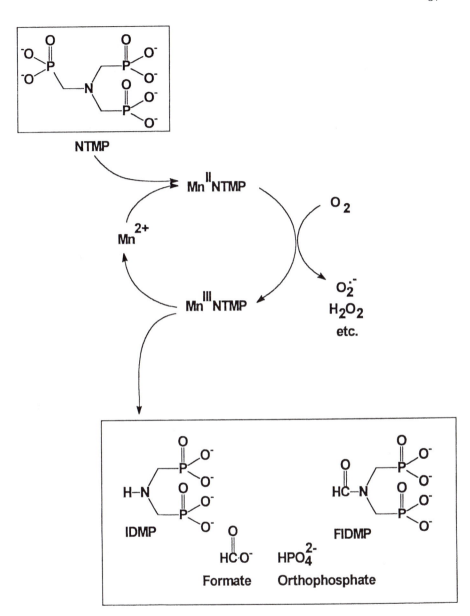

Figure 12. Mechanism of Mn^{II}-catalyzed NTMP autoxidation (adapted ref. (51)).

kinetics in the absence of added ligands is extremely slow below pH 8.5 (e.g. ref. (53)). The NTMP autoxidation reaction occurs at an appreciable rate across a wide pH range (4 < pH < 8.5), indicating that complexation by NTMP (i) broadens the pH range where Mn^{II} reaction with O_2 is thermodynamically favorable, and (ii) activates coordinated Mn^{II} towards reaction with O_2. As far as the first point is concerned, it is important to note that complexation by NTMP very likely lowers the reduction potential for the Mn^{III}/Mn^{II} half-reaction, since the logK for complexation of Mn^{III} is larger than the logK for complexation of Mn^{II}. As far as the second point is concerned, relatively little is known about how the identity of coordinated ligands and composition of the aqueous medium affect rates of Mn^{II} oxidation by O_2 (54). In one study, EDTA and pyrophosphate were found to greatly inhibit the reaction at pH 10 (55). Whether or not this same result would be obtained under neutral or acidic conditions was not investigated.

Mn^{III}-chelating agent complexes can be synthesized in a variety of ways, and the intramolecular oxidation of the coordinated (amino)carboxylate chelating agents has been extensively studied (e.g. ref. (56-59)). Thus, once $Mn^{III}NTMP$ is formed by the autoxidation reaction, NTMP breakdown is quite reasonable. We can speculate that phosphonate-containing ligands might exist that promote the reaction of Mn^{II} with O_2, but resist intramolecular oxidation by Mn^{III}. Soluble Mn^{III} complexes generated in this way would be strong oxidants in a thermodynamic sense, which might react quickly with other naturally-occurring or contaminant-derived chemical species.

DTPMP, EDTMP, and NTMP are all subject to Mn^{II}-catalyzed autoxidation, but HEDP and the breakdown products IDMP and FIDMP are not (51). Equilibrium speciation models indicate that the fraction of available Mn^{II} complexed by HEDP and IDMP is small, which may limit the rate of the forward reaction. Properties of the alpha-amino group may also be important; we are currently investigating whether phosphonates containing secondary amine versus tertiary amine groups exhibit different susceptibilities towards oxidation.

Of course, many more potential pathways for the chemical degradation of phosphonates need to be explored. The C-P bond in alkylphosphonates, for example, is weakened by electron-withdrawing substituents. Alpha-carbonyl phosphonates such as the antiviral pharmaceutical phosphonoformate are especially susceptible towards hydrolysis (60). Metal ion-catalyzed conversion of phosphonates into phosphonamides has been observed; nucleophilic attack of amine groups at the electrophilic phosphorus atom is believed responsible (61).

Conclusions

Synthetic organic compounds possessing a phosphonate group exist in multiple protonation levels, coordinate dissolved metal ions, and adsorb onto inorganic solid surfaces. During and subsequent to use, knowledge of the speciation of phosphonate-containing compounds is crucial for understanding chemical transformations and environmental effects.

Given the widespread use of a variety of phosphonate-containing chemicals, it is surprising how little is known about their thermodynamic properties (pK$_a$ and logK values), and about their reactions with the chemical constituents of natural waters. When a study has been performed on one chemical within a class, it is commonly assumed that the entire class with exhibit comparable behavior. This review indicates, however, that very small changes in structure (e.g., replacing one functional group with another) can markedly alter speciation, reaction pathways and reaction rates. As in most areas of environmental chemistry, phosphonate-containing synthetic organic compounds represent a rich new subject waiting to be explored.

Acknowledgements

U.S. Environmental Protection Agency, National Center of Environmental Research and Quality Assurance (Office of Exploratory Research) support for our work on the environmental chemistry of phosphonates under Grant R82-6376 is gratefully acknowledged.

Literature Cited

1. Williams, D. "Storing up trouble?", *Chem. In Brit.* **1998**, 48-50.
2. Kiss, T.; Lazar, I. "Structure and stability of metal complexes in solution", Chapter 9 in: "Aminophosphonic and Aminophosphinic Acids. Chemistry and Biological Activity", V.P. Kukhar, H.R. Hudson edits., Wiley, NY, **2000**.
3. Holm, R.H.; Kennepohl, P.; Solomon, E.I. "Structural and functional aspects of metal sites in biology", *Chem. Rev.* **1996**, *96*, 2239-2314.
4. Hancock, R.D.; Martell, A.E. (1989) "Ligand design for selective complextion of metal ions in aqueous solution", *Chem. Rev.* **1989**, *89*, 1875-1914.

5. Nash, K.L.; Horwitz, E.P. "Stability constants for europium(III) complexes with substituted methane diphosphonic acids in acid solutions", *Inorg. Chim. Acta* **1990**, *169*, 245-252.

6. Nash, K.L. "f-Element complexation by diphosphonate ligands", *J. Alloys Compd.* **1997**, *249*, 33-40.

7. Martell, A.E.; Smith, R.M.; Motekaitis, R. *NIST Critically Selected Stability Constants of Metal Complexes Database*, National Institute of Science and Technology, Gaithersburg, MD, **1998**.

8. Hagele, G.; Szakacs, Z.; Ollig, J.; Hermens, S.; Pfaff, C. "NMR-controlled titrations: characterizing aminophosphonates and related structures", *Heteroatom Chem.* **2000**, *11*, 562-582.

9. Sawada, K.; Araki, T.; Suzuki, T. "Complex formation of amino polyphosphonates. 1. potentiometric and nuclear magnetic resonance studies of nitrilotris(methylenephosphonato) complexes of alkaline earth metal ions", *Inorg. Chem.* **1987**, *26*, 1199-1204.

10. Gillard, R.D.; Newman, P.D.; Collins, J.D. "Speciation in aqueous solutions of di-ethylenetriamine-N,N,N',N'',N''-pentamethylenephosphonic acid and some metal complexes", *Polyhedron* **1989**, *8*, 2077-2086.

11. Sawada, K.; Kanda, T.; Naganuma, Y.; Suzuki, T. "Formation and protonation of aminopolyphosphonate complexes of alkaline-earth and divalent transition-metal ions in aqueous solution" *J. Chem. Soc. Dalton Trans.* **1993**, 2557-2562.

12. Sawada, K.; Miyagawa, T.; Sakaguchi, T.; Doi, K. "Structure and thermodynamic properties of aminopolyphosphonate complexes of the alkaline-earth metal ions", *J. Chem. Soc. Dalton Trans.* **1993**, 3777-3784.

13. Duan, W.; Oota, H.; Sawada, K. "Stability and structure of ethylenedinitrilopoly(methylphosphonate) complexes of the alkaline-earth metal ions in aqueous solution", J. Chem. Soc. Dalton **1999**, 3075-3080.

14. Sawada, K.; Duan, W.; Ono, M.; Satoh, K. "Stability and structure of nitrilo(acetate-methylphosphonate) complexes of the alkaline-earth and divalent transition metal ions in aqueous solution", *J. Chem. Soc. Dalton Trans.* **2000**, 919-924.

15. Sawada, K.; Araki, T.; Suzuki, T.; Doi, K. "Complex formation of amino polyphosphonates. 2. Stability and structure of nitrilotris(methylenephosphonato) complexes of the divalent transition-metal ions in aqueous solution", *Inorg. Chem.* **1989**, *28*, 2687-2688.

16. Imboden, D.M.; Lerman, A. "Chemical models of lakes", Chapter 11 In *Lakes: Chemistry, Geology, Physics*, A. Lerman, edit., Springer, NY, **1978**.

17. Stone, A.T.; Morgan, J.J. *Kinetics of chemical transformations in the environment*, Chapter 1 in: "Aquatic Chemical Kinetics", W. Stumm, edit., Wiley-Interscience, NY, **1990**.

18. Wilkins, R.G. *Kinetics and Mechanism of Reactions of Transition Metal Complexes*, 2nd edit., VCH, Weinheim, Germany, 1991.

19. Helm, L.; Merbach, A.E. (1999) "Water exchange on metal ions: experiments and simulations", *Coord. Chem. Rev.* **1999**, *187*, 151-181.

20. Nordin, J.P.; Sullivan, D.J.; Phillips, B.L.; Casey, W.H. "An ^{17}O-NMR study of the exchange of water on $AlOH(H_2O)_5^{2+}$(aq)", *Inorg. Chem.* **1998**, *37*, 4760-4763.

21. Hering, J.G.; Morel, F.M.M. "Kinetics of trace metal complexation: role of alkaline-earth metals", *Environ. Sci. Technol.* **1988**, *22*, 1469-1478.

22. Hering, J.G.; Morel, F.M.M. "Slow coordination reactions in seawater", *Geochim. Cosmochim. Acta* **1989**, *53*, 611-618.

23. Hering, J.G.; Morel, F.M.M. "Kinetics of trace metal complexation: ligand-exchange reactions", *Environ. Sci. Technol.* **1990**, *24*, 242-252.

24. Nowack, B.; Xue, H.; Sigg, L. "Influence of natural and anthropogenic ligands on metal transport during infiltration of river water to groundwater", *Environ. Sci. Technol.* **1997**, *31*, 866-872.

25. Xue, H.; Sigg, L. Kari, F.G. "Speciation of EDTA in natural waters: exchange kinetics of Fe-EDTA in river water", *Environ. Sci. Technol.* **1995**, *29*, 59-68.

26. Bedsworth, W.W.; Sedlak, D.L. "Sources and environmental fate of strongly complexed nickel in estuarine waters: the role of ethylenediamine-tetraacetate", *Environ. Sci. Technol.* **1999**, *33*, 926-931.

27. Heineke, D.; Franklin, S.J.; Raymond, K.N. "Coordination chemistry of glyphosate: structural and spectroscopic characterization of bis(glyphosate)metal(III) complexes", *Inorg. Chem.* **1994**, *33*, 2413-2421.

28. Agger, R.M.; Hedrick, C.E. "The effect of carbonate on the chromium(III)-EDTA reaction. An example of inorganic catalysis", *J. Chem. Educat.* **1966**, *43*, 541-542.

29. Buergisser, C.S.; Stone, A.T. "Determination of EDTA, NTA, and other amino carboxylic acids and their Co(II) and Co(III) complexes by capillary electrophoresis", *Environ. Sci. Technol.* **1997**, *31*, 2656-2664.

30. McArdell, C.S.; Stone, A.T.; Tian, J. "Reaction of EDTA and related aminocarboxylate chelating agents with $Co^{III}OOH$ (heterogenite) and $Mn^{III}OOH$ (manganite", *Environ. Sci. Technol.* **1998**, *32*, 2923-2930.

31. Weyh, J.A.; Hamm, R.E. "Iminodiaceto, methyliminodiacetato, and 1,3-propanediaminetetraacetato complexes of chromium(III)", *Inorg. Chem.* **1968**, *7*, 2431-2435.

32. Westall, J.; Hohl, H. "A comparison of electrostatic models for the oxide/solution interface", *Adv. Colloid Interface Sci.* **1980**, *12*, 265-294.

33. Stone, A.T.; Torrents, A.; Smolen, J.; Vasudevan, D.; Hadley, J. "Adsorption of organic compounds possessing ligand donor groups at the oxide/water interface", *Environ. Sci. Technol.* **1993**, *27*, 895-909.

34. Laiti, E.; Ohman, L.-O.; Nordin, J.; Sjoberg, S. "Acid/base properties and phenylphosphonic acid complexation at the aged γ-Al_2O_3/water interface", *J. Colloid Interface Sci.* **1995**, *175*, 230-238.

35. Laiti, E.; Ohman, L.-O. "Acid/base properties and phenylphosphonic acid complexation at the boehmite/water interface", *J. Colloid Interface Sci.* **1996**, *183*, 441-452.

36. Persson, P.; Laiti, E.; Ohman, L.-O. "Vibration spectroscopy study of phenylphosphonate at the water-aluminum (hydr)oxide interface", *J. Colloid Interface Sci.* **1997**, *190*, 341-349.

37. Barja, B.C.; Tejedor-Tejedor, M.I.; Anderson, M.A. "Complexation of methylphosphonic acid with the surface of goethite particles in aqueous solution", *Langmuir* **1999**, *15*, 2316-2321.

38. Hiemstra, T.; van Riemsdijk, W.H.; Bolt, G.H. "Multisite proton adsorption modeling at the solid/solution interface of (hydr)oxides: a new approach. I. Model description and evaluation of intrinsic reaction constants", *J. Colloid Interface Sci.* **1989**, *133*, 91-104.

39. Herbelin, A.; Westall, J.C. "FITEQL 3", Report 94-01, Department of Chemistry, Oregon State University, Corvallis, OR, **1994**.

40. Hiemstra, T.; van Riemsdijk, W.H. "A surface structural approach to ion adsorption: the charge distribution (CD) model", *J. Colloid Interface Sci.* **1996**, *179*, 488-508.

41. Nowack, B.; Stone, A.T. "Adsorption of phosphonates onto the goethite-water interface", *J. Colloid Interface Sci.* **1999**, *214*, 20-30.

42. Nowack, B.; Stone, A.T. "The influence of metal ions on the adsorption of phosphonates onto goethite", *Environ. Sci. Technol.* **1999**, *33*, 3627-3633.

43. Schindler, P.W. *Co-adsorption of metal ions and organic ligands: formation of ternary surface complexes*, Chapter 7 in: Mineral-Water Interface Geochemistry, M.F. Hochella and A.F. White, edits., Mineralogical Society of America, Reviews in Mineralogy Vol. 23, Washington, DC, **1990**.

44. Bordas, F.; Bourg, A.C.M. "Effect of complexing agents (EDTA and ATMP) on the remobilization of heavy metals from a polluted river sediment", *Aquat. Geochem.* **1998**, *4*, 201-214.

45. Nowack, B. (1998) "Behavior of phosphonates in wastewater treatment plants of Switzerland", *Water Res.* **1998**, *32*, 1271-1279.

46. Kari, F.G.; Hilger, S.; Canonica, S. "Determination of the reaction quantum yield for the photochemical degradation of Fe(III)-EDTA: Implications for the environmental fate of EDTA in surface waters", *Environ. Sci. Technol.* **1995**, *29*, 1008-1017.

47. Matthijs, E.; DeOude, N.T.; Bolte, M.; Lemaire, J. "Photodegradation of ferric ethylenediaminetetra(methylenephosphonic acid) (EDTMP) in aqueous solution", *Water Research* **1989**, *23*, 845-851.

48. Schowanek, D.; Verstraete, W. "Hydrolysis and free radical mediated degradation of phosphonates", *J. Environ. Qual.* **1991**, *20*, 769-776.

49. Holm, T.R.; George, G.K.; Barcelona, M.J. "Fluorometric determination of hydrogen peroxide in groundwater", *Anal. Chem.* **1987**, *59*, 582-586.

50. Steber, J.; Wierich, P. "Properties of aminotris(methylenephosphonate) affecting its environmental fate: degradability, sludge adsorption, mobility in soils, and bioconcentration", *Chemosphere* **1987**, *16*, 1323-1337.

51. Nowack, B.; Stone, A.T. "Degradation of nitrilotris(methylenephosphonic acid) and related (amino)phosphonate chelating agents in the presence of manganese and molecular oxygen", *Environ. Sci. Technol.* **2000**, *34*, 4759-4765.

52. Stumm, W.; Morgan, J.J. *Aquatic Chemistry*, 3rd edit., Wiley-Interscience, NY, 1996.

53. Davies, S.H.R.; Morgan, J.J. "Manganese(II) oxidation kinetics on metal oxide surfaces" *J. Colloid Interface Sci.* **1989**, 129, 63-77.

54. Coleman, W.M.; Taylor, L.T. "Dioxygen reactivity-structure correlations in manganese(II) complexes", *Coord. Chem. Rev.* **1980**, *32*, 1-31.

55. Bilinski, H.; Morgan, J.J. "Complex formation and oxygenation of manganese(II) in pyrophosphate solutions", Abstract presented before the Division of Water, Air, and Waste Chemistry **1969**, American Chemical Society National Meeting, Minneapolis, Mn.

56. Yoshino, Y.; Ouchi, A.; Tsunoda, Y.; Kojima, M. "Manganese(III) complexes with ethylenediaminetetraacetic acid", *Canad. J. Chem.* **1962**, *40*, 775-783.

57. Schroeder, K.A.; Hamm, R.E. "Decomposition of the ethylenediaminetetraacetate complex of manganese(III)", *Inorg. Chem.* **1964**, *3*, 391-395.

58. Hamm, R.E.; Suwyn, M.A. "Preparation and characterization of some aminopolycarboxylate complexes of manganese(III)", *Inorg. Chem.* **1967**, *6*, 139-145.

59. Klewicki, J.K.; Morgan, J.J. "Kinetic behavior of Mn(III) complexes of pyrophosphate, EDTA, and citrate", *Environ. Sci. Technol.* **1998**, *32*, 2916-2922.

60. Freedman, L.D.; Doak, G.O. "The preparation and properties of phosphonic acids", *Chem. Rev.* **1957**, *57*, 479-523.

71. Matczak-Jon, E.; Kurzak, B.; Sawka-Dobrowolska, W.; Lejczak, B.; Kafarski, P. "Zinc(II) complexes of phosphonic acid analogues of aspartic acid and aspargine", *J. Chem. Soc. Dalton Trans.* **1998**, 161-169.

72. Lacour, S.; Deluchat, V.; Bollinger, J.-C.; Serpaud, B. "Complexation of trivalent cations (Al(III), Cr(III), Fe(III)) with two phosphonic acids in the pH range of fresh waters", *Talanta* **1998**, *46*, 999-1009.

Chapter 5

The Effects of Soil/Sediment Organic Matter on Mineralization, Desorption, and Immobilization of Phenanthrene

S. B. Soderstrom, A. D. Lueking, and W. J. Weber, Jr.*

Environmental and Water Resources Engineering, Department of Civil and
Environmental Engineering, The University of Michigan,
Ann Arbor, MI 48109

The bioavailability, desorption profiles, and extent of immobilization of phenanthrene were assessed using three geosorbents that had been previously classified based on their relative sorption behaviors and the chemical characteristics of their associated organic matter. Contaminants sorbed to amorphous, geologically young sorbents were found to desorb at a faster rate and thus be more readily bioavailable than contaminants sorbed to geologically mature kerogens. However, the final extent of mineralization in biologically active systems was observed to be comparable for all three geosorbents tested regardless of geological age. Analysis of the geosorbents at the conclusion of the mineralization experiments implies that biological activity affects contaminant immobilization through both sequestration and transformation. Dimensional analysis of the experimental system with independent measures of phenanthrene degradation and mass transfer aided in interpretation of the mineralization data by providing a quantifiable parameter. The resulting dimensional parameter not only changes with experimental design, but also was found to change as experimental conditions changed within a single system.

Introduction

Sorption, bioavailability, and sequestration are interrelated phenomena affecting the transport and ultimate environmental fate of organic contaminants in subsurface systems. An important, yet poorly defined, condition that influences these phenomena is the physicochemical character of the sorbent, particularly that of its associated natural organic matter. Attempts to correlate bioavailability to soil type have been relatively unsuccessful (*1,2,3*), mainly due to the classification of the sorbent based solely on its macroscopic physical properties rather than the physicochemical character of its organic matter. A potentially useful way in which to classify sorbents for bioavailability studies is through use of the distributed reactivity concepts introduced in the early 1990's (*4*) and recently summarized and reviewed by Weber, et al (*5*). According to these concepts, soil organic matter (SOM) can be modeled as a macromolecular structure that ranges from completely amorphous to completely condensed. Geologically "young" soils contain primarily highly amorphous organic matter (e.g. fulvic acids) referred to as "soft-carbon" organic matter; whereas diagenetically altered soils contain a high degree of more condensed, "hard-carbon" organic matter (e.g. anthracite coals) (*5*). These "older" soils exhibit more nonlinear, slower, and only partially reversible sorption of hydrophobic organic compounds (HOCs) and have greater organic-carbon-normalized sorption capacities for such contaminants. Conversely, younger soils typically exhibit nearly linear, faster, and reversible HOC sorption, and lower organic-carbon-normalized sorption capacities.The distributed reactivity concept has been successfully used to explain the hysteresis effect in soils as well (*6*), and therefore may be a useful means for correlating bioavailability to soil type.

The bioavailability of a contaminant refers to the ability of a biotic species to access it. A common measure of bioavailability is the microbial mineralization profile; this is based on the assumption that unavailability to microorganisms implies unavailability to higher life forms. A standard assumption is that degradation occurs in the aqueous phase only; therefore, desorption is expected to play a key role in bioavailability. Bioavailability measurements have been shown to depend upon: the microbial ecology (*7*); the solid to solution ratio (*8,9*); and the reactor configuration (*10*). For these reasons, it is not expected that measurements of bioavailability in the laboratory will quantitatively reflect *in situ* conditions, although qualitative relationships should extend to the field. Such correlations would provide insight and guidance for risk assessment and remediation design.

Sequestration is a process in which the reactivity of a contaminant is reduced due to complexation with other materials and/or to only quasi-reversible transfer to another phase. Transformation, on the other hand, involves a chemical alteration of the contaminant. For example, sequestration might involve non-Fickian diffusion into tightly-knit SOM matrices, whereas transformation would involve a chemical or biochemical reaction that yielding a different reaction product. Sequestration and transformation can each increasingly reduce the bioavailability of a contaminant with increased contact time. This "aging" phenomenon has been used to qualitatively explain the reduction in extractability and toxicity of contaminants (*1, 11 - 14*). In pre-exposed soils, freshly added contaminants are often bioavailable, whereas native contaminants are not (*2, 15, 16*). The effect of aging is not incorporated into any of the current bioavailability models. The extent and conditions of the aging period have been shown to affect bioavailability assessments, although in some of these cases apparent equilibrium may not have been reached (1, 10, 11).

The purpose of this chapter is to: (1) test the applicability of the distributed reactivity model as a basis for classifying soils and/or sediments for bioavailability assessments; (2) compare bioavailability for different aging periods; and, (3) use dimensional analysis to evaluate the results of a biologically active soil system by conducting independent measures of mass transfer and degradation. In these experiments, we have attempted to eliminate factors that may complicate bioavailability comparisons, namely parameters that may affect the aqueous concentration of the contaminant. Parameters that affect the aqueous concentration "seen" by microorganisms will drastically affect mineralization profiles. For example, different soils with varying sorption capacities will require different amounts of contaminant to establish constant initial aqueous concentrations in different experimental systems. In addition, non-equilibrium conditions may result in initial aqueous concentration reductions due to ongoing sorption by the soil rather than by desorption and/or biodegradation. We have employed methods previously developed in our laboratories to establish sufficient equilibration times and variations in total contaminant mass to insure constant contaminant flux at the onset of bioavailability studies (6).

Theory

Equilibrium sorption and desorption behavior in natural soil systems are often nonlinear, and commonly modeled well by the Freundlich sorption equation:

$$q_e = K_F C_e^{\,n} \tag{1}$$

where K_F and n are the Freundlich capacity parameter and linearity factor, respectively, and, q_e and C_e, are the solid-phase and aqueous-phase equilibrium solute concentrations. In addition to being non-linear, natural geosorbents often exhibit some degree of sorption-desorption hysteresis, meaning that the partitioning between the solid-phase and aqueous phase may be history dependent. In abiotic systems, this is often due to some degree of sequestration of the contaminant within the soil. A hysteresis index was defined by Huang and Weber (6) to quantify this phenomenon:

$$Hysteresis\ Index = \frac{q_e^d - q_e^s}{q_e^s}\bigg|_{T,C_e} \tag{2}$$

where q_e^s and q_e^d are solid-phase solute concentrations for the single-cycle sorption and desorption experiments, respectively, and the subscripts T and C_e specify conditions of constant temperature and residual equilibrium solution phase solute concentration.

The distributed reactivity model explanation for the biphasic rate behavior commonly observed for desorption of HOCs from soils is that the soft-carbon sorbed, or "labile" fraction of the contaminant desorbs readily and reversibly, whereas the hard-carbon sorbed, or "resistant" component is released much more slowly. The slow desorption step has been attributed to non-Fickian diffusion into a tightly-knit SOM, polymerization, or entrapment within the SOM matrix. The rate model found in comparative analyses to be the most appropriate for description of such behavior (17) is a two-phase release model which couples first-order rate equations for both the slow, resistant, ϕ_s, and rapid, labile fractions, $\phi_r (= 1- \phi_s)$:

$$\frac{q(t)}{q_o} = \phi_s e^{-k_s t} + (1 - \phi_s)e^{-k_r t} \tag{3}$$

where $q(t)$ is the solid-phase sorbate concentration at a given time t, q_o is the initial solid-phase sorbate concentration, , and k_s and k_r are the first-order rate constants for the resistant (slow) and labile (rapid) fractions (day^{-1}), respectively.

Materials and Methods

Phenanthrene was chosen as the probe hydrophobic organic contaminant because of its important role in environmental systems and its extensive prior use in our laboratory. The preparation and characterization of the soils used in this study, as well as phenanthrene sorption behavior, have been described in previous work (6). In brief, Michigan Peat is a relatively young soil from which phenanthrene readily desorbs, resulting in a low hysteresis index; Chelsea soil is a high organic, muck-type top soil with an intermediate hysteresis index; Lachine Shale is a geologically old soil with a high hysteresis index. The soils were air dried, sieved with a 2-mm sieve, riffle split, and then stored in airtight containers until use. The soils were not sterilized in order to keep the chemical nature of the SOM intact. The sorption isotherms were fit to Equation 1 by Huang (6); these parameters are summarized in Table 1.

When preparing the soils for subsequent studies, equilibration times in excess of 60 days were used based on previous results (6). All systems were equilibrated and aged under anoxic conditions in an attempt to reduce biodegradation by indigenous microorganisms. For each aging period and soil type, five identical soil systems were used: (A) one to measure phenanthrene concentrations to assess biotic degradation at the conclusion of aging; (B) one to provide soil for desorption experiments; (C) two per soil to be used for the bioavailability experiments; and (D) one to serve as an abiotic control in the bioavailability study. A sixth soil-free system was used to measure abiotic

Table 1: Freundlich parameters of soil and phenanthrene loading

Soil Type	Freundlich parameters		Total phenanthrene mass to achieve Ce=800 µg/L	Resulting soild phase loading
	Log K_F	n	M_T (mg)	q_e (µg/g)
Michigan peat	1.08	0.89	45	4611
Chelsea soil	0.64	0.72	5.5	537
Lachine shale	2.37	0.53	80	8103

losses during aging. All systems were aged in 125-mL amber bottles sealed with screw caps, Teflon-lined silicone septa, and silver foil. A constant soil to solution ratio of 10 g to 100 mL was used. The average headspace in the bottles was approximately 10-15 mL. The background solution consisted of 0.015 M NaCl to control the ionic strength; 100 mg/L NaN$_3$ was added to the abiotic controls. In an attempt to achieve a constant solid to solution flux, the mass of non-radiolabeled phenanthrene was varied to achieve a constant aqueous phase concentration of 800 µg/L (Table 1). Radiolabeled phenanthrene (Sigma chemicals, 14.0 mCi/mmol, >98% purity), diluted in methanol, was added to achieve a final activity of 5 x 10^5 DPM. The samples were horizontally mixed at 25° C for a period of two and four months.

At the conclusion of aging the ^{14}C activity of each phase was quantified for use in subsequent studies. The radioactivity of the aqueous phase of each flask was determined on a LKB Wallace (Model 1219 RackBeta) liquid scintillation counter using ScintiSafe+ Liquid Scintillation Cocktail (Fisher Chemicals) and automatic quench correction. Triplicates of every sample were counted, and the average disintegrations per minute (DPM) were recorded after correcting for background activity. The aqueous phase of system (A) was used to confirm that this radioactivity was indeed phenanthrene and not possible metabolites formed during aging. The phenanthrene concentration was assessed by using reverse-phase HPLC (column: ODS, 5 µm, 2.1 x 250 mm; instrument: Hewlett-Packard Model 1090) with both a diode array detector (250 nm wavelength) and a fluorescence detector (Model HP1046A). System (B) was used as a representative measure of the solid-phase concentration. The soil was filtered with a 0.2 µm cellulose acetate filter, and the radioactivity was measured by combustion at 900 °C using a Biological Oxidizer (Model OX500). The soil was then stored at 4 ± 1 °C until use in the desorption experiments.

Measurements of desorption rate were conducted with a procedure similar that developed by Carroll, et al. (18). Sodium azide was added to the sample to prevent biological activity. Tenax TA polymer (Alltech Associates) was used as a model infinite sink. Prior to the desorption experiments, Tenax was soaked for 24 hours in a 25 ml glass centrifuge tube containing 10 mL of biological mineral media with 100 mg/L NaN_3. The dry to wet ratio of the resulting soil suspension was measured in parallel with the desorption experiments. Desorption experiments were commenced by adding 0.1 g of the aged soil (wet weight) to vials containing 10 mL of mineral media, Tenax, and sodium azide. At specified times, the soil, aqueous, and Tenax phases were separated by centrifugation and the concentration of phenanthrene in each phase was measured as described above. The radioactivity of the aqueous phase throughout the experiment was measured to test the infinite sink assumption. The desorption data were fit to Equation 3 using nonlinear regression.

Pseudomonas cepacia CRE7 was used in the mineralization studies to eliminate lag times. CRE7 was acclimated to phenanthrene as its sole carbon source in a biological mineral media. The ability of CRE7 to degrade phenanthrene was determined in soil-free systems using five concentrations (100-1000 μg/L) of phenanthrene. A high concentration of cells was used, ~10^7 CFU/mL (quantified using Tryptone agar plates), to eliminate growth during the experiment. A headspace of approximately 150-mL was used to insure sufficient oxygen supply. The phenanthrene concentration was measured as a function of time; at sampling, the cell solution was diluted in methanol, centrifuged and filtered to remove biomass, and then stored at 4 °C prior to HPLC analysis. Control experiments containing 100 mg/L NaN_3 were run in parallel to determine abiotic losses. Phenanthrene concentration data were fit to the Michaelis-Menten equation using nonlinear regression.

The mineralization studies were conducted in bioreactors (Ace Glass) that consisted of 250 ml glass Erlenmeyer flasks with a side arm. The aged soil samples from system (C) were quantitatively transferred to the bioreactors, while system (D) was transferred to the flasks to serve as an abiotic control. CRE7 was passed through a glass frit to remove phenanthrene crystals and concentrated to 10^9 CFU/ml for inoculation of the bioreactors; a sterile mineral media solution was added to the abiotic controls. A concentrated mineral nutrient solution was then added to the flasks. The volume of the mineral media and the cell culture was kept small (2 mL and 1 ml, respectively) to avoid dilution of the aqueous phase which could lead to desorption of equilibrated phenanthrene. During the course of the mineralization experiment, oxygen (>99.6%, Cyrogenic gases, 10ml) was added through the side arm to maintain aerobic conditions. $^{14}CO_2$ was captured in a septum-sealed center well containing NaOH. A small hole (~3mm) in the center well allowed gas transfer with the main flask. For sampling, all NaOH was removed from the center well and replaced with fresh solution. To minimize gas leakage, the center well septum was replaced after sampling. The radioactivity measured in the sodium hydroxide was converted to total CO_2 produced using the ratio of labeled to nonlabeled phenanthrene added to the system. To account for the unequal amounts of non-radiolabeled phenanthrene added to each soil type, the total CO_2 was normalized to the total carbon dioxide produced if all phenanthrene (both labeled and unlabeled) were completely mineralized.

After completion of the mineralization studies, the activities of the aqueous and solid phases were analyzed as described above. In addition to combustion at 900 °C, soil samples were extracted with methanol to recover sequestered ^{14}C-organics in the soil. The reactor components were either combusted or extracted in an effort to close the ^{14}C mass balance.

Results and Discussion

Equilibration and Aging

The soil-free abiotic controls showed that the mass balance during aging was nearly complete (>98%), and therefore no corrections were made to subsequent analysis. HPLC analysis of phenanthrene in system (A) showed that there were no detectable metabolites formed during aging. Therefore, the anoxic design of the experiment was sufficient to eliminate phenanthrene degradation during aging. The soil loading measured from system (B) corresponded to that of a soil in equilibrium with an aqueous concentration of 800 $\mu g/L$. However, the actual aqueous concentration was higher than expected, possibly due to the sorbance of phenanthrene to dissolved humic matter. Previous studies done by Huang (6) used a soil to solution ratio several orders of magnitude lower than this study, and therefore did not see this effect. Although the initial aqueous concentration was not constant for the different geosorbents, the soil loading corresponded to the constant flux condition.

Desorption Profile

The ^{14}C activity of the aqueous supernatant at each sampling time was negligible, confirming the ability of Tenax polymer to maintain an infinite-dilution condition. Due to the strong affinity of the Tenax for the phenanthrene, the desorption experiments mimicked a system in which biodegradation was instantaneous relative to the desorption rate. The phenanthrene desorption profiles from the two month aged soils are shown in Figure 1. The two and four month aged soils showed no distinguishable differences in desorption profiles, thus only the two month profiles are shown here.

102

The desorption profiles agreed qualitatively with that predicted from the soil organic matter classification. The geologically younger geosorbents, Chelsea Soil and Michigan Peat, showed faster initial desorption and a greater extent of desorption when compared to the geologically older Lachine Shale. The desorption data were fit to the two-phase desorption model of Equation 3, and are plotted along with the experimental data in Figure 1. The model parameters indicate that the labile fraction is highest in the Chelsea Soil, lower in the Michigan Peat and significantly lower in the Lachine Shale as shown in Table 2. The desorption rate constants for the labile fraction follow the same pattern.

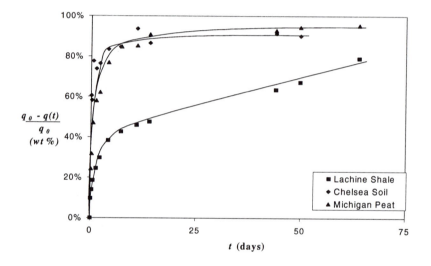

Figure 1: Desorption profile and two-phase desorption model regression for the three soils after two month aging period.

Soil Type and Mineralization Profile

The extent of phenanthrene mineralization parallels that expected based on the hysteresis index and SOM classification (6). Michigan Peat and Chelsea Soil are somewhat comparable, which is supported by the similar hysteresis indices and sorption capacity found by Huang (6). Initially, Lachine Shale has a mineralization profile that is considerably lower than the other two soils, but after extended time periods it approaches that of the other two soils. The mineralization profiles for two and four month equilibration times are shown in Figure 2.

Table 2: Desorption profile regression for the soil systems aged for two months

	Michigan Peat	Chelsea Soil	Lachine Shale
Labile Fraction			
ϕ_r	0.503517	0.693418	0.361826
k_r (1/d)	3.90E+00	1.58E+01	1.95E+00
Slow Desorption Fraction			
ϕ_s	0.496483	0.306582	0.638174
k_s (1/d)	1.41E-01	1.06E-01	2.40E-02

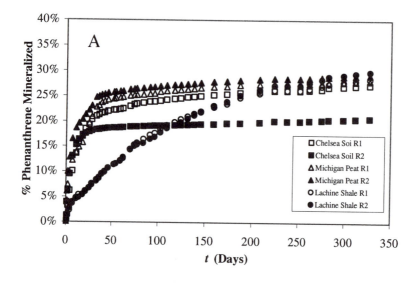

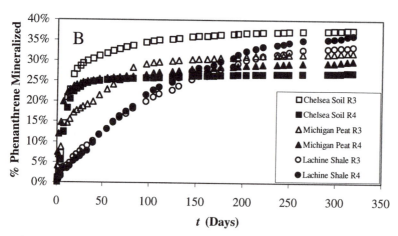

Figure 2: Carbon dioxide production versus time for three soils after two months aging (A) and four months aging (B).

Control flasks for Chelsea Soil and Lachine Shale showed no measurable mineralization over the course of the experiment. In the control flasks for the Michigan Peat, however, carbon dioxide was detected after a delayed period as shown in Figure 3. Due to the similarity of this delayed profile and the other profiles, it was concluded that the carbon dioxide production was not due to abiotic losses, but rather biotic losses. This establishes that sodium azide does not sterilize a system, but rather acts to inhibit the growth of new cells. Therefore, since the soil was never sterilized nor were the controls inoculated with CRE7, the delayed biodegradation is likely due to the acclimation of the indigenous microorganisms to phenanthrene. This is supported by the nature of Michigan Peat: it is a topsoil and is likely to have an active indigenous microbial population. The rate and extent of degradation for the indigenous microorganisms were comparable to those of the inoculated organisms.

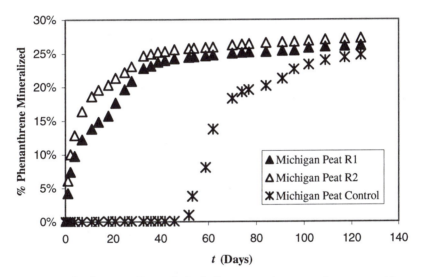

Figure 3: Once acclimated, the indigenous microorganisms present in Michigan Peat were able to mineralize phenanthrene.

Evaluation of the mineralization data from the bioreactors is not necessarily straightforward. First, not all phenanthrene degradation results in the formation of CO_2; partially degraded phenanthrene metabolites may react with the SOM and become integrated in the sorbent matrix. This effect is expected to be more pronounced in the younger soils that have a high degree of humic matter. Secondly, the humic matter may compete with phenanthrene as a carbon source. Although the CRE7 was acclimated to phenanthrene, the humic matter may be a higher-energy substrate, and, thus, may preferentially be degraded over phenanthrene. It is difficult to quantify this effect due to the competing mechanisms of desorption and substrate inhibition. For these reasons, a leveling off of $^{14}CO_2$ production does not necessarily indicate a reduction in microbial activity, but perhaps the beginning of the mineralization of alternative carbon sources.

Aging Period and Mineralization Profile

There is no significant difference in the mineralization profiles between two and four months aging for all three soils (Figure 2). This is likely because our experiments were designed to have longer equilibration times that many other studies reported in the literature. They were also designed to provide for a constant flux from the soil at the onset of the bioavailability study. The systems had a relatively large mass of soil, 10 g, which could lead to system heterogeneities and difficulties in achieving complete mixing in all systems; resulting in the variations of phenanthrene mineralization between systems of the same soil/sediment. This contrasts somewhat with the results found by other researchers. For example, Chung and Alexander (1) reported a statistically significant difference between 60 and 120 day aging period in eleven out of sixteen soils studied. Kelsey, et al. (11) found the extent of mineralization to be reduced by over 4% in phenanthrene aged in Lima Loam for 54 and 124 days.

Adsorption equilibration is a function of both contact time and experimental setup. Previous studies have shown that aging beyond three weeks to several months has negligible effect on desorption-resistant fractions or release rates (19). Johnson, et. al. have shown that initial contaminant loading has a significant effect on desorption profiles, but that the aging effect is small if aging is longer than two months for phenanthrene/geosorbent systems (17). Aging flasks with a high soil-to-water ratio are likely to take a longer time to reach an apparent equilibrium due to the lack of mobility of the contaminant. Incubation of the soil to apparent equilibrium is essential for accurate interpretation of mineralization profiles.

Immobilization

Combustion of the soil at the conclusion of the mineralization experiments revealed that a portion of the ^{14}C became immobilized within the soil, as shown in Figure 4. Immobilization may be due to either sequestration or transformation. Recall that sequestration is due to complexation with other materials and/or not readily reversible transfer to another phase; transformation involves a chemical alteration of the contaminant. The amount of ^{14}C-organics that are extractable from the soil gives an indication as to the portion of the contaminant that has become sequestered; for example, this portion of phenanthene may have diffused into tightly woven organic matter that was released upon Soxhlet extraction with methanol. The portion that was not extractable, however, possibly represents ^{14}C metabolites that have become chemically associated with the soil matrix.

The difference between the extractable and combustable ^{14}C-organic material is most pronounced for Michigan Peat, and generally follows the classification by hysteresis index and diagenetic alteration of the soil. The amorphous structure of the younger geosorbents is more reactive with both phenanthrene and any metabolites. This increased reactivity leads to greater sequestration and transformation as shown by the increased non-extractable fraction. This is also supported by the decreased CO_2 production in the younger sorbents when compared to the continued production of CO_2 from Lachine shale. We hypothesize that in the Chelsea Soil and Michigan Peat systems, the phenanthrene readily desorbed and a fraction was mineralized while another fraction was partially degraded and then underwent humification with the reactive SOM. On the other hand, in the Lachine Shale systems, the phenanthrene slowly desorbs and is mineralized due to the lack of reactive sites for humification and incorporation into the SOM.

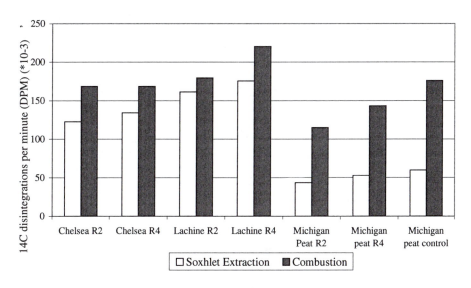

Figure 4: Methanol extraction and combustion results show that the amount of extractable organic material varies with the sorbent type at the conclusion of the mineralization experiments.

Mass Balance Evaluation

After completion of the experiments, the mass balance for each system was evaluated by summation of the mineralized $^{14}CO_2$, the aqueous phase ^{14}C organics, the extractable and immobilized ^{14}C organics, and the radioactivity associated with the reactor components. Glass reactor components were extracted with methanol. Other components were combusted in order to determine associated radioactivity. In all cases, the total radioactivity associated with the aqueous phase and the reactor components was between 3-5% of the initial radioactivity. In each system, there was a fraction of unrecovered radioactivity approximately equal to the amount mineralized. Experiments were conducted using ^{14}C labeled sodium bicarbonate to test the NaOH capture efficiency; this was found to be above 98%. Therefore, we hypothesize that the unrecovered radioactivity was lost through volatile metabolites that were not captured by the NaOH filled center well. In this case, the unrecovered ^{14}C would also represent bioavailable phenanthrene.

Dimensional Analysis

Analysis of the mineralization results is complicated by the competing effects of mass transfer, biodegradation, sequestration, and transformation, all of which are specific to our experimental conditions. This makes classification and generalization of the results difficult, at best. Previous attempts to generalize bioavailability measures have focused on developing either numerical models or correlations based on soil type. These models often focus on either biodegradation or mass-transfer, and make simplifying assumptions about the other phenomenon.

A simplified dimensional analysis can aid in interpreting our results. Dimensionless numbers are often used to generalize, scale-up, and compare results from one system to another. In the context of this study, a useful parameter is the bioavailability number, N_B (20), which is simply the ratio of the rate of mass transfer to the rate of biodegradation:

$$N_B = \frac{Mass\ Transfer\ Rate}{Biodegradation\ Rate} \qquad (4)$$

A bioavailability number less than unity indicates that the biological transformation of a contaminant in a system is limited by its mass transfer from the soil/sediment. Conversely, a bioavailability number greater than unity indicates that biodegradation itself is the rate-limiting step; i.e., the intrinsic mass transfer flux of sorbate to solution is greater than the sink flux attributable to degradation. In natural soil systems, the aqueous concentration is typically low and biodegradation of organic contaminants associated with soils and sediments is pseudo-first order (20). Hence, the bioavailability number can be expressed as the ratio of first-order volumetric rate coefficients for the two processes

$$N_B = \frac{k_{desorption}}{r_{max} / K_M} \qquad (5)$$

where $k_{desorption}$ (1/day) is the rate of mass transfer from the soil; and r_{max} (µg/(L-day) and K_M (µg/L) are the biokinetic parameters from the Michaelis-Menten equation.

In the field, parameters used in the bioavailability number may be difficult to quantify due to mixed populations of microorganisms, heterogeneous soil samples, and variable moisture conditions and exposure times. However, under controlled experimental conditions, this simplified analysis allows quantification and a relative measure of the effects of experimental variables, e.g. soil type and aging conditions.

To proceed with this analysis, we have made independent measurements of the biological degradation rate in soil-free systems and coupled this with the abiotic mass transfer rates found from the desorption studies. In other words, in an effort to understand the total system, we have separated two of the key factors in the ultimate fate of a contaminant, desorption and biodegradation, and measured each in the absence of the other. The biodegradation rate can be extracted from the intrinsic biokinetic parameters for the active degraders. In our case, we will employ the Michaelis-Menten parameters measured in soil-free systems for the inoculated culture CRE7. For an initial cell concentration of 1.94 x 10^7 CFU/mL, the biokinetic parameters for CRE7 were as follows: r_{max}=7.10 x 10^{-11} µg/L/day; K_M=883 µg/L. These parameters were coupled with both the labile and resistant desorption constants fit to the desorption profiles to calculate a bioavailabilty number which is summarized in Table 3.

This analysis indicates that the rate of mass transfer is the rate-limiting step for each soil type. This explains why the desorption and mineralization profiles were qualitatively similar. The difference in time scales may likely be due to the experimental design: the mineralization experiment only detected complete mineralization and did not detect the initial biodegradation to intermediate metabolites while the desorption experiment measured all phenanthrene that desorbed. The bioavailability number analysis shows that for each soil, the desorption resistant fraction has a bioavailability number that is several orders of magnitude lower than that of the labile fraction. This implies that although a contaminant may be initially biodegradable, slow desorption and mass transfer limitations decrease its availability to microorganisms with time.

Table 3: Desorption profile regression and bioavailability number

	Two month aging		
	Michigan Peat	Chelsea Soil	Lachine Shale
Labile Fraction			
ϕ_r	0.503517	0.693418	0.361826
k_r (1/d)	3.90E+00	1.58E+01	1.95E+00
N_B	2.89E-02	1.17E-01	1.45E-02
Slow Desorption Fraction			
ϕ_s	0.496483	0.306582	0.638174
k_s (1/d)	1.41E-01	1.06E-01	2.40E-02
N_B	1.05E-03	7.86E-04	1.78E-04

The bioavailability number is not necessarily constant for a non-steady state system, but is in fact a dynamic parameter that changes as both biological and mass transfer conditions change. For example, transformation of a contaminant and alteration of the soil organic matter may change mass transfer rates; likewise, evolution of a microbial community may change rates of biological degradation. Indeed, we have observed this latter case in the control experiment for Michigan Peat in which the indigenous microorganisms began to degrade phenanthrene. As the indigenous microorganisms became acclimated to phenanthrene, the overall biodegradation rate changed. Had we cultured the microorganims before and after acclimation, we would likely have seen a change in our determination of the biokinetic parameters. The first 50 days in the control experiment were likely indicative of a bioavailability number less than unity, while the behavior after fifty days was similar to the other systems which were mass transfer limited.

Of course there are several factors not explored in the bioavailability analysis, for example the soil to solution ratio. Previous investigators have shown that a soil to solution ratio of 1:10 is optimal for microbial degradation (*8,9*). In addition, the use of a microorganism that had been acclimated to phenanthrene clearly favored the case of a mass-transfer limited system. However, the use of the dimensionless bioavailability number will allow comparison for different cases in future work.

Conclusions

The physicochemical nature of soil organic matter has a marked influence on bioavailability, desorption, and immobilization of phenanthrene. Lachine Shale which contains SOM comprised primarily by hard carbon domains, exhibits slower desorption and mineralization rates, and a lower extent of phenanthrene immobilization. Two younger sorbents, Michigan Peat and Chelsea Soil containing SOMs comprised primarily by soft carbon domains, exhibit more rapid desorption and mineralization rates as well as a greater extent of immobilization. Our experiments were designed to have longer equilibration times that many other studies reported in the literature. They were also designed to provide for a constant flux from the soil at the onset of the bioavailability study. Under these conditions, there was no observable difference in mineralization profiles for aging periods of two to four months. A dimensional analysis of mass transfer and biodegradation, and employment of a dimensionless bioavailability number, simplifies the analysis of the complex processes that occur in a biologically active soil system. This analysis is useful for generalization of bioavailability assessments and provides a quantifiable parameter that is useful for comparison between systems. However, the bioavailability number is a dynamic parameter that may change with variations in the microbial community and/or alteration of the soil matrix.

Acknowledgments

The authors would like to thank former and present members of the Weber research group at Michigan, Dr. Weilin Huang (currently at Drexel University), Dr. Martin Johnson (currently at Dow Chemical Co.), and Minsun Kim for their invaluable research contributions and Michael Keinath for assistance in reviewing the manuscript. We also thank Dr. Hap Pritchard of the Naval Research Laboratories for helpful comments and suggestions in the experimental phases of the work. The work was funded in part by the US Environmental Protection Agency, Office of Research and Development, through grant GR825962-01-0 and grant R-819605 to the Great Lakes & Mid-Atlantic Center (GLMAC) for Hazardous Substance Research. Partial funding of the research activities of GLMAC is also provided by the State of Michigan, Department of Environmental Quality. Partial support was also provided in the form of an EPA STAR Graduate Fellowship award to Sara B. Soderstrom.

References

1. Chung and Alexander, M. *Environ. Sci. Technol.* **1998**, *32*, 855-860.
2. Carmichael, L.M.; Pfaender, F.K. *Environ. Toxicol. Chem.* **1997**, *16*, 666-675.
3. Pavlostathis, S.G.; Mathavan, G.N. *Environ. Sci. Technol.* **1992**, *26*, 532-538.
4. Weber, W.J., Jr.; McGinley, P.M.; Katz, L.E. *Environ. Sci. Technol.* **1992**, *26*, 1955-1962.
5. Weber, W.J., Jr.; LeBoeuf, E.J.; Young, T.M.; Huang, W. *Wat. Res.* **2001**, *35*, 853-868.
6. Huang, W.; Weber, W.J., Jr. *Environ. Sci. Technol.* **1997**, *31*, 2562-2569.
7. Guerin, W.F.; Boyd, S.A. *Appl. Environ. Microbiol.* **1992**, *58*, 1142-1152.
8. Carmichael, L.M.; Christman, R.F.; Pfaender, F.K. *Environ. Sci. Technol.* **1997**, *31*, 126-132.
9. Milhelcic, J.R.; Luthy, R.G. *Environ. Sci. Technol.* **1991**, *25*, 169-177.
10. Govind, R.; Fu, C.; Yan, X.; Gao, C.; Pfanstiel S. *Biotechnol. Prog.* **1997**, *13*, 43-52.
11. Kelsey, J.W.; Kottler, B.D.; Alexander, M. *Environ. Sci. Technol.* **1997**, *31*, 214-217.
12. Hatzinger, P.B.; Alexander, M. *Environ. Sci. Technol.* **1995**, *29*, 537-545.
13. Fu, M.H.; Mayton, H.; Alexander, M. *Environ. Toxicol. Chem.* **1994**, *13*, 749-753.
14. Landrum, P.F. *Environ. Sci. Technol.* **1989**, *23*, 588-595.
15. Erickson, D.C.; Loehr, R.C.; Neuhauser, E.F. *Wat. Res.* **1993**, *27*, 911-919.
16. Steinberg, S.M.; Pignatello, J.J., Sawhney, B.L. *Environ. Sci. Technol.* **1987**, *21*, 1201-1208.
17. Johnson, M.D.; Keinath, T.M.; Weber, W.J., Jr. *Environ. Sci. Technol.* **2001**, *35*, 1688-1695.
18. Carroll, Kenneth M.; Harkness, Mark R.; Bracco, Angelo A.; Balcarcel, Robert R. *Environ. Sci. Technol.* **1994**, *28*, 253-258.
19. Pignatello, J.J. *Environ. Sci. Technol.* **1990**, *9*, 1117-1126.
20. Bosma, T.N.P.; Middeldorp, P.J.M.; Schraa, G.; Zehnder, A.J.B. *Environ. Sci. Technol.* **1997**, *31*, 248-252.

Chapter 6

Reductive Transformation of Halogenated Aliphatic Pollutants by Iron Sulfide

Elizabeth C. Butler[1] and Kim F Hayes[2]

[1]School of Civil Engineering and Environmental Science, University of Oklahoma, Norman, OK 73019
[2]Department of Civil and Environmental Engineering, University of Michigan, Ann Arbor, MI 48109

Studies in the last decade show that iron sulfide minerals are very reactive in the reductive transformation of chlorinated aliphatic pollutants. These minerals, present in sulfate-reducing anaerobic environments, likely contribute to *in-situ* transformation of chlorinated aliphatic pollutants and have potential application in remediation technologies. Solution pH, the presence of organic co-solutes with functional groups representative of natural organic matter, and the thermodynamic or molecular properties of the halogenated aliphatic pollutant all influence the rates and/or products of pollutant transformation.

Halogenated aliphatic compounds, used widely in degreasing, dry cleaning, agriculture, and numerous synthetic, industrial, and manufacturing processes, are among the most common ground-water pollutants in the United States and other industrialized countries (e.g., *1, 2*). Under anaerobic conditions, these compounds are susceptible to *in-situ* transformations by microbial or abiotic processes, including mineral-mediated reductive dehalogenation. Iron sulfide minerals such as FeS, present in sulfate-reducing environments (*3, 4*), are very reactive in the reductive transformation of halogenated aliphatic pollutants (*5-*

22). This chapter reviews the literature to date on the reductive transformation of halogenated aliphatic pollutants, particularly chlorinated aliphatics, by iron sulfide minerals.

Occurrence of Iron Sulfide Minerals

Iron monosulfide, FeS, is produced in soils and sediments primarily through dissimilatory microbial reduction of sulfate to sulfide, which subsequently reacts with available iron to precipitate FeS (*3, 4*). The mineral mackinawite, often in poorly crystalline form (*3, 23-25*), is the initial FeS precipitate in the transformation of iron minerals by sulfate-reducing bacteria (*3*). For example, when the sulfate-reducing bacterium *Desulfovibrio desulfuricans* was grown at pH 8 in cultures containing a Fe(II)/Fe(III) oxyhydroxide and synthetic geothite (FeOOH), mackinawite was the predominant iron sulfide phase present after six and nine months, respectively (*26*). Even at lower pH values, mackinawite was the only iron sulfide phase detected after two weeks of microbial activity, and still a minor phase after that.

Both mackinawite and amorphous FeS have been identified in natural waters, including the pore waters of anaerobic freshwater and marine sediments (*27-29*) and ground water (*29*). Although other iron sulfide minerals such as greigite (Fe_3S_4) and pyrite (FeS_2) eventually form upon interaction of aqueous polysulfides with the FeS surface (*24, 30*), mackinawite is a metastable phase under sulfate-reducing conditions (*3*), making it very relevant for *in-situ* reductive transformations.

Because the activation energy for the phase transformation of mackinawite to the thermodynamically more stable FeS phase hexagonal pyrrhotite is quite high (493 kJ/mol (*31*)), mackinawite is more susceptible to oxidation to other mineral phases such as greigite and pyrite than transformation to pyrrhotite under ambient conditions (*31*).

Mackinawite Structure

In mackinawite, each iron atom is bound to four sulfur atoms in an almost perfect tetrahedron and each sulfur is bound to iron in distorted one-sided four-fold coordination (*32, 33*). Tetrahedra share edges to form layers that are stacked and interact by van der Waals forces (*32*). These layers are randomly stacked in poorly crystalline mackinawite formed by precipitation from solution (*34*). Within mackinawite layers, each iron atom is in square planar coordination with four other iron atoms with a relatively short Fe-Fe distance of 2.60 Å (*32, 33*), suggesting significant *d* orbital overlap and Fe-Fe metallic bonding

(*32, 33, 35*). Consistent with this, Vaughan and Ridout (*36*) have proposed extensive delocalization of *d* electrons within mackinawite layers and electronic band structure calculations indicate that mackinawite is a metallic conductor with a conduction band of mainly *d* character (*37*). The Fe-Fe metallic bonding in mackinawite results in a Fe oxidation state somewhat less than +II, so a non-stoichiometric formula is necessary for electroneutrality. Certain studies indicate that mackinawite non-stoichiometry is due to periodic sulfur "vacancies" in the crystalline lattice, resulting in the formula FeS_{1-x}, where $x \approx 0.025$ (*32*), while others report evidence for the presence of excess Fe atoms in the mackinawite crystalline lattice, resulting in the non-stoichiometric formula $Fe_{1+x}S$, where x = 0.04-0.07 (*38*). A more recent investigation of mackinawite structure (*33*) has not resolved the question of sulfur vacancies versus excess iron atoms.

It is likely that the unique metallic and conducting properties of mackinawite contribute significantly to its reactivity in electron transfer reactions. Because the conductivity of mackinawite is in the plane of the stacked layers, even poorly crystalline mackinawite with "random stacking" of layers (*34*) is likely to have conducting properties that contribute to its reactivity.

Reductive Dehalogenation Reaction Pathways

Halogenated organic compounds are susceptible to a variety of reductive transformations in aquatic systems, including dihaloelimination and hydrogenolysis. Dihaloelimination of *vicinal* polyhaloalkanes, also called 1,2-dihaloelimination and reductive *β*-elimination, involves loss of two halogens from *vicinal* (adjacent) carbons, resulting in formation of a carbon-carbon double bond. Formation of carbon-carbon triple bonds in the dihaloelimination of *vicinal* polyhaloalkenes is also possible (*10, 12, 15, 20, 39-47*). Dihaloelimination of *geminal* polyhaloalkanes, also called reductive *α*-elimination, results in loss of two halogen substituents from the same carbon atom, leading to formation of highly reactive carbene intermediates (*48*). These carbene intermediates are susceptible to subsequent hydrogen migration and hydrolysis reactions (*49, 50*). Hydrogenolysis of halogenated aliphatic compounds involves replacement of halogen substituents with hydrogen. In the presence of transition metals, hydrogenolysis typically begins with an initial rate-determining step involving formation of an alkyl radical (*51*). Subsequent reactions include hydrogen atom abstraction (*52*), or a second electron addition followed by protonation.

Understanding the factors that lead to different reductive dehalogenation products is very relevant, since certain products of reductive dehalogenation are regulated pollutants that are more toxic than their parent compounds, while other products are relatively benign. For the transformation PCE and TCE by FeS, for example, parallel transformation by both *vicinal* dichloroelimination and hydrogenolysis occurs (15), with hydrogenolysis leading to more toxic products than dichloroelimination. Sequential hydrogenolysis of PCE leads to TCE and cis-DCE, and sequential hydrogenolysis of TCE leads to cis-DCE and VC (15). Vicinal dichloroelimination of both PCE and TCE leads to the relatively benign product acetylene via the short-lived intermediates dichloro- and chloroacetylene (40, 41, 43, 44, 45, 47). Consequently, reaction conditions that favor transformation of PCE and TCE by dichloroelimination versus hydrogenolysis are desirable. In addition, since microbial decay of PCE and TCE by sulfate-reducing bacteria typically yields hydrogenolysis products (e.g., 53), detection of acetylene in sulfate-reducing environments contaminated with PCE or TCE may be indicative of abiotic degradation by FeS or other soil minerals.

Dehalogenation Reactions Mediated by Iron Sulfide Minerals

Kriegman-King and Reinhard (5, 7) first studied the reductive dechlorination of carbon tetrachloride (CT) in the presence of the iron disulfides pyrite and marcasite and observed a kinetic pattern characteristic of a surface reaction. In another study (6), they found that the rate of transformation of CT in the presence of biotite and vermiculite increased when aqueous HS⁻ was added to the experimental system, possibly due to formation of a reactive iron monosulfide or disulfide phase on the clay surface. Since then, numerous studies have reported the transformation of CT by iron mono- and disulfides, including pyrite (8, 16), iron metal treated with Na_2S (8), commercial iron sulfide (8, 16, 19), and precipitated FeS (13, 16, 17). Other halogenated aliphatics reported to undergo reductive dehalogenation in the presence of iron mono- and disulfides include hexachloroethane (HCA) by iron-containing clay minerals treated with bisulfide (5) and precipitated FeS (14); tetrachloroethylene (PCE) by precipitated FeS (15) and pyrite (20); trichloroethylene (TCE) by commercial FeS (10-12), precipitated FeS (15, 22), pyrite (20, 21), and iron metal treated with NaHS (18); cis 1,2-dichloroethylene (cis-DCE) by commercial FeS (9) and pyrite (20); trans 1,2- and 1,1-dichloroethylenes by commercial FeS (9); vinyl chloride (VC) by pyrite (20); and tribromomethane by precipitated FeS (17) and commercial iron sulfide (19). One recent study reported the transformation of five other chlorinated aliphatic compounds (pentachloroethane, 1,1,2,2- and 1,1,1,2-tetrachloroethanes, and 1,1,1- and 1,1,2-trichloroethanes) by precipitated FeS (17), and another reported the transformation of six other halogenated methanes (bromotrichloromethane, tetrabromomethane, dibromodichloromethane, tribromofluoromethane, dibromochloromethane, and trichloromethane) by commercial iron sulfide (19).

Trends in Reactivity

Transformation of halogenated aliphatics has been shown to take place at the FeS surface and to follow pseudo-first-order kinetics under conditions where reactive surface sites are present far in excess of the organic substrate (*14*). The prime notation (') on certain rate constants discussed below indicates correction to account for partitioning of reactants between the sample aqueous and vapor phases (*15, 43, 54, 55*). These corrected rate constants represent those that would be observed in a headspace-free system.

Influence of pH

Solution pH has been shown to affect the rates of many redox reactions, often through pH-dependent equilibria among reductant species such as hydroquinones (*56, 57*). For FeS, rate constants for HCA and TCE reductive dechlorination are highly pH-dependent, with faster rates observed at higher pH values (*14, 22*). A similar dependence of rate constants on pH, although over a larger pH range, was observed in the transformation of TCE by pyrite (*21*). The pH dependence in the transformation of HCA and TCE by FeS has been explained by an acid-base equilibrium between FeS surface functional groups, with more deprotonated surface functional groups having greater reactivity in reductive dechlorination (*14*). The difference in reactivity between the protonated and deprotonated forms of a surface acid/conjugate base pair can be explained by the greater driving force for electron donation by the more deprotonated ligand (*58*). Reactive surface acid/conjugate base pairs that could form upon hydration and subsequent protonation or deprotonation of partially uncoordinated FeS surface atoms include species such as $\equiv FeOH_2^+$ and $\equiv FeOH$ or $\equiv FeSH_2^+$ and $\equiv FeSH$, where "$\equiv$" indicates the bulk solid (*14*). Since mackinawite inter-layer surfaces are coordinatively saturated, and are bound only by van der Waals forces (*32*), sites of coordinative unsaturation would most likely be on the edges of mackinawite layers.

Similar to FeS, the influence of pH on the rate of transformation of TCE by pyrite has been attributed to pH-dependent equilibria between pyrite surface functional groups (*21*). In performing cyclic voltammetry with FeS_2 (pyrite) and $Fe_{1-x}S$ (pyrrhotite) electrodes over a wide pH range, Conway et al. (*59*) observed an increase in peak currents with increasing pH. In a controlled-potential technique such as cyclic voltammetry, the peak current is proportional to the rate of electron transfer at the electrode-solution interface (*60*), so an increase in peak current with increasing pH implies a faster rate of electron transfer at higher pH values. One explanation offered by Conway et al. (*59*) for the pH-dependence of the peak current for the pyrite electrode was greater deprotonation of sulfide species in the electrode/water interfacial region with increasing pH, an explanation that is generally consistent with the models described above.

Kinetic analysis of a two-reactive-site model leads to the following rate law (*14, 61*):

$$k_{obs}' = \alpha_0 k_0' + \alpha_1 k_1'$$

where k_{obs}' is the observed pseudo-first-order rate constant, k_0' and k_1' are the pseudo-first-order rate constants associated with the protonated and deprotonated FeS surface functional groups, and the ionization fractions α_0 and α_1 are the fractions of the total reactive surface sites in the protonated and deprotonated forms, respectively. The parameters α_0 and α_1 are functions of both pH and K_a, where K_a is the acid dissociation constant for the protonated form of the reactive surface functional group. A plot of k_{obs}', $\alpha_0 k_0'$, and $\alpha_1 k_1'$ for reductive dechlorination of HCA by 100 g/L FeS is shown in Figure 1, illustrating good agreement between data and model.

Transformation of HCA by FeS leads primarily to PCE, regardless of pH (*14*). PCE then undergoes subsequent transformation to acetylene and TCE at significantly slower rates (*15*). For TCE (and likely other pollutants as well), however, solution pH influences not only overall rate constants (k_{obs}' values) for FeS transformation, but also the distribution of reaction products (*22*). This is very relevant because TCE reductive transformation leads to products that are toxic (*cis*-DCE) as well as those that are benign (acetylene) (*15*). For example, one study found that as the pH increased between 7.3 and 9.3, the rate of FeS-mediated TCE dichloroelimination increased significantly, while the rate of TCE hydrogenolysis did not change substantially, with the result that the relative concentration of the harmful product *cis*-DCE decreased significantly with increasing pH. The branching ratio, or the rate constant for TCE dichloroelimination divided by the rate constant for TCE hydrogenolysis, increased by a factor of 10 between pH 7.3 and 9.3 for this system, as illustrated in Figure 2 (*22*).

A pH-dependent equilibrium between the acid and conjugate base forms of FeS surface species, discussed above to explain the influence of pH on rate constants for FeS-mediated reductive dechlorination of HCA, might also explain the pH-dependence of the TCE product distribution (*22*). Other factors proposed to explain the product distributions of TCE and other halogenated compounds in related systems include the reduction potential of reactive surface species (*62*), electron availability at the mineral surface (*63, 64*), and differences in enthalpies for dichloroelimination versus hydrogenolysis (*46*).

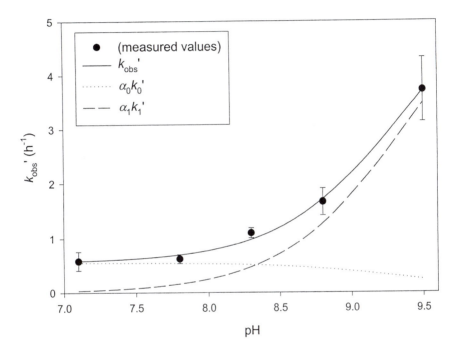

Figure 1. Observed pseudo-first-order rate constants for HCA reductive dechlorination versus pH; 100 g/L FeS. Error bars are 95% confidence intervals.
Adapted from Ref. *14*. American Chemical Society 1998.

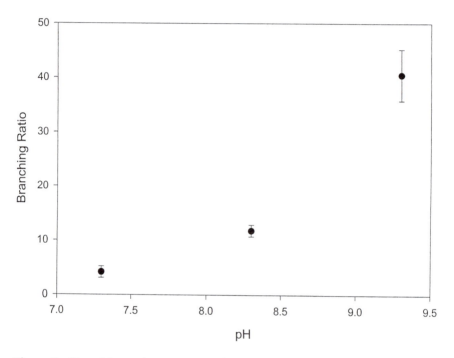

Figure 2. Branching ratios versus pH for TCE transformation by 10 g/L FeS. Error bars are 95% confidence intervals.

Influence of Natural Organic Matter

Organic molecules with strong affinities for the FeS surface may affect rate constants for surface electron transfer reactions. Obviously, understanding the influence of natural organic matter on reductive transformations is very relevant in predicting the rates and products of chlorinated pollutant transformation in natural systems. Studies to date indicate that organic co-solutes representative of the functional groups in natural organic matter can have no effect, or can increase or decrease observed rate constants (*14, 15, 22*).

In one study, numerous organic compounds with functional groups representative of natural organic matter were added to FeS aqueous slurries prior to reaction with HCA (*14*). These compounds included oxalate, succinate, hydroquinone, pyridine, 2,2'-bipyridine, 4,4'-bipyridine, 1,10-phenanthrolene, ethylenediamine, ethylenediaminetetraacetic acid, cysteine, and methionine. Oxalate, succinate, and hydroquinone, which contain phenolic and carboxylic functional groups, had no significant effect on the HCA transformation rate. Although the driving forces for succinate and hydroquinone adsorption to the FeS surface are not known, oxalate has a relatively large equilibrium constant for complex formation with Fe^{+2} (*65*), indicating that significant adsorption of oxalate to the FeS surface is likely, yet no significant effect on the reaction rate was observed when oxalate was present (*14*). These results suggest that either close interaction between HCA and FeS is not a prerequisite for electron transfer, or else that adsorption of these organic co-solutes is largely to non-reactive surface sites.

In contrast, certain aromatic nitrogen heterocyclic compounds were found to significantly affect the rate of HCA transformation by FeS. For example, addition of 1 mM 2,2'-bipyridine to FeS slurries caused a ten-fold increase in the rate of HCA transformation by FeS at pH 8.3 (*14*), as illustrated in Figure 3. Adsorption measurements indicated that the majority of the 2,2'-bipyridine in these experiments was associated with the FeS surface, suggesting that the rate increase in the presence of 2,2'-bipyridine is likely due to participation of a surface Fe(II)-2,2'-bipyridine complex in the electron transfer reaction. Addition of 1,10-phenanthrolene, which is structurally similar to 2,2'-bipyridine, to FeS slurries also dramatically enhanced the reaction rate. Since numerous natural products and biochemicals contain functional groups similar to 2,2'-bipyridine and 1,10-phenanthrolene, the presence of these compounds in natural systems may influence the rates of intrinsic abiotic transformations by FeS.

Unlike the case of HCA, 2,2'-bipyridine affected neither rate constants nor product distributions in the transformation of TCE by FeS at pH 8.3 (*22*). This lack of influence suggests that a close interaction between TCE and the FeS surface is not required for electron transfer, or else that surface sites for sorption of 2,2'-bipyridine are distinct from those where electron transfer to TCE takes place.

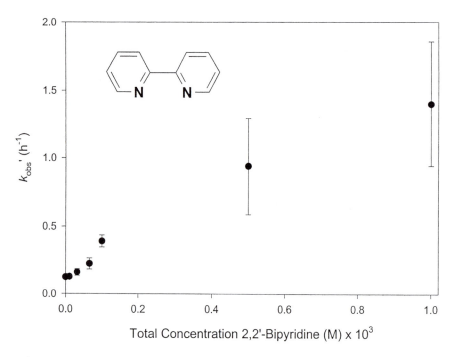

Figure 3. Observed pseudo-first-order rate constants for HCA reductive dechlorination versus total concentration 2,2'-bipyridine; 10 g/L FeS, pH 8.3. Error bars are 95% confidence intervals. Adapted from Ref. *14*. American Chemical Society 1988.

Addition of 1 mM cysteine, a thiol-containing amino acid, to aqueous FeS slurries reduced the rate of transformation of HCA by FeS by almost one half, and addition of 1 mM methionine, an alkyl sulfide amino acid, very slightly depressed the rate of HCA transformation (*14*). Addition of cysteine also significantly affected the rate of TCE reductive dechlorination by FeS (*15*). This rate reduction may be due to adsorption of cysteine or methionine to surface iron atoms in competition with HS⁻ or OH⁻, causing an energetic or steric barrier to electron transfer (*15*). Depending on structure and function, other thiol- and sulfide-containing natural products may also affect the rates of transformation of halogenated pollutants in aquatic systems where FeS is present.

Correlation Analysis

For the reaction of a homologous series of compounds with a common rate-limiting elementary reaction step, there is frequently a correlation between rate constants and overall free energies for the rate-limiting step, called a linear free energy relationship (LFER). A strong LFER may provide insight into the rate-limiting process involved in pollutant transformation reactions, while the absence of a strong LFER may suggest either the lack of a common reaction mechanism or the influence of other thermodynamic or molecular parameters on reaction rates.

Linear regression of one-electron reduction potentials (which are proportional to free energies for electron transfer reactions) with log k_{obs}' values for the transformation of 8 halogenated aliphatic compounds by FeS produced only a weak LFER (R^2=0.48) (*17*). The poor R^2 value and the small slope of the LFER were attributed to several factors, including the influence of thermodynamic or molecular parameters other than one-electron reduction potentials on reaction rates, different mechanisms of adsorption to the FeS surface for different pollutants, different reaction mechanisms at the FeS surface, and/or significant adsorption of certain pollutants to non-reactive FeS surface sites (*17*).

Empirical linear correlation analysis has also produced results that may be used in predicting the rates of FeS-mediated pollutant transformation based on pollutant thermodynamic or molecular properties. One study (*17*) performed linear correlation analysis of log k_{obs}' values for 8 halogenated aliphatic molecules with five molecular properties: one-electron reduction potentials, lowest unoccupied molecular orbital (LUMO) energies, free energies of formation of aqueous phase radicals formed upon one-electron reduction, gas-phase homolytic bond dissociation enthalpies, and aqueous solubilities. Of these parameters, homolytic bond dissociation enthalpies (D_{R-X} values) were best correlated with log k_{obs}' values for FeS reductive dechlorination (R^2=0.82). The correlation between log k_{obs}' and D_{R-X} is illustrated in Figure 4. Another parameter shown

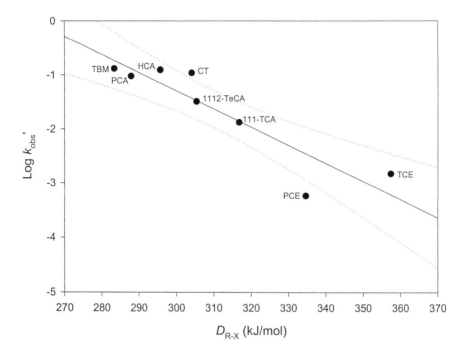

Figure 4. Log k_{obs}' versus D_{R-X}; 10 g/L FeS; pH 8.3. Dotted lines represent the 95% confidence interval. The linear equation is log $k_{obs}' = 8.8(\pm 4.9) - 3.4(\pm 1.6) \times 10^{-2} \times D_{R-X}$ ($R^2 = 0.82$). Units of k_{obs}' are h^{-1}. Chemical symbols not previously defined are TBM: tribromomethane; PCA: pentachloroethane; 1112-TeCA: 1,1,1,2-tetrachloroethane; and 111-TCA: 1,1,1-trichloroethane. Reproduced from ref. 17. American Chemical Society 2000.

to be reasonably correlated with log k_{obs}' values for FeS-mediated transformation of 8 halogenated methanes is the electron affinity of the halogenated methane (*19*).

Despite the limitations of existing correlations between log k_{obs}' values and thermodynamic or molecular parameters, LFERS or empirical linear correlations have the potential to predict at least order-of-magnitude trends in reactivity for chlorinated aliphatic compounds. These correlations may be useful in predicting the rates of *in-situ* natural attenuation based on molecular properties that are reported in the literature or that are readily computable using semi-empirical and quantum chemical computer models.

Application in Natural Attenuation and *In-Situ* Remediation

One current research challenge is to relate half-lives for laboratory experiments using pure FeS minerals to those for field-scale pollutant degradation reactions in sulfate-reducing environments (*19*). In the laboratory, however, FeS has been shown to degrade a range of halogenated aliphatic compounds with half-lives comparable to pollutant residence times in aquifers and sediments, i.e., hours to months, depending on available mineral surface area, pH, the presence of organic co-solutes, and the properties of the halogenated aliphatic compound. For many halogenated aliphatic pollutants, these half-lives are significantly shorter than those for hydrolysis, which can range from years to millennia (e.g., *66*). Consequently, FeS-mediated transformations must be considered when assessing the potential for natural attenuation of halogenated aliphatic pollutants in sulfate-reducing environments.

Since solution pH can significantly influence both reaction rates and the distribution of reaction products in FeS-mediated transformation reactions (*14, 15, 21, 22*), ground-water composition is likely to influence FeS reactivity, with even small differences in pH potentially affecting reaction rates or products. While FeS-mediated reaction rates are also significantly influenced by certain organic co-solutes, no organic co-solute has been found to completely inhibit the reductive dehalogenation reaction (*14*), indicating that FeS may be reactive even in waters with variable organic composition.

FeS may also be a useful alternative or addition to zero-valent iron metal for some *in-situ* reactive barrier applications (*11, 14*), particularly since FeS has been shown to be an order of magnitude more reactive per unit surface area than iron metal (*12*), and addition of pyrite and FeS to zero-valent iron has been shown to reduce the degradation time of CT (*8*). Factors affecting the long-term reactivity of iron sulfide minerals in field-scale remediation applications have yet to be explored, however. Generating FeS *in-situ* by enhancing the growth of sulfate reducing bacteria may also be a promising method for increasing or extending the reactivity of subsurface reactive barriers containing iron metal.

Pyrite has been added to iron metal in laboratory-scale reductive dehalogenation studies for the purposes of controlling pH (8, *44, 45, 67-73*) and removing dissolved oxygen (*73*). FeS has also been used as an additive to iron metal for controlling pH in laboratory-scale studies (*11*). A recent evaluation of the pilot-scale subsurface iron wall at Dover Air Force Base in Dover, Delaware indicated that use of a 10% pyrite/sand mixture in a pretreatment zone upgradient of the reactive iron portion of the wall was effective in removing dissolved oxygen from the ground-water passing through the pretreatment zone. The pyrite/sand mixture did not, however, result in significant pH control after the ground water passed into the reactive iron portion of the subsurface wall (*74*).

References

1. U. S. Environmental Protection Agency *Federal Register* **1985**, *50*, 46880 (Nov. 13, 1985).
2. U.S. Environmental Protection Agency "Common Chemicals found at Superfund Sites," (EPA 540/R-94/044), Office of Emergency and Remedial Response, Washington, D. C., 1994, as updated at URL: http://www.epa.gov/superfund/oerr/comtools/index/htm#generalpublications, April 1998.
3. Rickard, D. T. *Stockholm Contr. Geol.* **1969**, *26*, 49-66.
4. Freney, J. R. In *The Encyclopedia of Soil Science* Part 1, in *Encyclopedia of Earth Sciences*, Fairbridge, R. W. and Finkl, C. W., Jr., eds., Volume XII, Dowden, Hutchison and Ross, Inc., Stroudsburg, PA, 1979, pp. 536-544.
5. Kriegman-King, M. R.; Reinhard, M. in R. Baker, ed., *Organic Substances and Sediments in Water*, Vol. 2, Processes and Analytical, Lewis Publishers, Chelsea, Michigan, 1991, pp. 349-364.
6. Kriegman-King, M. R.; Reinhard, M. *Environ. Sci. Technol.* **1992**, *26*, 2198-2206.
7. Kriegman-King, M. R.; Reinhard, M. *Environ. Sci. Technol.* **1994**, *28*, 692-700.
8. Lipczynska-Kochany, E.; Harms, S.; Milburn, R.; Sprah, G.; Nadarajah, N. *Chemosphere* **1994**, *29*, 1477-1489.
9. Sivavec, T. M. U.S. Patent Number 5,447,639, 1995.
10. Sivavec, T. M.; Horney, D. P.; Baghel, S. S.; in *Emerging Technologies in Hazardous Waste Management VII*, American Chemical Society Special Symposium, September 17-20, 1995, Atlanta, GA, pp. 42-45.
11. Sivavec, T. M.; Horney, D. P.; Baghel, S. S. U.S. Patent Number 5,575,927, 1996.

12. Sivavec, T. M.; Horney, D. P.; *Preprints of Papers Presented at the 213th ACS National Meeting*, April 13-17, 1997, San Francisco, CA, Vol 37 (1), pp. 115-117.
13. Assaf-Anid, N.; Lin., K.-Y.; Mahony, J. *Preprints of Papers Presented at the 213th ACS National Meeting*, April 13-17, 1997, San Francisco, CA, Vol. 37 (1), pp. 194-195.
14. Butler, E. C.; Hayes, K. F. *Environ. Sci. Technol.* **1998**, *32*, 1276-1284.
15. Butler, E. C.; Hayes, K. F. *Environ. Sci. Technol.* **1999**, *33*, 2021-2027.
16. Devlin, J. F.; Müller, D. *Environ. Sci. Technol.* **1999**, *33*, 1021-1027.
17. Butler, E. C.; Hayes, K. F. *Environ. Sci. Technol.* **2000**, *34*, 422-429.
18. Hassan, S. M. *Chemosphere* **2000**, *40*, 1357-1363.
19. Kenneke, J. F.; Weber, E. J. *Preprints of Papers Presented at the 220th ACS National Meeting*, August 20-24, 2000, Washington, DC; American Chemical Society: Washington, DC, 2000; Vol 40 (2), pp 313-315.
20. Lee, W.; Batchelor, B. *Preprints of Papers Presented at the 220th ACS National Meeting*, August 20-24, 2000, Washington, DC; American Chemical Society: Washington, DC, 2000; Vol 40 (2), pp 338-340.
21. Weerasooriya, R.; Dharmasena, B. *Chemosphere* **2001**, *42*, 389-396.
22. Butler, E. C.; Hayes, K. F., manuscript in review, **2001**.
23. Berner, R. A. *J. Geol.* **1964**, *72*, 293-306.
24. Rickard, D. T. *Stockholm Contr. Geol.* **1969**, *26*, 67-95.
25. Lennie, A. R.; Fedfern, S. A. T.; Champness, P. E.; Stoddart, C. P.; Schofield, P. F.; and Vaughan, D. J. *American Mineralogist* **1997**, *82*, 302-309.
26. Rickard, D. T. *Stockholm Contr. Geol.* **1969**, *26*, 49-66.
27. Doyle, R. W. *Am J. Sci.* **1968**, *266*, 980-994.
28. Emerson, S. *Geochim. Cosmochim. Acta* **1976**, *40*, 925-934.
29. Davison, W. *Aquatic Sciences* **1991**, *53*, 309-321.
30. Pyzik, A. J.; Sommer, S. E. *Geochim. Cosmochim. Acta* **1981**, *45*, 687-698.
31. Lennie, A. R.; England, K. E. R.; Vaughan, D. J. *American Mineralogist* **1995**, *80*, 960-967.
32. Taylor, L. A. and Finger, L. W. *Carnegie Institution of Washington Year Book* **1970**, *69*, 318-322.
33. Lennie, A. R.; Redfern, S. A. T.; Schofield, P. F.; Vaughan, D. J. *Mineralogical Magazine* **1995**, *59*, 677-683.
34. Jellinek, F. In *Inorganic Sulphur Chemistry*; Nickless, G., Ed.; Elsevier: Amsterdam, 1968; p 720.
35. Kjekshus, A.; Nicholson, D. G.; Mukherjee, A. D. *Acta Chem. Scand.* **1972**, *26*, 1105-1110.
36. Vaughan, D. J. and Ridout, M. S. *J. Inorg. Nucl. Chem.* **1971**, *33*, 741-746.
37. Welz, D.; Rosenberg, M. *J. Phys. C: Solid State Phys.* **1987**, *20*, 3911-3924.

38. Clark, A. H. *Neues Jahrbuch fur Mineralogie Nonatshefte* **1966**(10), 300-304.
39. Belay, N. and Daniels, L. *Appl. Environ. Microbiol.* **1987**, *53*, 1604-1610.
40. Tezuka, M.; Yajima, T. *Denki Kagaku Oyobi Kogyo Butsuri Kagaku* **1991**, *59*, 517-518.
41. Nagaoka, T.; Yamashita, J.; Kaneda, M.; Ogura, K. *J. Electroanal. Chem.* **1992**, *355*, 187-195.
42. Hassan, S. M.; Wolfe, N. L; Cippolone, M. G. *Preprints of Papers Presented at the 209th ACS National Meeting*, April 2-7, 1995 Anaheim, CA, Vol. 35 (1), pp. 735-737.
43. Burris, D. R.; Delcomyn, C. A.; Smith, M. H.; Roberts, A. L. *Environ. Sci. Technol.* **1996**, *30*, 3047-3052.
44. Roberts, A. L.; Totten, L. A.; Arnold, W. A.; Burris, D. R.; Campbell, T. J. *Environ. Sci. Technol.* **1996**, *30*, 2654-2659.
45. Campbell, T. J.; Burris, D. R.; Roberts, A. L.; Wells, J. R. *Environ. Toxicol. Chem.* **1997**, *16*, 625-630.
46. Arnold, W. A.; Roberts, A. L. *Environ. Sci. Technol.* **1998**, *32*, 3017-3025.
47. Arnold, W. A.; Roberts, A. L. *Environ. Sci. Technol.* **2000**, *34*, 1794-1805.
48. March, J. *Advanced Organic Chemistry*, 3rd Ed.; Wiley: New York, 1985, p 873.
49. Castro, C. E.; Kray, W. C., Jr. *J. Amer. Chem. Soc.* **1966**, *88*, 4447-4455.
50. Fennelly, J. P.; Roberts, A. L. *Environ. Sci. Technol.* **1998**, *32*, 1980-1988.
51. Castro, C. E.; Kray, W. C., Jr., *J. Amer. Chem. Soc.* 1963, *85*, 2768-2773.
52. Vogel, T. M.; Criddle, C. S.; McCarty, P. L. *Environ. Sci. Technol.* **1987**, *21*, 722-736.
53. Bagley, D. M.; Gossett, J. M. *Appl. Environ. Microbiol.* **1990**, *56*, 2511-2516.
54. Picardal, F. W.; Kim, S.; Radue, A.; Backhus, D. In Emerging Technologies in Hazardous Waste Management 7; Tedder, Pohland, Eds.; Plenum Press: New York, 1997; pp 81-90.
55. Semadeni, M.; Chiu, P.-C.; Reinhard, M. *Environ. Sci. Technol.* **1998**, *32*, 1207-1213.
56. Wolfe, N. L. and Macalady, D. L. *J. Contam. Hydrol.* **1992**, *9*, 17-34.
57. Curtis, G. P.; Reinhard, M. *Environ. Sci. Technol.* **1994**, *28*, 2393-2402.
58. Fallab, S. Agnew. Chem. Internat. Edit. **1967**, 6, 496-507.
59. Conway, B. E.; Ku, J. C. H.; Ho, F. C. *J. Colloid Interface Sci.* **1980**, *75*, 357-372.
60. Wang, J. *Analytical Electrochemistry*; VCH, New York, 1994; p 2.
61. Millero, F. J. *Geochim. Cosmochim. Acta* **1985** *49*, 547-553.
62. Balko, B. A.; Tratnyek, P. G. *J. Phys. Chem. B.* **1998**, *102*, 1459-1465.
63. Pecher, K.; Haderlein, S. B.; Schwarzenbach, R. P. *Preprints of Papers Presented at the 213th ACS National Meeting*, April 13-17, 1997, San

Francisco, CA; American Chemical Society: Washington, DC, 1997; Vol 37 (1), pp 185-187.

64. Pecher, K.; Kneedler, E. M.; Tonner, B. P. *Preprints of Papers Presented at the 217th ACS National Meeting*, March 21-25, 1999, Anaheim, CA; American Chemical Society: Washington, DC, 1999; Vol 39 (1), pp 292-294.

65. Smith, R. M.; Martell, A. E. *Critical Stability Constants*; Plenum Press: New York, 1976.

66. Jeffers, P. M.; Ward, L. M.; Woytowitch, L. M.; Wolfe, N. L. *Environ. Sci. Technol.* **1989**, *23*, 965-969.

67. Burris, D. R.; Campbell, T. J.; Manoranjan, V. S. *Environ. Sci. Technol.* **1995**, *29*, 2850-2855.

68. Campbell, T. J.; Burris, D. R. *Preprints of Papers Presented at the 209th ACS National Meeting*, April 2-7, 1995, Anaheim, CA; American Chemical Society: Washington, DC, 1995; Vol 35 (1), pp 775-777.

69. Holser, R. A.; McCutcheon, S. C.; Wolfe, N. L. *Preprints of Papers Presented at the 209th ACS National Meeting*, April 2-7, 1995, Anaheim, CA; American Chemical Society: Washington, DC, 1995; Vol 35 (1), pp 778-779.

70. Harms, S.; Lipczynska-Kochany, E.; Milburn, R.; Sprah, G.; and Nadarajah, N. *Preprints of Papers Presented at the 209th ACS National Meeting*, April 2-7, 1995, Anaheim, CA; American Chemical Society: Washington, DC, 1995; Vol 35 (1), pp 825-828.

71. Allen-King, R. M.; Burris, D. R.; Specht, J. A. *Preprints of Papers Presented at the 213th ACS National Meeting*, April 13-17, 1997, San Francisco, CA; American Chemical Society: Washington, DC, 1997; Vol 37 (1), pp 147-149.

72. Allen-King, R. M.; Halket, R. M.; Burris, D. R. *Environ. Toxicol. Chem.* **1997**, *16*, 424-429.

73. Cipollone, M. G.; Wolfe, N. L.; Anderson, J. L. *Preprints of Papers Presented at the 213th ACS National Meeting*, April 13-17, 1997, San Francisco, CA; American Chemical Society: Washington, DC, 1997; Vol 37 (1), pp 151-152.

74. Remediation Technologies Development Forum. *Summary of the Remediation Technologies Development Forum Permeable Reactive Barriers Action Team Meeting*, February 16-17, 2000, Melbourne, FL, (http://www.rtdf.org/public/permbarr/minutes/).

Chapter 7

Non-Particle Resuspension Chemical Transport from Stream Beds

Louis J. Thibodeaux, D. D. Reible, and K. T. Valsaraj

Gordon A. and Mary Cain Department of Chemical Engineering, Louisana State University, Baton Rouge, LA 70803

The soluble fraction chemical release rates from contaminated stream bed sources to overlying water is more significant than once thought. This fact is supported by field measurements on the transport coefficients in rivers during time-periods when particle resuspension is absent. The numerical values of soluble release coefficients, k_f (cm/d), for PCBs in the Hudson, Grasse and Fox Rivers display similar magnitudes and an annual cyclic pattern. The lower values, 3 to 10 cm/d, occur in the winter while much higher values, 20 to 40 cm/d, occur in the summer. Candidate theoretical transport processes with characteristics capable of quantifying the observed magnitude and behavioral pattern of k_f were selected from the literature and reviewed. The bed-side process of bioturbation driven biodiffusion coupled to a water-side benthic boundary layer resistance was selected as the theoretical transport algorithm. With flow, temperature and other data from the Fox River numerical coefficient calculations were made using a PCB congener. Based on these results and other evidence it was shown that the proposed theoretical algorithm predicted the correct magnitudes of k_f and was capable of mimicking its cyclic behavior.

The bed sediments of many streams, estuaries, and harbors retain large quantities of organic and metal pollutants long after their discharges to water have been eliminated. As a consequence, water quality remains a problem in some of these streams due to bed-residing chemicals being transported back to the water column. In many cases bed sediment remediation is being proposed as a solution to the problem. Typically, the types of remediation being proposed include monitored natural recovery, in situ containment or treatment and excavation followed by containment or treatment of the dredged material. Prior to selecting one or more of these remedial option extensive studies are sometimes made of the natural recovery processes in the stream-bed and the associated response of the aquatic and biota ecosystem. The primary tool used in these studies are mathematical models of the stream system and the processes therein. These include stream geometry, flow hydraulics, particle behavior, chemodynamics and finally biotic uptake of chemical pollutants, These stream process models are supported by field and laboratory measurements and are used to project the future behavior of stream media concentrations to both active and passive events including remediation and storms. Recently published studies on the Fox (1,2), Grasse (3) and Hudson Rivers (4,5,6,) highlight the efforts underway.

Presently these models are undergoing rapid refinement as a consequence of advances in the scientific knowledge on chemical and biological processes in streams and similar aquatic environments. The availability of large sets of high quality chemical data and has aided these advances as well. These things happened within the last decade. This paper focuses on the advances that have occurred with understanding transport processes at the sediment-water interface. The processes that occur here commence chemical mobility and regulate its release rate from the bed. Traditionally the chemodynamic modeling of hydrophobic organics and metals has been tied primarily to tracking particle resuspension and deposition at the sediment interface of streams. Large mass quantities of polychlorinated organics (PCBs) are appearing in the water column under low stream-flow conditions, apparently without any particle resuspension occurring (2-6). The current chemical fate and transport (CFaT) models being used for these rivers are employing empirically derived algorithms to account for the release rate of this soluble fraction from the bed. The objective of the paper is to develop a theoretically sound framework for the transport phenomena that quantifies the release rate of the soluble chemical fraction from the bed in the absence of particle resuspension. In addition it will be used to explain and quantify the empirically derived transport algorithms.

Prior to developing the theoretical framework the basic principles of riverine CFaT models will be presented and the empirical evidence obtained from the three rivers will be reviewed. Next candidate transport process selected from the published literature will be reviewed. The alogorithm representing the theoretical framework will then be presented and applied to obtain numerical results transport coefficients for the Fox River. Finally, a discussion and comparison of field derived and theoretical coefficients will be made.

Chemical Fate and Transport Models

The state-or-the-art of chemical riverine modeling is captured in the applications ongoing with PCB in the Hudson River (4,5,6,). This site has received extraordinary study over several decades by federal and state authorities as well as the General Electric Company. The basic elements of the CFaT models presented here are derived from these sources using conventional nomenclature employed by practitioners in the field.

Water column model. A comprehensive chemical mass balance in the water column should account for mass change with time, advection and dispersion, particle deposition, soluble release, particle resuspension from the bed, evaporation to air and degradation. Over a differential distance x in the direction of flow (L) these processes are.

$$\frac{\partial C_T}{\partial t} = \frac{\partial}{\partial x}(UC_T) - \frac{D_S}{hm_w}(f_pC_T) + \frac{k_f}{h}(C_S - f_dC_T) + \frac{R_s}{hm_s}(f_{Ps}C_{TS}) - \frac{K_L}{h}\left(f_dC_T - \frac{C_a}{H}\right) \cdot S_x \quad (1)$$

The dispersion term is absent since dividing the reach into Δx completely mixed segments accomplishes dispersion numerically. In equation 1 t is time (t), C_T is soluble, particulate, and colloidal, concentration (M/L^3), U is average water velocity (M/t), D_s is particle deposition flux (M/L^2t), h is water column depth (L), m_w is suspended solids concentration (M/L^3), f_p and f_d are fractions chemical on particles and in solution, k_f is the soluble fraction bed release mass-transfer coefficient (L/t), C_s is the total, soluble and colloidal, concentration at the sediment-water interface (M/L^3), R_s is particle resuspension flux $(M/L^2 t)$, m_s is the particulate chemical concentration in the surface sediment (M/L^3), $f_{ps} C_{TS}$ is the fraction on particles and total chemical concentration in the surface sediment (M/L^3), K_L is the evaporation mass-transfer coefficient (L/t), C_a is chemical vapor concentration in air (M/L^3), H is Henry's constant (L^3 / L^3) and S_x is the chemical lost by reaction $(M/L^3 t)$. It is conventional to use the local or instantaneous equilibrium theory to quantify the dissolved fraction, f_d, particulate fraction, f_p, and colloidal fraction, f_{DOM}, in both the water column and bed. The equations needed to quantify these fractions appear elsewhere (4, 5, 6) and are omitted here for brevity.

Sediment bed model. Both particles and chemicals move across the sediment-water interface. This is reflected in the processes of deposition, dissolution and resuspension quantified above in the water column mass balance. It is conventional to segment the bed vertically in the z-direction. The transient equation describing the vertical distribution of chemicals within the bed should include an accumulation term, a Fickian-like dispersion formulation for chemical migration vertically both on particles and in solution, movement by porewater advection and burial by particle deposition of both solid and soluble fractions. The mass balance yields

$$\frac{\partial C_{TS}}{\partial t} = \frac{\partial}{\partial z}\left(E_p \frac{\partial p}{\partial z} + E_d \frac{\partial C}{\partial z}\right) - \frac{\partial}{\partial Z}(U_Z C + W_b C_{TS}) \qquad (2)$$

where E_p and E_d are the particle and soluble fraction including colloids "dispersion" coefficients (L^2/t), U_z is the inward Darcian porewater velocity (L/t), W_b in the bed accretion velocity (L/t), p is chemical concentration on particles (M/L^3) and C the soluble plus colloidal concentration in the porewater. Whereas porewater advection moves the soluble and the colloidal fractions the burial process captures all the phases. To be consistent with Eq. 1 $p \equiv f_{ps}C_{TS}$ and $C=(f_{ds}+f_{DOMS})C_{TS}$ in Eq. 2.

Together equations 1 and 2 are the CFaT portion of the riverine model. It is dependent open a solids mass balance module for determining the suspended solids concentration plus the particle deposition and resuspension fluxes. The hydrodynamic, particle balance as well as the biota up-take modules are beyond the scope of this study. In addition to the noted coupling to other models, equations 1 and 2 are coupled at the sediment-water interface through the flux expressions.

The quasi-steady state assumption. The riverine models described above are solved simultaneously in the transient case. A simpler approach will now be used. Although over long time-periods the accumulation process is necessary in both equations for shorter time-periods the quasi-steady-state (QSS) assumption applies (8). For a large bed-source mass the small chemical releases will not decrease it in the weeks to month time-period so that a steady-state conditions is closely approximated. For a stream of width w (L) of volume section whx and area A=wx (L^2), Equation 1 with the particle process terms omitted can be transformed to

$$\frac{-d}{dA}(Qf_dC_T) = k_f(C_s - f_dC_T) - K_L\left(f_dC_T - \frac{C_a}{H}\right) \qquad (3)$$

where Q is the volumetric flow rate (L^3/t) of the stream. Under the same QSS assumptions Equation 2 becomes

$$n = E_p\frac{dp}{dz} + E_d\frac{dC}{dz} - U_zC \qquad (4)$$

where the net outward flux of the chemical from the bed, $n(M/L^2t)$, is constant and equal to all the terms in Eq.3. Equations 3 and 4 are coupled mathematically at the sediment-water interface and the parameters k_f, E_p and h_s the depth of the surface mixed layers, are calibrated to chemical data. Together with the concentration state variables they capture the essence of the non-particle resuspension soluble fraction chemical release flux from the stream bed. In the next section field methodologies and evidence based on the soluble release processes will be reviewed. The more traditional particle resuspension approach will be reviewed also.

Field Evidence of Soluble Release

Traditional reverine modeling approaches that emphasize the particle process of erosion and deposition do not contain the k_f term in Eq. 3 or the two vertical dispersion coefficient terms in Eq. 4. However they do contain soluble release process element (1,7). It is assumed to be some form of solute diffusion in the porewater channels of the bed and the flux is quantified by $(D/h_s)(C_s-C_w)$. Where D is the effective diffusion coefficient in the porewater. Typically the D/h_s used is in the range of 0.1-1.0 cm/d (1,7). The magnitude of the chemical release contributed by this process is insignificant as will be demonstrated later. In order for these traditional models to accommodate the significant non-particle resuspension mass quantities released during the low-flow and mid-flow regimes of the river various calibration methodologies associated with the particle resuspension process were adopted. This accommodation was done in the CFaT PCB model for the Fox River (1, 2). In this instance the PCB water column data, for example, was used as a particle tracer to adjust "gross settling and resuspension" velocities. Termed "background" resuspension velocities they ranged from 0.0015 to 0.10 mm/d and were inadvertently used to quantify the soluble release fraction. The molecular diffusion soluble contribution was estimated to be at least one order of magnitude smaller than the background resuspension (1). A refined model for the Fox River retained the background resuspension scheme but also incorporated the soluble release rate shown in equation 3 to account for the exchange of this fraction between the water column and the top surficial sediment segment(2). The authors note that the k_f rates used were significantly higher than molecular diffusion. For the River above DePere Dam a value of 0.20 cm/day was used from December through March when ice cover is present and 39 cm/day for the remainder of the year. The authors note that the required higher rates during no-ice conditions may be due to the wave-driven exchange and higher boating and biological activity. For the River below DePere Dam the ice and no-ice calibrated values were 4.0 and 30 cm/d respectively. No direct information was available for determining particle mixing rates, E_p, or depth h_s. The model was therefore calibrated with a value of 5.1E-8 cm^2/s in the top sediment layer which was 1.25 to 10 cm in thickness.

More recent riverine modeling approaches attempt to disconnect the particle and chemical data calibration steps. Specifically, chemical data is not used to adjust parameters for the solids mass balance module (6). What follows is a review of the approaches taken by recent CFaT modelers and the results obtained using Eqns. 3 and 4 for calibrating the release of soluble PCB fractions on the Hudson and Grasse Rivers.

During low flow conditions the water in of the river is clear with very low total suspended solids (TSS) levels. During these periods the particle resuspension from the bed is assumed to be absent or negligible. A section of the stream is isolated for study using upstream (us) and downstream (ds) measurements of concentration and flow. This plus data on entering side-streams, bottom sediment porewater concentration, point discharges etc., was used to quantify or otherwise account for all other inputs into the section. Using $C=f_d C_T$ and solving Eq. 3 for k_f yields the following relationship for estimating the soluble release mass- transfer coefficient.

$$k_f = \frac{(QCl_{ds}-QCl_{us})}{AC_s} + \frac{K_L C}{C_S} \qquad (5)$$

It was assumed that the soluble concentration in porewater, C_S, is much larger than the average concentration in the water column, C, and that C_a/H is much smaller than C. These are very good assumptions in the case of the PCBs level in these rivers. Typically k_f is quantified using the flow-concentration data only with the evaporative contribution, the second group on the right side of the equality, omitted. Typically the evaporative contribution is estimated to be about 10% or less.

Sufficient data was collected around the Thompson Island pool (TIP) located between the Thompson Island Dam and Fort Edwards on the Hudson River, New York. The TIP is 5.9 miles in length, has an average velocity of 0.2 m/s, an average depth 2.5 m and average width of 216 m. Data on PCBs containing three or more chlorine atoms (PCB_{3+}) displaying k_f vs time-of-year is shown in Fig. 1 (6). During the winter months k_f is approximately 3 cm/day. It increases in early spring and peaks between 10 and 14 cm/day in the late spring to early summer and declines through the summer into fall. The timing of peak and subsequent decline suggest that it may be linked to increased flow or to biologically mediated mixing in the surface sediments (4, 6). On the basis of model calibration, a value of $E_p=10^{-7} cm^2/s$ was used for both cohesive and non-cohesive sediment. A mixing depth of 3 to 10 cm was used which was determined by calibration as well. Although both E_p and h_s were determined by calibration both are supported by the composition of the benthic community and vertical concentration profiles of ^7Be and PCBs observed in high resolution sediment cores (4).

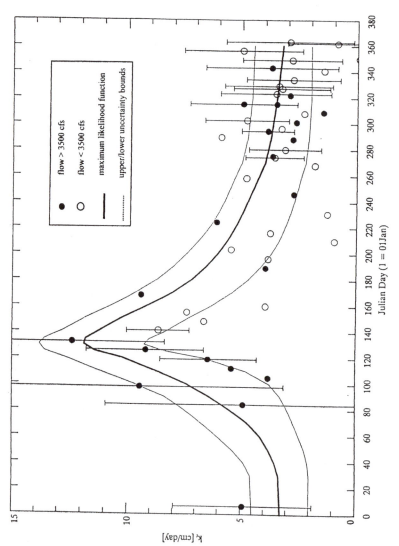

Figure 1. Sediment-water column PCB mass-transfer coefficient for the Hudson River (Reprinted with permission from reference 6. Copyright 2000 American Chemical Society)

Another study performed around the TIP for 1993-1997 data gave similar results (5). The seasonally-variable low flow effective mass-transfer coefficient, k_f, was observed on PCB_{3+} load gains across TIP under non-resuspending conditions. The k_f values were distinctly higher in the summer months relative to most of the year. From early May to mid-June k_f increases from about 10 to 25 cm/d and declines to about 10 cm/d at the end of August. The authors note that the peak value occurs in June and causal factors are poorly understood since water column temperature peaked in August. However, the peak does coincide with the time of the highest groundwater influx rates. Sediment particle mixing was determined on the basis of observed sediment core depth profiles, judgements on distributions of biological activity and model calibration to long term concentration trajectories. A value of 1.2E-6 cm²/s was used for E_p in the top sediment layer. Particle mixing depths, h_s, were 6 to 10 cm and a function of sediment cohesiveness (5).

The lower portion of the Grasse River, NY contains PCBs in its bed sediments as well. The contamination covers a six-mile stretch from the Massena Power Dam to the St. Lawerence River. Data collected by ALCOA over the time period 1997-98 were used along with model studies to understand the fate and transport of PCBs in the river. The river has relatively low water velocities; it has steep banks with water depths ranging from 10 to 15 feet and widths of 400 to 600 ft. Flow and concentration measurements were used with Eq. 5 to calculate k_f values. A seasonally varying pattern resulted. The general trend displayed an early summer peak between 6 and 7 cm/day a tail-off in autumn and a sustained low value throughout the winter months of 1-2 cm/day. The investigators described the mass-transfer coefficient as an empirical parameter that incorporates a number of processes occurring in the surface sediment, including molecular diffusion, turbulence, bioirrigation, bioturbation and advection. Since the lower Grasse River is predominantly colonized by aquatic worms (oligochaetes) and midges (chironomids) some degree of mixing must be occurring within the surficial sediments. Based on ^{210}Pb cores and model calibration, a partially mixed surficial sediment depth of 10 cm was chosen to represent conditions in the bed. Although the authors note that the magnitude of the particle mixing coefficient, E_p, is uncertain, a value of 5E-7 cm²/s was used for the upper 5 cm or the bed (3).

Riverine CFaT model applications were reviewed in this section with the emphasis on process algorithms used for quantifying the soluble fraction being released from the bed during non-particle resuspension flow conditions. It was shown that the more traditional models use adjustments to the solids mass balance resuspension algorithms to accommodate for what (apparently) is the soluble fraction. More recent process developments disconnect the solids and chemical mass balances by providing algorithms to accommodate it directly into the latter. Abandoning the very slow molecular diffusion release equation the new algorithm uses a calibrated transport parameters, k_f, for the processes that are occurring on both sides of the sediment-water interface. Some of the calibrated parameters display a seasonal pattern to which flow rate, biological activity and other factors ongoing in the aquatic environment are offered as possible forcing variables. In the next section a theoretical framework will be presented to support the soluble release processes characterized by k_f both qualitatively and quantitatively.

Theoretical Framework

What follows is a theoretical underpinning of the non-particle resuspension, soluble release processes represented by Eq. 3 and 4. A broad approach will be taken in that numerous transport processes will be considered. These will include the individual processes plus several combined processes that operate in series or parallel on both sides of the interface.

Particle Resuspension

Although clearly not a soluble transport mechanism, the particle resuspension process will be covered as well. Its use in traditional models was highlighted above and its inclusion here is so as to contrast it with known soluble release mechanisms. Theory and correlations are available to estimate these and the final algorithms used are based in-part on tests performed with bed-sediment core samples. The particle resuspension process term appears in Eq. 1. It is the fourth term on the right side. The portion that represents the so-called "resuspension velocity" is R_s/m_s and has the dimensions of L/t. It can be viewed as a solid particle mass-transfer coefficient. Calibrated values for it appear in Figure 2 for the Fox River(1). The values at and below 0.1 mm/d are background resuspension velocities since they were based on PCB concentrations and not suspended solids concentrations. Generally, for the Fox River water flow rates below 183 m^3 / s did not result in particle resuspension (2). So, it is highly likely that these low resuspension velocities represent the soluble fraction release from the bed.

Molecular Diffusion

This transport mechanism operates to move solute molecules through the water filled pore spaces between the sand, silt and clay clumps or grains that form the bed. The solids pieces are fixed thereby providing a torturous and partially blocked pathway for solute molecules driven by a higher concentration in the pore water at depth h_s than at the interface. As mention above this process

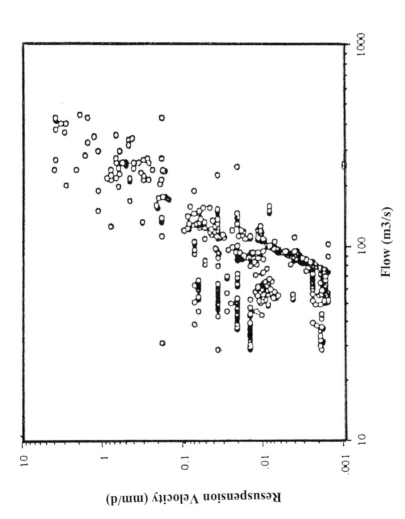

Figure 2. Calibrated resuspension velocities versus flow in the lower Fox River (Reprinted with permission of the International Association of Great Lakes Research, reference 1.)

is always included as a soluble release mechanism in traditional models even though it contribution to the chemical release rate is normally insignificant. Part of the solute in the pore water is associated with the dissolved organic carbon (DOC). The term DOC is used as an operational analytical chemistry designation to characterize colloidal organic matter in water. These carbon-rich fine particles are highly loaded with sorbed quantities of hydrophobic material and move toward the sediment-water interface by Brownian diffusion. Unlike solute molecular diffusion these particles respond to a DOC gradient in the bed (10, 12). Combined with the truly dissolved fraction the DOC fraction typically contributes an additional amount to the diffusive release form bed sediment (9, 12). A factor of 10 increase is typical for pore waters with 20 mg/L DOC (10). Algorithms for estimating both colloidal and solute diffusivities are readily available (9, 11).

Biodiffusion

Enhanced chemical transport in the upper sediment layer by the process called bioturbation is termed biodiffusion. It moves both the particle bound and soluble chemical fractions due primarily to the activities of macroinvertabrates and is not dependent on a chemical concentration gradient. The parameters E_p and E_d in Equations 2 and 4 are termed the particle and pore water biodiffusion coefficients respectively (L^2/t).

A recent review of the literature on the subject of bioturbation covers the early period of its discovery, through the development of theoretical models to describe the process in mathematical terms (13). Several theoretical approaches including box models (14), non-local transport (15), levy-flight (16), and Fickian diffusion formalisms (17) have been proposed to describe the process features and quantify the important aspects. Although the non-local and levy flight theories mimic animal activities more closely, in practice the Fickian diffusion approach dominates in the area of model applications. This approach is supported by data on in-bed particle reworking profiles derived form radioisotope studies interpreted with diffusion models. The transition form bioturbation of particles to soluble chemical fraction occurred with the focus on seabed minerals and nutrients (18). This was later followed by application to sparingly soluble organics where it was demonstrated that particle bound PCBs behaved similarly (19, 12). The activities and consequently the bioturbational effects of deposit feeding organisms are largely restricted to a thin surficial zone in marine sediments with a "worldwide, environmentally invariant mean of 9.8 cm depth and with a standard deviation of 4,5 cm," (20). This same layer thickness has been observed in freshwater sediments (9). Both particles and pore water are moved from depth to sediment /water interface and back down again in subsequent activities. This particle process provides a rapid mechanism of transporting bound contaminants to the bed surface. For strongly sorbed chemicals p >> C in Eq. 4 so that when $E_p \approx E_d$ the biodiffusion process for the particle-bound fraction is the much larger one. Typically however, the numerical values of E_d are 5 to 10 times greater than E_p. A correlation for E_p as a function of biomass density has been developed (22) as well as particle burial velocities, W_b (23). Recently a few laboratory measurements using three hydrophobic organic chemicals with oligochaetes in the bed support the E_p vs biomass density correlation (24).

The Benthic Boundary Layer

The region of the water column where the velocity is influenced by the sediment surface is referred to as the benthic boundary layer (BBL). At the sediment-water interface the water velocity must approach zero so that a slow moving layer exist here. Recent measurements reveal that the diffusive sublayer thickness for dissolved oxygen is 0.12 to 1.2 mm and decreases with increasing flow (21). Several correlations are available for estimating the soluble chemical mass-transfer coefficients, β, in the BBL (L/t). Values in the range of 0.02 to 2.5 cm/hour have been reported applicable to the marine environment (25). Apparently none have been measured in freshwater streams but flume experiments aimed at simulating river hydraulics yielded values in the range of 11.8 to 14 cm/hour (9). These cover both flat bed sediment surfaces (21,25) and surfaces with sand waves (9). When used in its flux equation form it replaces k_f in Eq. 3. Because of the data difficulties involved in measuring C_s field estimates of k_f likely include β plus some bed-side transport resistances as well. Therefore a distinction is made in the two coefficients.

Other Transport Processes

Only five individual processes are considered in detail in this paper. The five are generally operative at all sites and include the well known and significant ones for chemical release from bed-sediment to the water column. Porewater advection, which was mentioned above, can operate in both direction. When the water flow is downward and into the bed it retards the chemical release process. During the wet and dry seasons the groundwater surface elevation changes, with respect to the stream water surface elevation, and this causes changes in the advective flow directions. Normally so-called seepage meters can be used to quantify the magnitude and direction of the advective flow in sediment beds. Measurements conducted on the Grasse River indicate it is not significant at that site (3). Because of a lack of measured data for the other rivers it will not be included in this theoretical treatment, however it has the potential of being as significant process in some locations.

Localized porewater advection occurs through sand waves on the bottom as well as in and around other objects that protrude above the bed surface creating localized pressure gradients. Gas formation and bubbles released from the bed have been observed and numerically reported to be a possibly significant release mechanism. Conceptually the process directly moves a volatile and semi-volatile chemical mass in the gas phase. In addition the bubbles cause fine particles to be injected into the water column. Indirectly the bubbles may enhance transport by mixing sediment, bringing contaminated particles from depth to the surface. Some macro-inverterbrates have been observed injecting fine particles into the BBL as well. Other groups of inverterbrates residing in constructed tubes within the bed move porewater by a pumping action. The process is termed bioirrigation of burrows. The secretion of mucus by some benthic organisms results in the binding together of particles. When deposited on the surface a degree of armoring may result that retards particle resuspension and even soluble release. More research is needed before all of these processes, except porewater advection, can be quantified exactly and ranked as to their significance. Once that occurs, they can be included as chemical release processes from the bed.

Particle Concentration Normalized Coefficients

Including the particle resuspenion velocity, the details of five individual mass-transfer coefficients were presented above. These appear in Table I normalized to the chemical concentration on solid particles in the bed. Although the particle resuspension and particle biodiffusion transport coefficients remain unchanged the ones defined with solute concentrations require the $m_s K_D$ term for conversions to the equivalent particle concentration form. The K_D is the particle-to-porewater chemical partition coefficient (L^3/M). In doing so all can be compared on the same numerical basis.

Table I Particle Concentration Normalized Transport Coefficients (L/t)

Process	Coefficient
Particle resuspension velocity	R_s/m_s
molecular diffusion in the bed	$D/h_s m_s K_D$
Particle biodiffusion in the bed	E_p/h_s
porewater biodiffusion in the bed	$E_d/h_s m_s K_D$
Benthic boundary layer	$\beta/m_s K_D$

The resistance-in-series concept. The outward movement of a solute molecule under non-particle resuspending conditions is potentially controlled by transport resistances on both sides of the interface. Water is the continuous phase from the particle surface where the local equilibrium assumption (LEA) applies, through the bed porewater pathways and then through the interface to the water column. Derived elsewhere (9) the resistance-in-series flux algorithm is shown below. Using a concentration in solution difference it is

$$n = \frac{(C_S - C)}{\left(\dfrac{1}{\beta} + \dfrac{h_S}{(m_s K_D E_p + E_d)} \right)} \quad (6)$$

The LEA relates the chemical particle and porewater concentrations using K_D, the partition coefficient. All terms have been defined previously. It is seen that the particle and porewater biodiffusion processes terms are in parallel. They are added directly as conductances while the BBL coefficient β enters as an added resistance term.

The overall mass-transfer coefficient. As shown above the resistance-in-series concept provides a means for combining transport processes on either side of the interface. Defined as the ratio of the flux to the concentration on particles, K_S, the overall sediment soluble release mass-transfer coefficient (L/t) is

$$\frac{1}{K_S} = \frac{m_s K_D}{\beta} + \frac{h_S}{\left(E_p + \dfrac{E_d}{m_s K_D} \right)} \quad (7)$$

where the definition is $K_s \equiv n/\Delta p$. With this result it is possible to use theoretical or laboratory measured transport parameters to make quantitative prediction of the soluble release coefficient at a particular site in the riverine system. Other processes, such as molecular diffusion, can be included by adding an appropriate conductance term in the bracket.

The purpose of this section was to review the known or suspected individual processes for moving soluble chemicals near and within bed sediments. From this set those that: 1) are supported by data, 2) have a sound theoretical foundation and 3) have algorithms for making computations were selected as the candidate theoretical processes. They are listed in Table I. The well known and accepted resistance-in-series concept was used to connect processes on either side of the interface and to represent the combined contributions as the overall mass-transfer coefficient. Some coefficients are normally defined using a solute concentration basis and some are defined using a particle concentration basis. The particle concentration form was chosen to express all the coefficients so that numerical values can be compared directly.

Theoretical calculations representing a site on the Fox River will be used next to illustrate some numerical results.

Fox River Numerical Application

The bed-sediment of the Fox River contains PCBs and is under intensive study. CFaT models are being used to estimate the present and future releases to Green Bay (1, 2). Site conditions and model parameters for the Fort. Thompson Turning Basin will be used in this numerical study of transport coefficients.

Physical-chemical data representative of the site appear in Table II. A particular PCB congener, 2, 4, 2', 4'-tetrachlorobiphenyl was used to specify a molecular diffusivity, and value of the biodiffusion coefficient representative of Lake Huron sediments was used in the absence of one for the Fox River. The latter was corrected for temperature; a factor of 2 increase for each 10°C increase was assumed (14). A correlation developed for stream-beds was used to estimate β as a function of flow rate (9). The results of these computations appear in Table II.

Table II. Fox River Physical-Chemical Parameter

Parameter	Value	Source
Particle density (kg/m^3)	2600	1
Bed porosity (cm^3/cm^3)	0.78	1
PCB average K_D(m^3/kg)	13.4	1
Biodifusion depth (m)	0.05	20
Molecular diffusion depth (m)	0.05	20
Molecular diffusivity (cm^2/s)*	5.52 E-6	-
Biodiffusivity of particles (cm^2/s)		
$\quad$ 0°C	0.3 E-7	-
$\quad$ 20°C	1.2 E-7	9
$\quad$ 25°C	1.7 E-7	-
Benthic BL Coefficient (mm/d)		
$\quad$ Flow 50 m^3/s	.047	-
$\quad$ Flow 200 m^3/s	.096	-
$\quad$ Flow 500 m^3/s	.160	-

*2, 4, 2^1, 4^1-tetrachlorobiphenyl used.

The data in Table II was used along with the theoretical transport equations given for the individual process coefficients in Table I to make predictions for the Fox River. These calculated results appear in Table IIIb. Equation 7 was used to combine coefficients for a resistance-in-series computation. In addition to these calculated coefficients Table IIIa contains those obtained from field modeling studies on the Fox and the two additional rivers. Based on a particle density and an average PCB partition coefficients for each sediment the product $m_S K_D$ was determined for each river. They were 4150, 8640 and 7680 for the Hudson, Grasse and Fox Rivers respectively. The k_f values reviewed previously for each were divided by this $m_S K_D$ product to convert each to a PCB concentration on particle basis. These coefficients are listed in Table IIIa as a range of minimum and maximum values. Also listed in the table are the particles and background resuspension velocities for the Fox River. In all calculations the PCB concentration in the water column was assumed to be zero. Compared to the elevated particle concentration on the bed surface this is a realistic assumption. Calculations for more than one PCB congener should be performed for a more robust analysis.

Discussion and Conclusions

The objective of this paper was to develop a theoretically sound process framework for quantifying the release rate of the soluble chemical fraction from bed-sediment in the absence of particle resuspension. Both Equations 6 and 7 represent the final algorithm of this discussion. First, a brief review of what has gone before.

Table III. Mass-Transfer Coefficients for the Soluble* PCB Fraction (mm/d)

River	Coefficient	Source
a. Field Evidence		
Fox River		
Particle resuspension $(R_s/m_s)^+$	.20 to 4.3	1
Background resuspension (R_s/m_s)	.0015 to .010	1
Above DePere Dam (k_f)	.00026 to .051	2
Below DePere Dam (k_f)	.0052 to .039	2
Hudson River (k_f)		
Study 1	.0072 to .096	4
Study 2	.024 to .060	5
Grasse River (k_f)	.0012 to .0081	3
b. Theoretical Calculations		
Molecular diffusion	.00088	na
MD and BBL in series	.00085	na
Biodiffusion of particles	.0052 to .030	na
Benthic boundary layer	.047 to .16	na
Biodiffusion and BBL in series	.0049 to .024	na

*Coefficient based on PCB concentration on particles.
+This category is not a soluble transport one.
na- not applicable

CFaT riverine models were presented for both the water column and bed sediment. They were then simplified to focus onto the non-flow resuspension soluble fraction using the quasi-steady state assumption to isolate the key water-side and sediment-side process elements. Field evidence of soluble release based on CFaT model derived data was reviewed for three rivers. Both the traditional particle background resuspension process and more recent soluble fraction process algorithms data interpretation were covered. Numerical field calibrated resuspension velocities and soluble mass-transfer coefficients were presented. Candidate water-side and sediment-side transport processes, selected from the literature were reviewed. Those that provided the best theoretical explanation and contained laboratory and/or field data support were selected. Finally, the flux and the overall transport coefficient which captures the essential features of the framework were presented. Following this the theoretical mass-transfer coefficients were applied to a site on the Fox River below De Pere Dam. Numerical calculations were made for the transport coefficients for both individual and combined processes.

The focus of the discussion will be on the summary data shown in Table III. With one exception the data displayed in part a of the table are soluble fraction mass-transfer coefficients. Under the field evidence section of the table the first line contains particle resuspension velocities. There is little doubt that these large values are reflective of solids particles being eroded from the bed. For high stream flow and/or storm-like flow conditions these large values are not un-expected during these time-periods when the particle resuspension process is the dominant chemical transport process from the bed to the water column. The second line contains resuspension velocities that are much smaller in magnitude. These so-called background resuspensions values are in the range of the soluble k_f values listed in the following two lines of the table. These calibrated velocities were based on PCB concentrations in the water column and not on the suspended solids concentration in the water column. So, it is not suprising that they reflect the soluble release coefficients. Information as to the seasonal variation of the background resuspension velocities were absent in the study.

One very interesting aspect concerning k_f, the soluble release mass-transfer coefficient, is its seasonal behavior. Typically this variation is one with low values in the winter months; rapidly increasing ones in the spring, reaching a maximum in early summer then gradually returning to low values again late in the fall. Another interesting aspect is that the ranges of the k_f values for these rivers are similar. The values for the Grasse River are generally lower. However, this may be due to the Grasse being a very low velocity stream. So slow it becomes thermally stratified and takes on the characteristics of a lake. It is not altogether clear why the k_f ranges of the three rivers should be so similar. Equation 7 may provide a partial explanation.

Assuming that the observed k_f values are a measure of the BBL transport coefficient β in Eq. 7, then they should be a function of stream velocity. Correlations exist based on laboratory flume experiments that clearly show β increasing with flow. This may explain the spring-time k_f increase as due to the increase stream flow which normally occurs after snow melt. For example, it also explains why the values of k_f for the Grasse River, which stratifies like lake water, are lower than the values for the flowing streams. However, a strict interpretation of k_f as being equal to β is problematic. The way it is extracted from field data suggest it may contain other transport resistances. Equation 6, used to quantify k_f contains the concentration in the surficial sediment porewater, C_S. Ideally, if the field measured values of k_f are equal to β then the soluble and colloidal concentration is that at the sediment-water interface. The practicalities of obtaining a surficial sediment sample and extracting the porewater for measuring C_S and to assume that it is the interface concentration in nearly impossible. Sampling the top one centimeter of the surface sediment for C_S is possible but this is not the interface. Therefore, due to the sampling and analysis limitations, the k_f evidence appearing for the rivers in Table IIIa likely contains chemical transport resistance from both sides of the interface.

The overall transport coefficient K_s in Eq. 7, contains a particle biodiffusion coefficient, E_p and a particle mixing depth, h_s. In comparison the porewater contribution, E_d, is much smaller for hydrophobic organics and will be neglected in this discussion on PCBs. Data were presented in the field evidence section of the manuscript on the calibrated E_p and h_s values used for each of the three rivers. These values ranged from 0.5 E-7 to 12.5 E-7 cm^2/s for E_p and 1.25 to 10 cm for h_s. The individual values comprising these ranges are likely reflective of the variations of in upper sediment particle biomixing for the various rivers. The available theory on biodiffusion reviewed above suggest that population density and temperature are important independent variables. Population density is in turn dependent on bed properties such as fraction organic matter and particle size. So, it appears that the range of model calibrated E_p/h_s ratio may be such that its value dominates the magnitude soluble release coefficient k_f. Likely the coefficients ranges given in Table III are in part controlled by transport processes on either side of the interface as related by Eq. 7.

A key parameter appearing in Eq. 7 is K_D, the particle-to-water chemical partition coefficient. This parameter should reflect chemical desorption from the particles rather than absorption. There is considerable evidence in the published literature which demonstrates they are different, numerically. In Eq. 7 as K_D becomes large then $K_s \sim \beta/m_s K_D$ so that the BBL or the water-side resistance controls the magnitude of the overall transport coefficient. Recent work by the authors supports this prediction (26), however the concept is counter intuitive. Never-the-less the theoretical behavior of the K_D effect displayed in Eq. 7 suggest that as chemical binding or sequesteration potential to the particle increases the water-side resistance controls the transport coefficient magnitude. Additional research is needed to support this predicted effect.

Theoretical calculations based on Eq. 7 were performed for a site on the Fox River below DePere Dam and are summarized in part b of Table III. The predicted numerical values may be compared to the field evidence k_f values in part a of the table. The first value shown is for the process of solute molecular diffusion in the bed. It is 0.00088 mm/d and on the low range of the reported k_f values. The next line includes the BBL resistance. However, this coefficient is large, as seen by the entries in line four of the table, so it contribute little to the overall series resistance which is 0.00085 mm/d. Clearly, the molecular diffusion process is too slow to explain the large coefficients on the upper end of the k_f range in Table IIIa. Using an E_p value published for a lake located nearby, along with suggested average mixing depths and temperature variations for E_p, produces the range of biodiffusion mass-transfer coefficients given in line three of Table IIIb. Slight reductions in magnitude occur when BBL resistance is added; these appear in line five. The range of these theoretical coefficients are in good agreement with the field evidence on k_f shown in line four of Table 3a. The addition of the BBL resistance lowers the theoretical values slightly as shown in the last line in Table III.

Closure. This completes the discussions of the theoretical framework proposed for the non-particle resuspension soluble release transport from bed sediment. It provides a basis for explaining the numerical range and some of the behavioral character of the field observations. Currently riverine CFaT models are fitted with calibrated parameters for quantifying the soluble release. Whereas these more recent approaches are mathematically correct they lack a theoretical basis. The presented theoretical framework and associated algorithms provide this link. With this theoretical backing a means exist for estimating the magnitude of the numbers and for incorporating variations in the process parameters. Having available the theoretically correct process and associated algorithms provides a high degree of confidence in future predictions and extrapolations of the CFaT model. Based on the above theoretical framework a sound means now exist by which some CFaT model process parameters can be adjusted in order to accommodate different locations or changes in the riverine system at a fixed location.

References

1. Velleaux, M.; Endicott., D.; J. Great Lakes Res. **1994,** 20 (2) 416-434.
2. Limno-Tech, Inc. Development of an Alternative Suite of Models for the Lower Fox River. Prepared for the Fox River Group of Companies. Ann Arbor, MI. 1999, pp 29-31 and addendum p 21.
3. Alcoa. A Comprehensive Characterization of the Lower Grasse River. 1999, p 10-11.
4. QEA, LLC. PCBs in the Upper Hudson River. Vol. 2-A Model for PCB Fate, Tansport, and Bioaccumulation for General Electric, Albany NY. 1999, pp 4-24 to 4-26.
5. TAMS Consultants, Inc. Hudson River PCBS Reassessment RI/FS, Volume 2D- Book 1 of 4 Fate and Transport Models. 2000. Report for U.S. Environmental Protection Agency Region 2 and U.S. Army Corps of Engineers, Kansas City District.

6. Connolly, J.P.; Zahakos, H.A.; Benaman, J.; Zigler, C.K.; Rhea, J.D.; Russell, K. Eviron. Science Technol. **2000**, 34, 9.

7. Farley, K. J. etal. An Integrated Model of Organic Chemical Fate and Bioaccumulation in the Hudson River Estuary. Prepared for the Hudson River Foundation. Environmental Engineering Department, Manhattan College, Riverdale, NY. 1999.

8. Valsaraj, K. T.; Thibodeaux, L. J.; Reible, D. D. Environ. Toxicol. Chem. **1997**, 16, 391-396.

9. Thibodeaux, L. J. *Environmental Chemodynamics.* 2nd ed. John Wiley, N.Y. 1996, Ch. 5.

10. Thoma, G.J. et al. *Organic Substances in Sediments and Water*; Baker, R.A., Ed.; Vol. 1, Humics and Soils, Lewis Publishing: Chelsea, MI, 1991.

11. Chapra, S. C.; K. H. Reckhow. *Engineering Approaches for Lake Management* Vol. 2; Mechanistic Modeling. Butterworth Pub., Woburn, A., 1983, p. 184-185.

12. Reible, D.D., Valsaraj, K.T.; Thibodeaux, L.J. *Hbdk of Environmental Chemistry*, Hutzinger, O., Ed.; Ch. 3, Vol 2 Part F, Reactions and Processes, Springer-Verlag Gmbh. Germany; 1991, p 185-228.

13. Thoms, S. Models for Alteration of Sediments by Benthic Organisms; Project 92-NPS-2; Water Environmental Research Foundatioin; Alexandria, VA, 1995.

14. Berner, R. A. *Early Diagnesis: Theoretical Approach*, Princeton U. Press, Princeton, N. J. 1980

15. Boudreaux, B. P. Am. J. Sci. **1986**, 1, 286, p 161-198.

16. Mohany, S. 1997. MS thesis, Louisiana State University, Baton Rouge, LA, 1997.

17. Guinasso, N. L.; Schink, D.R. Jo. Geophysical Res. **1975**, 80, 21, p 3032-3043.

18. Aller, R. C. Ph. D thesis, Yale University, New Haven, CN, 1997.

19. Thibodeaux, L. J., et al. A Theoretical Evaluation of the Effectiveness of Capping PCB Contaminated New Bedford Harbor Bed Sediment. Final Report, Hazardous Waste Research Center, Louisiana State University; Baton Rouge, LA, 1990.

20. Boudreaux, B. P. Limnol. Oceanography. **1998**, 43, 3.

21. Steinberger, N; Hondoz, M. Journal of Environmental Engineering. **1999**, 125, 192-200.

22. Matisoff, G. *Animal Sediments Relations*; by McCall, R. L.; Tevasz, M. I. S, Eds.; Mathematical Models of Biochemistry, Ch. 8, Plenum Press, New York, 1982.

23 Boudreaux, B. P. *Diagenetic Models and Their Implementation.* Springer, NY. 1997, pp 138-141.

24. Reible, D. D.; Popov, V.; Valsaraj, K. T.; Thibodeaux, L. J., Lin, F.; Dikshit, M.; Todaro, A.; Fleeger, J. A. Water Res. **1996**, 30, 3.

25 Boudreaux, B. P. 1997. Op cit. p. 178-186.

26. Thibodeaux, L. J.; Valsaraj, K. T.; Reible, D. D. Environmental Engineering Science. **2001**, (accepted 19 Dec. 2000).

Chapter 8

Development of the Speciation-Based Metal Exposure and Transformation Assessment Model (META4)

Application to Copper and Zinc Problems in the Alamosa River, Colorado

Allen J. Medine[1], James L. Martin[2], and Elizabeth Sopher[1]

[1]HydroQual, Inc., 900 Valley Lane, Boulder, CO 80302
[2]Engineer Research and Development Center, Waterways Experiment Station, U.S. Army Corps of Engineers, 3909 Halls Ferry Road, Vicksburg, MS 39180

The Metal Exposure and Transformation Assessment (META4) Simulation Program is a generalized metals transport, speciation, and kinetics model developed for application to a variety of receiving waters. Algorithms for the simulation of crucial metal transformation processes, such as aqueous speciation, sorption/desorption, chemical precipitation/ dissolution, pH dynamics and kinetics were added to the basic structure of the Water Analysis Simulation Program (WASP, Version 4.32), resulting in the META4 model. The modeling of metal dynamics in the Alamosa River downstream of the Summitville Mine Site (CO) was performed to support the Use Attainability Analysis and ultimate goals for remedial actions. Following calibration, the model was used to hindcast metal concentrations for pre-mining conditions, prior to 1873, and conditions prior to heap-leach mining beginning in 1984 (pre-Galactic). Results demonstrate much improved water quality conditions in the Alamosa River during both eras. Analysis of data collected after significant metal load reductions at the Summitville mine site (1998) provided a verification of model framework and configuration for the original analysis.

Aquatic resource management strategies for toxic metals, including waste allocation, remedial action (restoration) or total maximum daily loading, are best evaluated by an estimation of the impacts of processes affecting metals concentrations in the water and sediments. These types of analyses are complicated by the fact that metal behavior is non-conservative in aquatic systems and that the transport, transformations, and attenuation depend upon the particular forms of metal present and chemical attributes of the environmental system. A chemical equilibrium simulation routine, based upon a solution approach similar to MINTEQ, was developed as a submodel to the generalized Water Analysis Simulation Program (WASP) to support these analyses. The META4 (Ver. 3.0) submodel addresses metal speciation and kinetics for the metal reactions and uses a solution approach similar to that of MINTEQA2 (Ver. 3.11) developed by, and distributed by EPA (1).

Physical and chemical processes that affect the transport of metals are taken into account in the model including advection, dispersion, erosion, sedimentation, and chemical reaction (speciation, adsorption, desorption, precipitation, and dissolution). Algorithms for the simulation of crucial metal transformation processes are thoroughly described in the User Manual (2). The modeling procedure is illustrated in Figure 1.

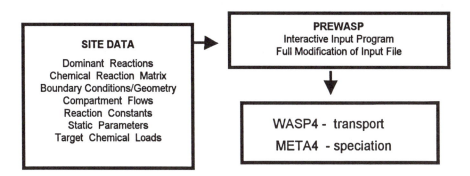

Figure 1. The procedure used for modeling with WASP4/META4.

Some of the advantages of WASP4/META4 are that it can represent 1, 2 or 3-dimensional environments (streams, rivers, reservoirs, multiple benthics), sequential deposition or scouring of benthic bed layers, transient storage, constant or variable pH, numerous point and non-point loads, and multiple metal and major ion reactions, including individual aqueous species. WASP4/META4 was used to model metal dynamics in the Alamosa River downstream of the Summitville Mine (3)

Site Background

The water quality of the Alamosa River has been impacted by naturally mineralized bedrock in the watershed and by mining activities in the basin that began in the late 19th century and continued through most of the 20th century. Gold was initially discovered at Summitville by early Spanish explorers and then developed by settlers in the San Luis Valley in the 1870s (4). Silver, copper and lead have also been produced at the site. The first amalgamation mill was erected in 1875 and, by 1883, the district was the largest gold producer in the state, with nine mills. More recent operations at the site include copper exploration in the late 1960s, gold and copper exploration in the late 1970s, and further gold prospecting in the early 1980s. In 1968, significant disturbances were evident at the base of the mountain (12 acres). Wightman Fork, the stream leaving the site, was an orange-red color downstream. By 1980, the additional disturbed area increased to 33 acres.

In 1984, Galactic Resources Limited (SCMCI) obtained a permit to operate a heap leach gold mining operation. Construction of the leaching pad and operation began in the summer of 1986. Twenty million tons of rock were removed from an open pit in the 3 years of operation. Development of the open pit resulted in the increase of copper concentrations emanating from the Reynold's Adit from approximately 25 mg/L to a high of 650 mg/L (5); inaccurate water-balance predictions and pump breakdowns caused direct release of cyanide and copper contaminated fluids; and placement of sulfide rich waste rock piles caused acid drainage and metals contamination in Wightman Fork. The disturbed area at the site increased to 633 acres. Numerous operating, environmental, and financial problems led to bankruptcy of the company in 1992. The USEPA initiated Emergency Response Actions in December, 1992 to control contaminant releases from the site.

Alamosa River Study Area

The study area for this report includes the Alamosa River and tributaries upstream of Terrace Reservoir. Major tributaries that have significant impacts on water quality in the Alamosa River include Wightman Fork, Iron Creek, Alum Creek and Bitter Creek. Wightman Fork contributed the greatest mass of copper and zinc to the river. For example, during 1993, copper and zinc loadings were, in general, 100X greater from Wightman Fork compared to other tributaries (6). Other tributaries generally provide dilution water that tend to improve overall water quality (Figure 2).

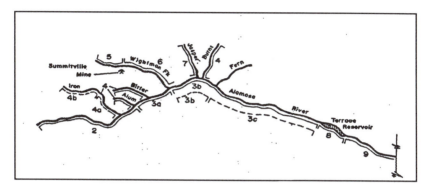

Figure 2. Study area for the Alamosa River modeling showing Colorado State River Segments (Source:6).

The Use Attainability Assessment (UAA) as well as recent USEPA monitoring indicated that water quality standards in the basin were not being met due to a combination of the natural geologic conditions and metals contamination due to mining. The UAA also discussed historic aquatic life uses in the basin (6). Anecdotal evidence from local persons and a variety of other Forest Service, Division of Wildlife, and USEPA data are reported to indicate that the Alamosa River supported at least a limited fishery until 1990. While there is evidence of mining activity in the upper Alamosa River, it was determined to be a minimal influence of stream water quality.

Conceptual Model for Metals Dynamics in the Alamosa River

The Conceptual Model for contaminant source, transport, transformations, fate, and resource impacts indicates that the primary mechanisms responsible for metal transport in surface waters are advection and dispersion. Advection is typically the dominant process responsible for metal transport in dynamic surface waters such as streams. However, in the stream sediment layer, or benthos, where water velocities are much lower, dispersion may become a significant mechanism for chemical release and transport. Additional dilution and chemical input from adjacent tributaries and alluvial groundwater contribute to the observed concentrations of contaminants in the Alamosa River.

Analysis of existing data indicates that the Alamosa River upstream of Station 45.5 (above Wightman Fork) and Wightman Fork are the major contributors to metal degradation in Alamosa River below the confluence with Wightman Fork (Colorado State Segments 3b and 3c). Chemical sinks within the Alamosa River include coprecipitation and sorption losses to the bed region, alluvial system transfers, and settling of particulate metals.

Adsorption of dissolved species is the most significant attenuation process for copper; zinc adsorption is limited when pH is less than 6.5. In many surface waters affected by mining, including the Alamosa River, it is generally accepted that sorption of metals to iron oxyhydroxides is the dominant chemical process regulating the dissolved concentrations of metals in waters and sediments (7). Chemical precipitation of Fe and Al is very rapid under pH conditions greater than 4 while pH's greater than 6.0 accentuate the sorption properties of suspended and bed sediments for copper. While adsorptive processes are very effective in removing metals from the water column, the equilibration of the sediment porewater with the contaminated sediments will generally result in the contaminated porewater. Stream flow changes that result in an increase in sediment resuspension can lead to significant releases of dissolved metals to the overlying water column during such transient events.

Modeling Alamosa River Water Quality

The modeling of the Alamosa River basin included the area from Wightman Fork downstream to Terrace Reservoir in southern Colorado (Figure 2). The objective of the modeling was to evaluate the likely pre-mining (1873) and pre-Galactic (1984) metal concentrations in the Alamosa River and potentially achievable environmental conditions. A pre-mining and pre-Galactic analysis of both the Alamosa River upstream of Wightman Fork (AR45.5) and Wightman Fork (WF0.0) was needed to serve as point sources (upper boundary condition) to the model segments downstream of Wightman Fork. Evaluation of current and historical water quality data was performed to provide an estimate of seasonal conditions in Wightman Fork prior to SCMCI's operations and prior to mining (*3*), and these estimates were then used as input to the META4 model. The model input at AR45.4 (upstream boundary) for the calibration was based on a September, 1995 synoptic sampling conducted by the USGS while the model inputs developed for pre-SCMCI and pre-mining for copper, zinc, iron, manganese and aluminum are shown in Tables 1 and 2.

Table 1. Pre-Mining concentrations in the Alamosa River below the mouth of Wightman Fork (AR 45.4)

	Concentrations (μg/L)							
	Winter		Spring		Summer		Fall	
	Mean	85th-ile	Mean	85th-ile	Mean	85th-ile	Mean	85th-ile
Cu (Diss)	36.9	35.1	11.7	30.5	7.58	19.1	13.7	22.1
Zn (Diss)	105	91.8	22.4	60.0	26.5	41.2	44.8	82.2
Fe (Total)	6,680	7,790	4,410	10,900	2,010	5,100	5,710	8,940
Mn (Diss)	565	557	126	378	124	207	282	544
Al (Diss)	3,550	4,370	422	2,380	112	215	1,440	2,340

Table 2. Pre-Galactic concentrations in the Alamosa River below the mouth of Wightman Fork (AR 45.4)

	Concentrations (μg/L)							
	Winter		Spring		Summer		Fall	
	Mean	85th-ile	Mean	85th-ile	Mean	85th-ile	Mean	85th-ile
Cu (Diss)	211	233	88.6	179	141	154	136	158
Zn (Diss)	127	137	57.2	94.0	82.6	131	147	268
Fe (Total)	6,920	8,290	4,590	11,100	2,540	6,000	6,210	9,780
Mn (Diss)	580	647	189	503	224	405	475	929
Al (Diss)	3,460	4,420	456	2,440	197	334	1,540	2,500

Modeling Framework for the Alamosa River

Based on the Conceptual Model, the relevant, interacting physical-chemical processes that affect overall water quality were included in the modeling framework. The water quality conditions used for the inputs to the model represented the worst case seasonal water quality observed in the Alamosa River above Wightman Fork and as estimated for Wightman Fork at the mouth. The fall and winter periods represent the greatest degradation of water quality for copper and zinc in the Alamosa River downstream of Summitville. Of these two seasons, the best monitoring data for modeling was available for only the fall period; consequently, the model simulations were based on fall flow conditions.

Modeling Approach

The META4 model simulates transport of flows and loads through compartments in the water column and underlying sediment as equilibrium chemistry is calculated for each segment, including the interactions with adjacent compartments. The modeling process involved the evaluation of site data as shown in Figure 1. Model input data included volumetric flows, metal loading rates, and inflow metal concentrations. Other data needs included calculation time step, model compartmentalization, cell volume, channel geometry, volumetric flow rates, and metal-sorbent reaction constants. Channel geometry and compartments volumes were estimated from time of travel information for the fall monitoring data collected by the USGS (September, 1995).

More complex sediment modeling, as used within META4, required a delineation of the importance of dominant mechanisms which affect accumulation of metals in sediments and potential releases of metals from sediments. The accumulation of metals in sediments generally includes five major mechanisms: 1) adsorption onto fine-grained materials, including oxides, 2) precipitation of trace element compounds, 3) coprecipitation with Fe, Al and Mn oxides/carbonates, 4) association with organic matter, and 5) incorporation in crystalline minerals. Due to the dominance of the sediment sorption reactions by iron precipitates, only adsorption to iron was simulated in the model. The model configuration included copper and zinc speciation as affected by the composition of major water quality variables. The major parameters used to describe the speciation of metals, and the corresponding effects on sorption and desorption by the sediments, include H^+, CO_3^{2-}, Ca^{2+}, SO_4^{2-}, Mg^{2+}, Cu^{2+}, Zn^{2+} and three classes of adsorbing solids: Solid1 (iron oxides), Solid2 (aluminum oxides), and Solid3 (residual source/sink surfaces).

Chemical Reactions

The reactions to be included in the META4 submodel for WASP4 were determined after a series of chemical speciation calculations using MINTEQA2 (Versions 3.11). Using a variety of water quality conditions, including estimates of future conditions, only the reactions that represented a significant percentage of each individual metal were included in the model. Generally, reactions were excluded if they comprised less than 0.2 percent of the dissolved metal concentration. The two-layer sorption model (diffuse layer or double layer model) as implemented in the geochemical model MINTEQA2 is regarded as the most representative approach for modeling metal sorption to iron oxides and was used for the modeling. Constants for hydrous iron oxide sorption of copper and zinc to strong and weak sites were based on literature (7).

Model Calibration

The modeling for Segments 3b and 3c (Wightman Fork to Fern Creek: Fern Creek to Terrace Reservoir) of the Alamosa River includes 14 surface water compartments along the main channel along with the corresponding 14 benthic compartments for each water column compartment (Figure 3). Model compartments were developed from information concerning the physical and chemical characteristics of stream reaches (i.e. slope, hydrology, sediment type) and locations for major loads to the system.

The specification of overall model details including control parameters, system configuration, hydrology, boundary conditions, reactions, loadings and other model parameters were developed for the model input file. The model was calibrated to the September, 1995 monitoring data. The initial calibration activity, following the balancing of flows and travel time, included the simulation of a conservative anion (to check on flow balancing) and total suspended solids (TSS represented by particulate iron) within the Alamosa River from Wightman Fork to Terrace Reservoir. After solids were calibrated, subsequent steps included the combined calibration of benthic metal concentrations and total/dissolved metal in the water column of each compartment (3).

Sediment conditions for the adsorbed fraction of zinc and copper were determined from Terrace Reservoir data (8). The data showed that the operationally defined labile zinc and copper in the sediments were about 49.2 ± 5.26 % and 23.1 ± 3.62 % of the total sediment burden of these metals. The iron oxide fraction for iron was approximately 21.8 ± 7.10 % of the total iron content. The conditions were used to describe the partitioning of copper and zinc to the sediments as reflected in double layer sorption model used in WASP4/META4.

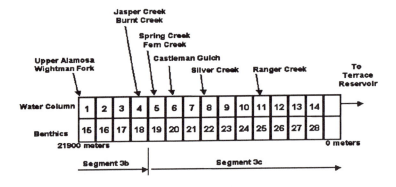

Figure 3. Model compartmentalization, upper boundary and tributary locations.

The results of the calibration for water column total zinc and total copper are shown in Figures 4 and 5, respectively (dissolved copper and zinc were nearly equal to total metals and are not shown). Both total and dissolved metals in the water column compartments show excellent agreement with observed data and represent the combined effects of loading, dilution, partitioning and attenuation. Three observed data points were available for comparison at AR43.6, AR41.2 and AR34.5 for the September, 1995 modeling. The relative percent difference for each paired data for observed and modeled metal concentration is generally well below 5%. Following calibration, a Monte Carlo simulation was used to describe the potential pre-Galactic and pre-mining water quality.

Monte Carlo Simulation of Pre-Galactic and Pre-Mining Chemistry

The model simulations performed for pre-Galactic and pre-mining included calculations of both dissolved and total zinc and copper in the various model compartments from Wightman Fork downstream to Terrace Reservoir. From a modeler's perspective, addressing uncertainty in the application of numerical modeling to Summitville was central to sound decision-making. Sensitivity analysis followed by uncertainty analysis provided a greater ability to make informed decisions. One of the more basic aspects of uncertainty analysis was statistical in nature and was based on a comparison of the calibrated model with the observed distributions of metal concentrations as described below (Calibration Prediction Error). Obviously, this comparison is not possible for the pre-mining and pre-Galactic scenarios. To address the model uncertainty associated with input parameters, it was decided that Monte Carlo simulation, widely used in water quality modeling, would best reflect this uncertainty.

After a number of model sensitivity runs, it was determined that the uncertainty could be best reflected by varying copper and zinc concentrations in the Alamosa River (above Wightman Fork) and in Wightman Fork. Input files for 100 simulations were randomly constructed from the statistical distribution of the original data sets from WF0.0 and AR45.5 for the upstream boundary condition. After the input files were assembled to describe the pre-Galactic and pre-mining chemical inputs in the first model compartment (as described for Tables 1 and 2), the model was run and the data entered into a database management program for analysis.

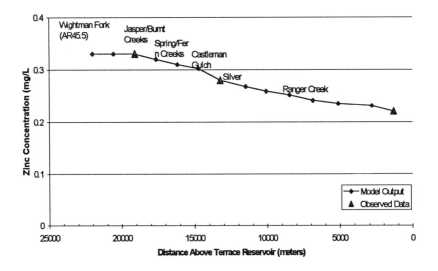

Figure 4. Model calibration for total zinc during September, 1995.

Pre-Galactic Water Quality

The pre-Galactic water quality for copper and zinc during fall (Figures 6-7) likely represent the worst-case seasonal values in the Alamosa downstream of Wightman Fork. Copper concentrations show a pronounced effect of the iron oxide partitioning in the water column and well as reduced transport of copper due to retention by stream sediments.

160

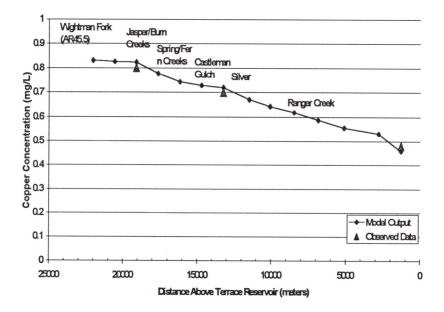

Figure 5. Model calibration for total copper during September, 1995.

Concentrations of dissolved and total zinc were nearly identical and indicated the limited partitioning of zinc to the iron oxides at pH conditions in the Alamosa River (<7.0) for the pre-Galactic scenario. While the concentration of total copper is reduced by 82% from Wightman Fork to Terrace Reservoir through combined adsorption (to bed sediments) and dilution, the zinc concentration is reduced by 47%. Zinc and copper are both removed to the bed sediments, although zinc is removed less efficiently. If pH were in the range of 7.5-8.0 in the system, both zinc and copper adsorption would be enhanced and the associated dissolved metal in the water column would be decreased further.

Pre-Mining Water Quality

The pre-mining water quality for copper and zinc during fall time is shown in Figures 8-9 and, as for copper, would represent the highest seasonal concentrations for both dissolved and total metals in the Alamosa River downstream of Wightman Fork. Total copper concentrations decrease by 66%

while total zinc concentration decreases by only 16%. The main difference between the efficiencies of removal is again related to system pH. Modeling indicates that during the pre-mining era the dissolved copper and zinc concentrations would be below 4 µg/L and 50 µg/L, respectively, and would not result in significant impacts to fisheries. Pre-Galactic concentrations are higher and were found to be below 30 µg/L and 130 µg/L, respectively, in the lower portion of the Alamosa River below Silver Creek. The water quality for other seasons is expected to have been better than these estimates and would have likely permitted fisheries to be established as far upstream as Spring and Fern Creek under the pre-Galactic modeling scenario.

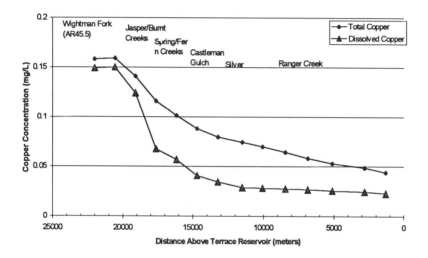

Figure 6. Simulated Pre-Galactic copper concentrations during fall.

Summary and Conclusions

This modeling study evaluated the likely pre-mining (prior to 1873) and pre-Galactic (prior to 1984) concentrations in the Alamosa River downstream of Wightman Fork. Following model calibration, Monte Carlo simulations estimated water quality after calculating the effects of adsorption, advection, settling, dilution, and dispersion. The Monte Carlo simulation provided an assessment of model uncertainty with respect to the boundary conditions. The combined effects of these processes indicate that in the pre-Galactic scenario, using worst-case fall conditions, 82% of the copper and 47% of the zinc were removed, and in the pre-mining scenario 66% of the copper and 16% of the

162

zinc were removed in the Alamosa River between Wightman Fork and Terrace Reservoir. Based on water quality standards and anecdotal evidence, the modeled concentrations in the Alamosa River would have been adequate to support a fishery in the river prior to mining, and in the pre-Galactic period, would likely have supported a fishery in the river downstream of Fern and Spring Creeks.

Acknowledgement

The authors would like to thank Mr. Ed Bates, Project Manager, USEPA, National Risk Reduction Management Laboratory, Cincinnati, for his foresight in funding development and application of the model.

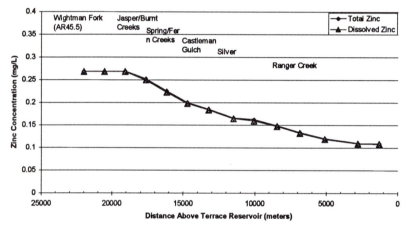

Figure 7. Simulated Pre-Galactic zinc concentrations during fall.

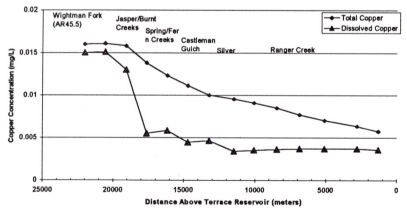

Figure 8. Simulated Pre-mining copper concentrations during fall.

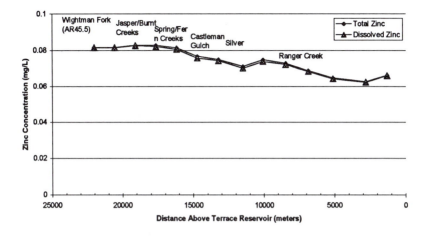

Figure 9. Simulated Pre-mining zinc concentrations during fall.

Literature Cited

1. Allison, J. D.; Brown, D. S.; Novo-Gradac, K. J. MINTEQA2/ PRODEFA2, Geochemical Assessment Model for Environmental Systems: Version 3.0 User's Manual, EPA/600/3-91/021, USEPA, 1991.

2. Martin, J. L.; Medine, A. J. META4 - Metal Exposure and Transformation Assessment Model, Model Documentation for Version 3, December, 1998.

3. Medine, A. J. Technical Assessment of Pre-Mining and Pre-Galactic Water Quality for Wightman Fork and the Alamosa River. USEPA, National Risk Management Research Laboratory, Cincinnati, OH, 1997.

4. Shriver, D. R. History of the Summitville Mining District. The San Luis Valley Historian, Volume XXII (1), 1990.

5. Pendleton, J. A.; Posey, H. H.; Long, M. B. Characterizing Summitville and Its Impacts: Setting the Scene. In: Proc.: Summitville Forum '95. CO Geol. Survey, Special Publication 38, Denver, Colorado, 1995.

6. CDPHE. Use Attainability Assessment, Alamosa River Watershed through 1996. Posey, H. H.; Woodling, J.; Campbell, A.; and Pendleton, J. A. USEPS and Colorado WQCC, July 12, 1996.

7. Dzombak, D. A.; Morel, F. M. M. *Surface Complexation Modeling, Hydrous Ferric Oxide.* John Wiley and Sons, New York, 1990.

8. Horowitz, A. J.; Robbins, J. A.; Elrick, K. A.; Cook, R. B. Bed Sediment-Trace Element Geochemistry of Terrace Reservoir, near Summitville, Southwestern Colorado. USGS, Open-File Report 96-344, 1996.

Chapter 9

Transport and Transformation Processes Affecting Organophosphorus Insecticide Loads in Surface Runoff

Joel A. Pedersen and I. H. (Mel) Suffet*

Environmental Science and Engineering Program,
University of California, Los Angeles, CA 90095

Surface runoff from agricultural fields represents an important route for the introduction of organophosphorus insecticides into aquatic ecosystems. Concentrations in surface runoff reflect the summation of numerous physical, chemical and microbial processes that alter the amount of organophosphorus compounds available for transport in overland flow and shallow interflow. Volatilization and transformation processes serve to reduce the amount of parent compound available for transport in runoff. Sorption of organophosphorus insecticides to inorganic and organic soil components determines their rate of transfer from soil solution to overland flow and their mode of transport in surface runoff. Dissolved insecticides are transferred into overland flow by molecular diffusion, raindrop impact induced turbulent diffusion and shear stress induced mass transfer. Sorbed insecticides are mobilized and transported by soil erosion processes. Processes controlling insecticide flux from the soil matrix into overland flow influence the magnitude and temporal variability of organophosphorus insecticide loads in surface runoff from cultivated lands.

Introduction

Organophosphorus insecticides are among the most widely used insect control agents in the United States. Due to their efficacy, reduced lipophilicity and decreased environmental persistence, they have largely replaced the organochlorine pesticides. Organophosphorus insecticides represent a class of compounds sharing a basic chemical structure and primary molecular mode of toxicity (i.e., acetylcholinesterase inhibition). Beyond these commonalities, organophosphorus insecticides vary greatly in their physicochemical properties. The general structure of these compounds is:

$$R_1O-\overset{\overset{\displaystyle O(S)}{\|}}{\underset{\underset{\displaystyle R_2}{|}}{P}}-X$$

where R_1 represents a methyl or ethyl group (except in a few early systemic insecticides); R_2 is an alkyl, aryl, alkoxy, amino, substituted amino or thioalkyl group; and X is a leaving group. The OR_1 and R_2 groups of most currently used organophosphorus insecticides are both methoxy or ethoxy moieties. The electronegativity of the sulfur or oxygen atom double bonded to the phosphorus atom affects the chemical and biochemical reactivity of organophosphorus compounds with the oxonate form (i.e., P=O) usually being substantially more reactive. Aside from the influence of the doubly bonded S or O, the physico-chemical properties and biochemical activity of organophosphorus insecticides are largely determined by the nature of the X group. Table I displays physico-chemical properties of representative organophosphorus insecticides. A comparison of the octanol-water partition coefficients, K_{ow}, reveals the role the X group plays in determining compound hydrophobicity. The relatively high K_{ow} value of chlorpyrifos (log K_{ow} = 4.70) is due primarily to the hydrophobic 3,5,6-trichloro-2-pyridinyl X group, while the low K_{ow} value of dimethoate (log K_{ow} = 0.70) is due to the polar 2-mercapto-N-methyl acetamide X group. Chlorpyrifos and dimethoate also differ in their R_1 and R_2 groups, but this has a relatively minor effect on the difference in K_{ow}. This can be illustrated by comparing octanol-water partition coefficients of chlorpyrifos (log K_{ow} = 4.70) and chlorpyrifos-methyl (log K_{ow} = 4.24). The variation in hydrophobicity, aqueous solubility and vapor pressure among organophosphorus insecticides results in profound differences in their transport, transformation and fate in the environment.

Surface runoff from agricultural lands represents an important route for the transport of organophosphorus insecticides into surface waters. Edge of field concentrations are necessarily site-specific due to variations in application

Table I. Physico-chemical Properties of Representative Organophosphorus Insecticides[a]

Compound		MW	$\log P^0$	$-\log C_w^{sat}$	$\log K_{ow}$
Acephate	O,S-dimethyl acetylphosphoramidothioate	183.2	-0.65[b]	-0.63	-0.89
Chlorpyrifos	O,O-diethyl O-(3,5,6-trichloro-2-pyridinyl) phosphorothioate	350.6	0.43	5.40	4.70
Chlorpyrifos-methyl	O,O-dimethyl O-(3,5,6-trichloro-2-pyridinyl) phosphorothioate	322.5	0.48	5.09[c]	4.24
Diazinon	O,O-diethyl O-[6-methyl-2-(1-methylethyl)-4-pyrimidinyl] phosphorothioate	304.3	1.08	3.57	3.30
Dichlorvos	O,O-dimethyl O-(2,2-dichlorovinyl) phosphate	221.0	3.32	1.44	1.90
Dimethoate	O,O-dimethyl S-methylcarbamoylmethyl phosphorodithioate	229.3	-0.60	1.11	0.70
Disulfoton	O,O-diethyl S-[2-(ethylthio)ethyl] phosphorodithioate	274.4	1.11	4.04	3.95
Malathion	O,O-dimethyl S-[1,2-bis(ethoxycarbonyl)ethyl] dithiophosphate	330.3	0.72	3.36	2.75
Methamidophos	O,S-dimethyl phosphoramidothioate	141.1	0.67	<0.15[c]	-0.80[c]
Methidathion	[(5-methoxy-2-oxo-1,3,4-thiadiazol-3(2H)-yl) O,O-dimethylphosphorodithioate	302.3	-0.60[c]	3.18	2.20
Trichlorfon	O,O-dimethyl (2,2,2-trichloro-1-hydroxyethyl)-phosphonate	257.4	-0.30	0.33	0.43[c]

[a] Physico-chemical property data from (5). All parameters measured at 25°C unless otherwise noted. P^0 = vapor pressure (mPa), C_w^{sat} = aqueous solubility (mol·L$_w^{-1}$), K_{ow} = octanol-water partition coefficient (mol·L$_o^{-1}$)·(mol·L$_w^{-1}$)$^{-1}$.
[b] 24°C.
[c] 20°C

amount and method, formulation, crop type and maturity, irrigation practices, meteorology, soil composition and moisture content, and management practices. Several published reviews summarize pesticide occurrence in surface runoff including the relatively few published reports on organophosphorus insecticides (*1 – 4*).

Organophosphorus insecticides are applied to plants and soils using a variety of methods and formulations. Because formulation and initial placement affect exposure of these compounds to transformation processes and their availability for transport in surface runoff, the influence of these factors must be understood. Formulation in particular may exert an important influence on organophosphorus insecticide loads in surface runoff. Organophosphorus insecticides are rarely applied alone, but are mixed with other substances to enhance their performance and safety. These formulation ingredients can make up to 99.5% of the applied pesticide product and include organic solvents, surfactants and polymers.

This chapter focuses on physical, chemical and microbial processes influencing the concentrations of organophosphorus insecticides discharged to receiving waters in surface runoff (Figure 1). Research on these processes is evaluated emphasizing areas with significant information gaps. Many of the processes described govern the transport and transformation of organic contaminants in general; however, particular emphasis is given to those affecting organophosphorus insecticides. The potential effect of formulation and spray adjuvants is also considered.

Application

Formulation and initial placement influence the susceptibility of organophosphorus insecticides to transport in surface runoff, as well as their degradation by abiotic and microbial processes. Formulation affects the kinetics of insecticide release into soil water and overland flow, as well as sorption to soil solids and plant surfaces. Spray adjuvants affect initial placement by influencing the amount of insecticide depositing on foliar and soil surfaces. Initial placement determines the relative importance of such processes as volatilization, photolysis, biodegradation, and leaching out of the zone of interaction with overland flow.

Formulations and Adjuvants

Formulations are designed to maximize product stability and availability to target pests, improve handling and minimize human exposure. Pesticide formulations consist of the active ingredient, carrier (e.g., organic solvent, clay),

168

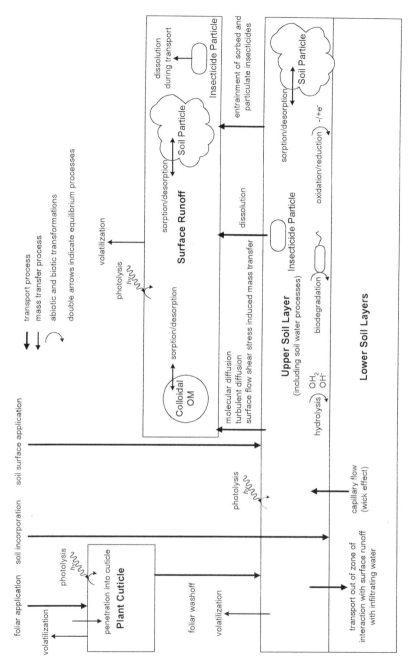

Figure 1. Transport and transformation processes affecting the movement of organophosphorus insecticides into surface waters.

surfactants and other ingredients (e.g., dyes, potentiators, stabilizers). In outdoor agricultural use scenarios, organophosphorus insecticides are usually applied as granules or sprays. For liquid sprays, additional chemicals (i.e., spray adjuvants) may be added before final application.

Granular formulations consist of an inert carrier (e.g., mineral clay) impregnated with the insecticide (6). Polymers are often added as binding agents and coatings. The composition and thickness of the granule coating can be designed to control insecticide release rate and increase persistence. Granular pesticides are applied dry and must be wetted before the active ingredient can diffuse into soil water. This increases insecticide persistence and delays availability for environmental transport and transformation processes.

Formulations applied as sprays include wettable powders, suspension concentrates, emulsifiable concentrates, encapsulated formulations and ultra-low-volume formulations. Wettable powders consist of finely divided pesticide particles combined with a finely ground dry carrier (e.g., synthetic silica, mineral clay) and surfactants (6). When the powder is mixed with water, a stable homogeneous particle suspension is formed. Suspension concentrates are particulate insecticides premixed with a liquid. When sprayed onto porous media, the water in particle suspensions penetrates the medium leaving the insecticide at the surface. On nonporous surfaces, the water evaporates leaving a deposit.

Emulsifiable concentrates consist of the active ingredient combined with organic solvents, surfactants and other enhancers (e.g., polymers as protective colloids and suspending agents). Insecticides applied as emulsions are able to penetrate porous media more readily than those in particle suspensions. Nearly all the insecticide is available for sorption to soil surfaces and is thus exposed to degradative processes. Microencapsulated formulations provide controlled release of insecticides applied as sprays. The composition and thickness of the polymeric capsule wall control the release rate. Microcapsule suspensions are often applied in conjunction with adhesive compounds to improve retention to foliar surfaces. Ultra-low-volume formulations consist of pure insecticides (e.g., malathion) or highly concentrated solutions. When diluents are used, vegetable oils or solvent mixtures are employed rather than water. The lower surface tension of oil carriers allows insecticide penetration through waxy leaf cuticles.

Prior to application of spray formulations, spray adjuvants are typically added to the insecticide mixture to enhance the efficacy of the active ingredient. Spray adjuvants include surfactants, compounds that impart adhesion and viscoelasticity to spray droplets (e.g., latex), compounds that provide protection from ultraviolet light and reduce volatilization, and activators. The co-application of these compounds affects organophosphorus insecticide dissipation and transfer to surface runoff. Some adjuvants and formulation ingredients are toxicologically significant themselves (e.g., nonylphenol ethoxylates).

The effect of formulation and spray adjuvants on insecticide efficacy has received considerable attention from the pesticide industry. However, few detailed mechanistic studies on the role these additives play in environmental fate processes have appeared in the open literature. Application of laboratory-derived process information to field scenarios is hindered by the fact that most laboratory investigations have used technically pure (unformulated) organophosphorus insecticides. Including the effects of formulation ingredients on such processes as volatilization and sorption to soil solids would allow laboratory studies to better predict the environmental behavior of these compounds.

Initial Placement

Depending on the pest control scenario, organophosphorus insecticides may be applied to foliar or soil surfaces, or incorporated into the soil matrix. For example, fenamiphos is applied as a spray or in granular form to the soil surface and incorporated postapplication (7). Incorporation into the soil reduces the importance of photolytic processes and may place a fraction of the applied chemical below the zone of interaction with overland flow. Emulsifiable concentrates are applied to plant and soil surfaces as ground and aerial foliar sprays and via chemigation (application through irrigation systems). Unless soil moisture is high, spray water and emulsifying solvents (if present) quickly evaporate when applied to soil surfaces, leaving a deposit of mostly pure active ingredient. Deposits formed from emulsion formulations are comprised of crystalline pesticide with a residue of surfactant which may enhance initial mobilization of the insecticide during irrigation and rain events (7). The extent of spray droplet retention on leaf surfaces affects the amount of insecticide depositing on soil and is enhanced by nonionic surfactants (e.g., organosilicones) (8).

Phase Transfer Processes

After organophosphorus insecticides are deposited on or incorporated into the soil matrix, they distribute themselves between the air, soil air, soil water and soil solid phases. Insecticides applied to foliar surfaces partition between the plant cuticle, air and any water present on the leaf surface (e.g., dew). Phase distribution affects the amount of organophosphorus insecticide available for transfer into overland flow, as well as the mode of transport in surface runoff.

Dissolution into Soil Water

In addition to the aqueous solubility of organophosphorus insecticides, dissolution into soil water depends strongly on formulation. Insecticides in emulsifiable concentrates partition between the carrier liquid and soil water phases. Insecticides applied in granular formulations must diffuse out of the granule into the soil water. During irrigation and precipitation events, dissolution of insecticide crystals and granules may limit the amount available for transport in surface runoff. The kinetic release of organophosphorus pesticides from formulations has not been extensively investigated (*3*). Davis et al. (*7*) found that the release of fenamiphos from emulsifiable concentrate deposits into flowing water was considerably more rapid than from granular formulations. Cryer and Laskowski (*9*) determined chlorpyrifos release rates from clay granules as a function of interstitial water velocity.

Volatilization

Volatilization of soil and foliar applied insecticides reduces the amount available for transport with runoff water. Volatilization depends on the vapor pressure and water solubility of the compound. Partitioning between the water and vapor phases is expressed by the Henry's law constant, $K_H = C_a/C_w$, where C_a and C_w are the air and water phase concentrations, both expressed in $mol \cdot L^{-1}$. In addition to the physico-chemical properties of the pesticide, the volatilization rate from soil also depends on soil moisture content, temperature, humidity, airflow rate over the surface and soil properties (e.g., organic matter content, clay content, porosity, density) (*10*).

Insecticide vapor density in soil air, and thus volatilization rate, depends on the concentration of pesticide dissolved in the soil water. More strongly sorbed compounds have lower soil water concentrations and lower vapor densities. As a soil dries, organophosphorus insecticides sorb onto the dry soil and become less available for volatilization. Temperature increases typically cause increased partitioning into the air phase, but may also result in drying of the upper soil layer with concomitant reduction in volatilization. Air flow rate and turbulence also influence pesticide flux into the bulk air (*10*). In a field study, Majewski et al. (*11*, *12*) observed that daytime chlorpyrifos and diazinon volatilization flux decreased rapidly during the first few hours after sunrise. The drying of the upper few millimeters of soil was likely responsible for the reduction in air concentrations as the day progressed. So-called "wick evaporation" can also affect organophosphorus insecticide volatilization (*13*). Tillage practices have been shown to impact volatilization rates of chlorpyrifos and fonofos (*14*).

Volatilization also represents an important loss process for foliar applied organophosphorus insecticides. Although volatilization generally occurs relatively rapidly (*15*), formulation can strongly influence loss rates. For example, significantly less chlorpyrifos volatilizes from tomato leaves when applied with mineral or cottonseed oil as compared to emulsifiable concentrates because oil formulations enhance penetration into the leaf cuticle (*16*).

Sorption to Solid Surfaces

Sorption to soil solids and plant cuticular material represents an important process influencing the chemodynamic behavior of insecticides, including their transport in surface runoff. Sorption phenomena affect the volatilization, hydrolysis, photolysis and microbial transformation of organophosphorus insecticides. Furthermore, species sorbed to soil particles are transported by erosion processes rather than as solutes in the water phase. Sorption to foliar surfaces reduces the amount of pesticide mobilized by washoff.

Sorption to Inorganic and Organic Soil Components

Soils represent structurally and compositionally heterogeneous mixtures of inorganic and organic components. Mineral domains potentially contributing to organophosphorus insecticide sorption include (i) exposed surfaces on particle exteriors and within macropores; (ii) internal mesopore and micropore interfaces; and (iii) interlayer surfaces of swelling clays (*17*). The reactivity of soil organic matter with nonionic organic contaminants depends on its origin and diagenetic history. Several researchers have found utility in distinguishing between two broad classes of organic matter that differ in sorption properties and diagenetic histories: expanded ("soft" or "rubbery") and condensed ("hard" or "glassy") organic matter (*18*).

Sorption of organophosphorus insecticides to soil particles depends primarily on compound hydrophobicity and the fraction of natural organic matter (NOM) in the soil (*19*, *20*). Mineral phases appear to exert more influence on sorption processes for organophosphorus insecticides than for more extensively investigated hydrophobic organic compounds (e.g., DDT). The sorption of most hydrophobic organic compounds is dominated by NOM when the fraction of soil organic matter, f_{om}, exceeds 0.002 (*21*). For organophosphorus insecticides, however, mineral surfaces appear to exert an influence on sorption at a higher organic matter content ($f_{om} < 0.02$) (*19*). A few studies suggest that soils contain two organic matter domains differing in their sorptive affinity for phosphorothioate insecticides (*15*). Systematic investigations of the roles of different inorganic and organic domains on the sorption of organophosphorus insecticides are currently lacking.

Numerous studies have examined the sorption of organophosphorus insecticides to natural soils. Experimentally derived organic carbon-water equilibrium distribution coefficients, K_{oc}, for the same organophosphorus compound vary greatly between studies and between soils used in the same study. This reflects variation in methodology, as well as differences in NOM composition. Methodological differences include drying and sieving procedures; pH, ionic strength and colloidal NOM concentration of the aqueous solution; solid and liquid phase separation method; and analytical methods employed to determine dissolved and sorbed concentrations (22, 23). Most studies conducted prior to 1990 failed to consider the effect of colloidal NOM on experimentally derived K_{oc} values.

Few studies have investigated the underlying physical and chemical mechanisms responsible for the interaction of organophosphorus insecticides with NOM. The structure of both the soil organic matter and the sorbate determine the extent of interaction. Several studies have shown the extent of hydrophobic organic compound sorption to humic substances correlates strongly with organic matter aromaticity (24 – 26). NOM aliphaticity may also play a role (27). Evidence also exists that the steric environment within NOM exerts an effect on hydrophobic organic chemical sorption (25, 28). The significance of bulk organic matter properties on organophosphorus insecticide sorption has not been studied. A more complete understanding of the influence of NOM composition and conformation on sorption processes will require the application of more sophisticated analytical techniques (29).

Several other aspects of organophosphorus insecticide sorption require further study. Although sorption-desorption hysteresis (20), aging and bound residue formation (30) have been noted, the molecular mechanisms responsible for these phenomena have not been elucidated. The kinetics of organophosphorus insecticide release from soil particles including the effects of biosurfactants (31) have not been extensively investigated, nor have the effects of formulation ingredients on sorption processes (32).

Sorption to Plant Cuticular Material

The absorption of organophosphorus insecticides by plant cuticles decreases the mass available for washoff by irrigation water, rain or dew. Washoff onto the soil surface or into surface runoff depends on the properties of the insecticide, plant surface and formulation ingredients. The physico-chemical properties of the aerial surfaces of crop plants are controlled primarily by the cuticle, a continuous thin layer ($< 0.1 – 10$ μm) of mainly lipid material (33). The cuticle comprises a crosslinked matrix of high molecular weight polyester cutins forming a three-dimensional network infiltrated with solid waxes. The identity and relative abundance of the molecules comprising the cuticle varies by species. The microstructure of the cuticle surface also influences insecticide deposition (*33*). Cuticle surface roughness varies considerably between species. The cuticle can be relatively smooth, ridged or papillose; or covered with microcrystalline epicuticular wax or trichomes (*33*). Understanding organophosphorus insecticide interaction with plant surfaces requires detailed knowledge of the chemical composition and physical properties of the cuticle.

The mass of organophosphorus insecticide available for washoff depends on the dissolution of foliar surface deposits and desorption from the cuticle matrix. The nature and concentration of formulation and spray adjuvants applied with an insecticide strongly influences its distribution at the plant surface. For example, chemigation of chlorpyrifos in soybean oil solution resulted in retention of three times more active ingredient on the foliar surface than an emulsion formulation (*34*). Anionic surfactants decrease absorption of insecticides through the plant cuticle and increase their resistance to washoff by irrigation water, rain or dew. Nonionic surfactants increase penetration through the plant cuticle, but may render insecticides subject to washoff (*35*). The kinetics of organophosphorus insecticide desorption from plant cuticular materials including the effect of adjuvants warrants further study.

Transformation Processes

Abiotic and biotic transformation processes reduce the amount of parent compound available for transport with runoff. Transformation products warrant concern, however, when they exhibit significant toxicity. Processes occurring on plant and soil surfaces are the most important for reducing insecticide mass available for runoff. Because the residence time of surface runoff on agricultural fields is typically short, transformations during overland transport are usually of less consequence. Organophosphorus insecticide transformations in simple aqueous media have been extensively studied. In contrast to this, relatively little is known about chemical transformations involving dissolved and particulate NOM. Many studies of organophosphorus insecticide dissipation in soil matrices have examined overall degradation rates and products (*15*), as well as their correlation with soil properties. The underlying mechanisms responsible for many of these processes have yet to be elucidated.

Hydrolytic decomposition represents an important transformation process for organophosphorus insecticides in soil and runoff water. Hydrolysis in homogeneous aqueous solutions has been extensively studied. Nucleophilic displacement (S_N2) occurs at either the phosphorus or carbon atom with an alcohol or diester being the leaving group (*21*). Hydrolysis of an organophosphorus insecticide (C) depends on the specific acid and base-catalyzed and neutral reactions: $-d[C]/dt = k_H[C] = k_A[H^+][C] + k_B[OH^-][C] + k_N[C]$, where k_H, k_A, k_B and k_N are the overall and specific acid, base and neutral hydrolysis rate constants. For organophosphorus compounds, OH⁻ is a substantially better nucleophile than H_2O. The base-catalyzed reaction generally results in dissociation of the best leaving group from the phosphorus atom. In most cases, acid-catalyzed hydrolysis is relatively unimportant unless the leaving group contains a basic functional group that enhances reactivity when protonated (e.g., the pyrimidine ring of diazinon) (21). The rate of phosphorothioate insecticide hydrolysis is typically less than that of the corresponding phosphate compounds because the greater electrophilicity of the oxon renders it more susceptible to nucleophilic attack (21). Association with particulate, colloidal or dissolved NOM significantly reduces the base-catalyzed hydrolysis rate of phosphorothioate ester insecticides, but has no effect on the rate of the neutral reaction (36, 37). The effect of association with NOM on acid-catalyzed hydrolysis is poorly understood (38). Hydrolysis of organophosphorus insecticides can be catalyzed by dissolved metal ions (particularly Cu(II)) (39) and constituent or exchangeable metal ions in aluminosilicate minerals (e.g., montmorillonite) (40).

Abiotic oxidation and reduction reactions may be important transformation processes for some organophosphorus insecticides under certain soil conditions. The nitroaromatic group of methyl parathion is reduced extremely rapidly to the corresponding aromatic amine in freshwater sediments (41). Functional groups in NOM similar to quinone-hydroquinone redox pairs are believed to participate in such reductive processes (42, 43). Iron porphyrins can also act as effective reducing agents for nitroaromatic compounds in the presence of a bulk reductant (e.g., sulfide) (43). Extracellular enzymes may be responsible for a significant fraction of the reductive activity of soil NOM (44). Further research is needed to elucidate the mechanisms by which NOM participates in the reduction of organophosphorus and other organic compounds.

Photolytic processes at soil and plant surfaces are not well understood. The primary factor limiting photolysis at soil surfaces is light attenuation by organic and inorganic chromophores (45). The photic zone of soils is confined to the top 0.5 mm (46). Upward movement of soil incorporated organophosphorus insecticides into the photic zone with evaporating water may affect the degree of photolysis (47). Because of competitive light absorption by soil components, indirect photolysis probably plays a greater role in the phototransformation of organophosphorus insecticides than direct photolysis (48).

The plant cuticle represents a complex surface containing compounds that may act as photosensitizers, quenchers and potential reaction partners. Most studies of the phototransformation of organophosphorus insecticides on leaf surfaces ignored the structure of the plant cuticle and employed surrogate surfaces which cannot provide dependable predictive information. Schynowski and Schwack (49) found a strong correlation between the olefinic content of fruit cuticles and photodegradation rate of parathion.

Microbial transformation represents a significant degradation process for some organophosphorus insecticides. In most cases, biodegradation appears to result from co-metabolism. However, a few bacterial strains capable of utilizing organophosphorus insecticides as their sole carbon source have been isolated (*50*, *51*). Microbial flora are also capable of oxidizing phosphorothioate insecticides to their corresponding oxons (*52*) and reducing the nitroaromatic group of parathion to the corresponding aromatic amine (*53*). Formulation and spray adjuvants can also affect biodegradation rates (*54*). In addition to bacteria, soil fungi (*55*), microalgae and cyanobacteria (*56*) have also been shown to degrade a variety of organophosphorus insecticides.

Mobilization of Dissolved and Sorbed Species

Irrigation and precipitation events result in the entry of organophosphorus insecticides into the hydrologic cycle. Organophosphorus compounds deposited on foliar surfaces, dissolved in soil water and sorbed to particles are mobilized and transported with surface runoff. The main mobilization processes are discussed below.

Foliar washoff

The washoff of foliar-applied pesticides has been investigated primarily in terms of amount washed off and correlations with rainfall intensity, rainfall amount and period between application and rainfall. The timing of irrigation or precipitation with respect to application is an important determinant of the amount of insecticide in foliar washoff. As time from application increases, not only does the amount available for washoff decrease due to volatilization and transformation, but resistance to washoff also increases (*57*). Increased resistance to washoff is thought to be due to sequestration of organophosphorus insecticides in the cuticular matrix or microscopic cracks on the leaf surface.

Formulation strongly affects the amount of organophosphorus insecticide washed off from foliar surfaces. For example, azinphos-methyl and phosmet formulated as emulsifiable concentrates were highly susceptible to washoff, while wettable powder formulations were not (*58*). Thin-walled fenitrothion microcapsules were less prone to washoff than the emulsifiable concentrate (*59*). The composition of spray adjuvants also affected the degree of washoff (*58*).

Rainfall intensity appears to be inversely correlated with insecticide concentration in washoff (*60*) and concentrations are typically highest in first few millimeters of washoff (*57*). Lower insecticide concentrations at higher rain intensities may be due to the greater thickness of the water layer on the leaf surface and the shorter "residence time" of a unit volume of water on the leaf.

Transport Away from Soil Surface with Infiltrating Water

With the onset of a rainfall or irrigation event, water penetrates the soil surface and begins to infiltrate the soil profile under gravity and capillary forces. Infiltrating water traveling through the soil column displaces soil water containing dissolved organophosphorus insecticides and transports them downward. Dissolved insecticides may also diffuse into the infiltrating water from immobile water in the soil matrix and desorb from soil solids. In addition to transporting dissolved chemicals, infiltrating water may also carry insecticides associated with soil colloids (61). The removal of organophosphorus insecticides from the soil surface zone results in depletion of the mass available for transport by overland flow. The rate of mass transport with infiltrating water depends on source strength, infiltration rate, aqueous solubility, sorption, and chemical reactions. More water-soluble compounds have the potential to contaminate groundwater, while more hydrophobic compounds have limited potential to be transported away from the soil surface with infiltrating rainfall.

Mass Transfer of Dissolved Pesticides from Soil Solution to Overland Flow

When rainfall intensity exceeds infiltration rate and surface-storage capacity has been reached, overland flow begins. The transfer of dissolved pesticides from the soil matrix to overland flow consists of several mechanisms: desorption from soil organic matter, mineral surfaces and plant residues; dissolution of insecticide crystals or granules; and diffusive and turbulent transport of dissolved insecticide from soil water into overland flow (2, 62). The relative importance of each process depends on the physico-chemical properties of the chemical, formulation, initial placement, soil properties, recent hydraulic history and vegetation (62).

Dissolved insecticides are transferred from soil solution to surface runoff through the concurrent mechanisms of molecular diffusion, raindrop impact induced turbulent diffusion, and shear stress induced mass transfer (63, 64). In addition, shallow interflow may contribute dissolved chemicals to surface runoff as it returns to the surface downslope or seeps into rills and furrows (65). Most studies of dissolved chemical transport into overland flow have employed inorganic tracers such as bromide, gypsum ($CaSO_4 \cdot 2H_2O$) and ^{32}P (66, 67). The behavior of organophosphorus insecticides, however, is considerably more complex due to association with particulate and colloidal natural organic matter.

Solute concentrations in surface runoff are typically significantly lower than those in soil water, even when measured just below the soil surface (68). Molecular diffusion of solutes occurs in response to the concentration gradient between the soil solution and runoff water. Although turbulent diffusion enhances the transfer of solutes from soil solution into overland flow, molecular diffusion can transfer a significant amount of chemical when the concentration gradient is steep (63).

Several researchers have postulated that raindrop impacts impart turbulence to soil water and produce large fluctuating pressure gradients resulting in convective mixing (63, 69). The degree of interaction between the soil solution and precipitation is greatest in the upper 0.2-0.3 cm and falls off rapidly with depth, although a small amount of soil water is transferred to overland flow from as deep as 2.0 cm (63). Solute transfer to overland flow increases with rainfall kinetic energy (63, 70), as well as with the hydraulic conductivity of the soil surface (68, 71) since raindrop induced mixing is greater and extends to greater depths. Accelerated transfer of solutes from depths much below the soil surface is believed to result from a pumping action caused by raindrop impacts (63).

As water flows over the soil surface, solute mass is transferred via mechanical dispersion from pore spaces into overland flow (64, 69). Shear stresses generated by overland flow accelerate removal of solutes from porous media into surface runoff (64, 72). The dispersion coefficient, D, resulting from overland shear flow increases in proportion to soil permeability, k, and the square of shear velocity, u^*: $D \propto k u^{*2}$ (64, 72). Flow regime is an important determinant of the mechanism responsible for interfacial frictional resistance between the porous medium and overland flow.

Mobilization of Particle-associated Pesticides

A significant fraction of hydrophobic organophosphorus insecticides is transported sorbed to suspended particles. The movement of these compounds in the environment depends on the erosion and transport of the particles to which they are sorbed. Soil erosion is a selective process resulting in the enrichment of clay-sized particles and NOM, and therefore hydrophobic organophosphorus insecticides, in the eroded material.

Particles are detached from the soil surface by boundary shear stress and raindrop impact. Forces resisting particle detachment are due to gravity (particle weight), interparticle frictional resistance, and interparticle and interaggregate cohesiveness (73, 74). Fluid flow generates two motive forces that may overcome these resisting forces: a lateral drag force acting at the surface and a hydraulic lift force (a consequence of the Bernoulli effect and the buoyancy exerted by the fluid) (74). When the total fluid force exceeds the frictional and gravity forces, erosion results. Although soil strengths are typically on the order of kPa, average overland flow shear stresses on the order of Pa cause particle detachment from upland soils. Velocity fluctuations associated with turbulent flow impart instantaneous and localized levels of shear stress that greatly exceed average shear stress and result in particle detachment (75). Numerous additional factors complicate this generalized description including particle size distributions, shape, spatial heterogeneity, aggregation of soil particles, and soil surface slope and roughness. Granular and crystalline insecticides can also be mobilized by particle detachment processes.

Raindrop impact on the soil surface results in particle detachment and splash. Although most of the soil splashed during a storm event is not transported from the field, detached soil particles are trapped in water-filled depressions and clog surface pores. This reduces infiltration and causes greater surface runoff and erosion (74). The erosivity of rainfall depends on its intensity, duration and energy; size distribution and terminal velocities of raindrops; slope direction and steepness; wind speed and direction; and surface roughness (74).

Transport in Surface Runoff

Once dissolved and particle-associated organophosphorus insecticides have been mobilized, they are transported with surface runoff toward the receiving water. Overland flow generally starts as sheet flow, but as its velocity increases and it gains erosive power, it scours the soil surface to form rills. Rills join to form increasingly higher order channels producing a dendritic pattern of converging branches (74).

Dissolved Insecticide Transport

Once dissolved pesticides have been extracted from the soil matrix into overland flow or reach the surface through shallow interflow, they are transported toward the field outlet with surface runoff. For one-dimensional overland flow, dissolved insecticide transport can be expressed as (76):

$$h\frac{\partial C}{\partial t} + q\frac{\partial C}{\partial x} = kC_s(t) - C \cdot f$$

where h is flow depth; q is unit discharge in the slope direction; k is the mass transfer coefficient; $C_s(t)$ is solute concentration at the soil surface; and $f(t)$ is the spatially uniform rainfall rate. Using mathematical simulations, Rivlin et al. (76) demonstrated that field outlet concentration distributions depended strongly on the kinematic behavior of overland flow during the early stages after runoff initiation. A similar dependence was expected for the falling stage of overland flow (i.e., after cessation of rainfall). The outlet concentration profile for the intervening time period depends on the rate of dissolved chemical transfer from the soil solution to surface runoff.

Discharge can be related to overland flow depth by the St. Venant equations (77). The kinematic wave form is the most commonly used simplification for sheet flow: $q = \alpha h^m$, where α is a conveyance factor and m is a constant depending on flow regime. Hydraulic relationships derived from sheet and larger channel flow do not hold for flow in actively eroding rills. Separate relationships between discharge, flow velocity and hydraulic friction must be applied (78).

Because overland flow depth and infiltration rate are interdependent processes, the transient, spatially variable infiltration rate must also be accounted for. The two-dimensional Richard's equation can be used to calculate the soil water flux at any position x along the slope:

$$\frac{\partial \theta}{\partial t} = \frac{\partial}{\partial x}\left[K(\psi)\frac{\partial \phi}{\partial x}\right] + \frac{\partial}{\partial z}\left[K(\psi)\frac{\partial \phi}{\partial z}\right]$$

where θ is the volumetric water content; $K(\psi)$ represents hydraulic conductivity; $\psi(x, z)$ is the matrix potential of the soil water; and ϕ is the hydraulic head. Infiltration affects the distribution, magnitude and timing of surface runoff.

During transport with surface runoff, organophosphorus insecticides redistribute themselves between the dissolved, colloidal and suspended particle phases. Such phase redistribution during overland transport has not been investigated and the common assumption of phase equilibrium at the field outlet has not been tested. The validation of physically based numerical models of pesticide transport in surface runoff will require careful laboratory and field experimentation that includes the effects of infiltration and sorption (77).

Transport of Sorbed and Particulate Insecticides

Sorbed and particulate (e.g., crystalline, granular) insecticides are transported by the same processes involved in soil erosion. Three types of particle transport can be distinguished: rolling/sliding at low sediment transport rates (analogous to bedload movement in streams), saltation (particle movement by irregular jumps) and suspended particle motion at high sediment transport rates (sediment eroded from surface and washed away). Stream power, Ω, is the best predictor of unit sediment load: $\Omega = \rho_w gsq$, where ρ_w is water density; g is gravitational acceleration; s is slope; and q is unit discharge (78).

Suspended particle transport is the result of dynamic equilibrium between particle detachment and deposition (sedimentation). A suspended particle will settle when the fluid shear stress drops below a critical level. Neglecting reaction, mass conservation of an insecticide associated with particles in settling velocity class i can be described by:

$$\frac{\partial(S_i P_i h)}{\partial t} + \frac{\partial(S_i P_i q)}{\partial x} = S_i\left(r_i + r_{ri} + r_{gi} - d_i\right) + K_{sw,i} P_i\left(S_i^* - S_i\right)h$$

where S_i is the insecticide concentration sorbed to particles in class i; P_i is the particle concentration; r_i is the entrainment rate; r_{ri} is the re-entrainment rate for previously deposited particles; r_{gi} is the rate of particle input due to gravity processes (e.g., slumping of rill walls, head cut collapses); d_i is the mass rate of deposition per unit area; $K_{sw,i}$ is the overall mass transfer coefficient between the sorbed and dissolved phases; and S_i^* is the concentration of sorbed insecticide in equilibrium with the water phase (73, 79, 80). The entrainment rate applies to the original cohesive soil, while the re-entrainment rate relates to detachment of particles in the cohesionless layer of deposited sediment.

Conclusion

While the occurrence of most processes affecting organophosphorus insecticide concentrations in surface runoff are well documented, many are relatively poorly understood on a mechanistic level. Key processes requiring further study include sorption to plant and soil surfaces, and transfer from soil water into overland flow. Sorption to foliar surfaces determines the amount of applied insecticide available for washoff onto the soil surface or into surface runoff. The kinetics of desorption from plant cuticular materials and the effect of adjuvants on foliar sorption processes need additional research.

Sorption of organophosphorus insecticides to inorganic and organic soil components determines their availability for transfer from soil water to overland flow and their mode of transport in surface runoff. The underlying physical and chemical mechanisms responsible for the interaction of organophosphorus compounds with soil organic matter require more thorough investigation, as does the role of clay mineral surfaces in organophosphorus insecticide sorption. Desorption kinetics, aging phenomena and the effect of formulation ingredients on sorption processes also warrant further study.

The transfer of dissolved and colloid-associated species from the soil surface zone to overland flow determines the magnitude and timing of organophosphorus insecticide flux in surface runoff. Previous research on the physics of solute transfer into overland flow focused on inorganic molecules. Future work should concentrate on moderately hydrophobic organic species (e.g., organophosphorus insecticides) and the role colloids may play in mass transfer into surface runoff. Careful laboratory and field experimentation will be required to understand the importance of individual transport processes for these compounds. Improved understanding of organophosphorus insecticide sorption and transport will enhance our ability to predict the fate of these compounds in agricultural systems.

References

1. Wauchope, R.D. *J. Environ. Qual.* **1978**, *7*, 459-472.
2. Leonard, R.A. In *Pesticides in the Soil Environment: Processes, Impacts, and Modeling*; Cheng, H.H., Ed.; Soil Science Society of America: Madison, WI, 1990; pp. 303-349.
3. Burgoa, B.; Wauchope, R.D. In *Environmental Behaviour of Agrochemicals*; Roberts, T.R.; Kearney, P.C., Eds.; John Wiley & Sons: New York, NY, 1995; pp. 221-255.
4. Larson, S.J.; Capel, P.D.; Majewski, M.S. *Pesticides in Surface Waters: Distribution, Trends and Governing Factors*, Ann Arbor Press: Chelsea, MI, 1997.
5. *The Pesticide Manual*, 11th ed.; Tomlin, C.D.S., Ed.; British Crop Protection Council: Surrey, U.K., 1997.
6. Matthews, G.A. *Pesticide Application Methods*, Longman Scientific & Technical: New York, 1992.
7. Davis, R.F.; Wauchope, R.D.; Johnson, A.W.; Burgoa, B.; Pepperman, A.B. *J. Agric. Food Chem.* **1996**, *44*, 2900-2907.
8. Reddy, K.N.; Locke, M.A. *Pestic. Sci.* **1996**, *48*, 179-187.
9. Cryer, S.A.; Laskowski, D.A. *J. Agric. Food Chem.* **1998**, *46*, 3810-3816.
10. Thomas, R.G. In *Handbook of Chemical Property Estimation Methods*, Lyman, W.J.; Reehl, W.F.; Rosenblatt, D.H., Eds.; McGraw-Hill, New York, 1982; pp. 16-1 to 16-50.
11. Majewski, M.S.; Glotfelty, D.E.; Seiber, J.N. *Atmos. Environ.* **1989**, *23*, 929-938.
12. Majewski, M.S.; Glotfelty, D.E.; Paw, K.T.; Seiber, J.N. *Environ. Sci. Technol.* **1990**, *24*, 1490-1497.
13. Spencer, W.F.; Claith, M.M. In *Fate of Pollutants in the Air and Water Environments*; Suffet, I.H., Ed.; John Wiley & Sons: New York, 1977; pp. 107-126.
14. Whang, J.M.; Schomburg, C.J.; Glotfelty, D.E.; Taylor, A.W. *J. Environ. Qual.* **1993**, *22*, 173-180.
15. Racke, K.D. *Rev. Environ. Contam. Toxicol.* **1993**, *131*, 1-150.
16. Veierov, D.; Berlinger, M.J.; Fenigstein, A. *Med. Fac. Landbouww Rijksuniv. Gent* **1988**, *53*, 1535-1541.
17. Luthy, R.G.; Aiken, G.R.; Brusseau, M.L.; Cunningham, S.D.; Gschwend, P.M.; Pignatello, J.J.; Reinhard, M.; Traina, S.J.; Weber, W.J., Jr.; Westall, J.C. *Environ. Sci. Technol.* **1997**, *31*, 3341-3347.
18. Young, T.M.; Weber, W.J., Jr. *Environ. Sci. Technol.* **1995**, *29*, 92-97.
19. Wahid, P.A.; Sethunathan, N. *J. Agric. Food Chem.* **1978**, *26*, 101-108.
20. Felsot, A.; Dahm, P.A. *J. Agric. Food Chem.* **1979**, *27*, 557-563.
21. Schwarzenbach, R.P.; Gschwend, P.M.; Imboden, D.M. *Environmental Organic Chemistry*; John Wiley & Sons: New York, 1993.

22. Freeman, D.H.; Cheung, L.S. *Science* **1981**, *214*, 790-792.
23. Gschwend, P.M.; Wu, S.-C. *Environ. Sci. Technol.* **1985**, *19*, 90-96.
24. Gauthier, T.D.; Seitz, W.R.; Grant, C.L. *Environ. Sci. Technol.* **1987**, *21*, 243-248.
25. Chin, Y.-P.; Aiken, G.R.; Danielsen, K.M. *Environ. Sci. Technol.* **1997**, *31*, 1630-1635.
26. Perminova, I.V.; Grechishcheva, N.Yu.; Petrosyan, V.S. *Environ. Sci. Technol.* **1999**, *33*, 3781-3787.
27. Chefetz, B.; Deshmukh, A.P.; Hatcher, P.G.; Guthrie, E.A. *Environ. Sci. Technol.* **2000**, *34*, 2925-2930.
28. Uhle, M.E.; Chin, Y.-P.; Aiken, G.R.; McKnight, D.M. *Environ. Sci. Technol.* **1999**, *33*, 2715-2718.
29. Hedges, J.I.; Eglinton, G.; Hatcher, P.G.; Kirchman, D.L.; Arnosti, C.; Derenne, S.; Evershed, R.P.; Kögel-Knabner, I.; de Leeuw, J.W.; Littke, R.; Michaelis, W.; Rullkötter, *J. Org. Geochem.* **2000**, *21*, 945-958.
30. Andréa, M.M.; Wiendl, F.M. *Pesq. Agropec. Bras.* **1995**, *30*, 695-700.
31. Mata-Sandoval, J.; Karns, J.; Torrents, A. *Environ. Sci. Technol.* **2000**, *34*, 4923-4930.
32. Sánchez-Camazano, M.; Arienzo, M.; Sánchez-Martin, M.J.; Crisanto, T. *Chemosphere* **1995**, *31*, 3793-3801.
33. Holloway, P.J. *Pestic. Sci.* **1993**, *37*, 203-206.
34. Wauchope, R.D.; Young, J.R.; Chalfant, R.B.; Marti, L.R.; Sumner, H.R. *Pestic. Sci.* **1991**, *32*, 235-243.
35. Marer, P.J.; Flint, M.L.; Stimmann, M.W. *The Safe and Effective Use of Pesticides*, University of California: Oakland, CA, 1988.
36. Macalady, D.L.; Wolfe, N.L. *J. Agric. Food Chem.* **1985**, *33*, 167-173.
37. Noblet, J.A.; Smith, L.A.; Suffet, I.H. *J. Agric. Food Chem.* **1996**, *44*, 3685-3693.
38. Macalady, D.L.; Tratnyek, P.G.; Wolfe, N.L. In *Aquatic Humic Substances: Influence on Fate and Treatment of Pollutants*; Suffet, I.H., MacCarthy, P., Eds.; American Chemical Society: Washington, D.C., 1989; pp. 323-332.
39. Smolen, J.M.; Stone, A.T. *Environ. Sci. Technol.* **1997**, *31*, 1664-1673.
40. Pusino, A.; Petretto, S.; Gessa, C. *J. Agric. Food Chem.* **1996**, *44*, 1150-1154.
41. Wolfe, N.L.; Macalady, D.L.; Kitchens, B.E.; Grundl, T.J. *Environ. Toxicol. Chem.* **1987**, *6*, 827-837.
42. Tratnyek, P.G.; Macalady, D.L. *J. Agric. Food Chem.* **1989**, *37*, 248-254.
43. Schwarzenbach, R.P.; Stierli, R.; Lanz, K.; Zeyer, J. *Environ. Sci. Technol.* **1990**, *24*, 1566-1574.
44. Schnoor, J.L.; Licht, L.A.; McCutcheon, S.C.; Wolfe, N.L.; Carreira, L.H. *Environ. Sci. Technol.* **1995**, *29*, 318A-323A.
45. Miller, G.C.; Zepp, R.G. *Residue Rev.* **1983**, *85*, 89-110.
46. Hebert, V.R.; Miller, G.C. *J. Agric. Food Chem.* **1990**, *38*, 913-918.

184

47. Donaldson, S.G.; Miller, G.C. *Environ. Sci. Technol.* **1996**, *30*, 924-930.
48. Wolfe, N.L.; Mingelgrin, U.; Miller, G.C. In *Pesticides in the Soil Environment: Processes, Impacts, and Modeling*; Cheng, H.H., Ed.; Soil Science Society of America: Madison, WI, 1990; 103-168.
49. Schynowski, F.; Schwack, W. *Chemosphere*, **1996**, *33*, 2255-2262.
50. Ou, L.-T.; Sharma, A. *J. Agric. Food Chem.* **1989**, *37*, 1514-1518.
51. Mallick, K.; Bharati, K.; Banerji, A.; Shakil, N.A.; Sethunathan, N. *Bull. Environ. Contam. Toxicol.* **1999**, *62*, 48-54.
52. Alexander, M. *Biodegradation and Bioremediation*, Academic Press: San Diego, CA, 1999.
53. Gottschalk, G. *Bacterial Metabolism*, Springer-Verlag: New York, 1986.
54. Charnay, M.-P.; Tarabelli, L.; Beigel, C.; Barriuso, E. *J. Environ. Qual.* **2000**, *29*, 1618-1624.
55. Al-Mihanna, A.A.; Salama, A.K.; Abdalla, M.Y. *J. Environ. Sci. Health* **1998**, *B33*, 693-704.
56. Megharaj, M.; Madhavi, D.R.; Sreenivasulu, C.; Umamaheswari, A.; Venkateswarlu, K. *Bull. Environ. Contam. Toxicol.* **1994**, *53*, 292-297.
57. Willis, G.H.; McDowell, L.L.; Smith, S.; Southwick, L.M. *J. Environ. Qual.* **1994**, *23*, 96-100.
58. Nord, J.C.; Pepper, W.D. *J. Entomol. Sci.* **1991**, 26, 287-298.
59. Ohtsubo, T.; Tsuda, S.; Takeda, H.; Tsuji, K. *J. Pestic. Sci.* **1991**, *16*, 609-614.
60. Willis, G.H.; Smith, S.; McDowell, L.L.; Southwick, L.M. *Arch. Environ. Contam. Toxicol.* **1996**, *31*, 239-243.
61. Nelson, S.D.; Letey, J.; Farmer, W.J.; Williams, C.F.; BenHur, M. *J. Environ. Qual.* **1998**, *27*, 1194-1200.
62. Wallach, R.; Jury, W.A.; Spencer, W.F. *Trans. ASAE* **1989**, *52*, 612-618.
63. Ahuja, L.R. *Soil Sci. Soc. Am. J.* **1990**, *54*, 312-321.
64. Richardson, C.P.; Parr, A.D. *J. Hydraul. Eng.* **1991**, *117*, 1496-1512.
65. Ahuja, L.R.; Ross, J.D.; Lehman, O.R. *Water Resour. Res.* **1981**, *17*, 65-72.
66. Ahuja, L.R.; Lehman, O.R.; Sharpley, A.N. *Soil Sci. Soc. Am. J.* **1983**, *47*, 746-748.
67. Zhang, X.C.; Norton, L.D.; Nearing, M.A. *Water Resour. Res.* **1997**, *33*, 809-815.
68. Ahuja, L.R.; Lehman, O.R. *J. Environ. Qual.* **1983**, *12*, 34-40.
69. Parr, A.D.; Zou, S.; McEnroe, B. *J. Environ Eng.* **1998**, *124*, 863-868.
70. Ahuja, L.R.; Sharpley, A.N.; Lehman, O.R. *J. Environ. Qual.* **1982**, *11*, 9-13.
71. Sharpley, A.N.; Ahuja, L.R.; Menzel, R.G. *J. Environ. Qual.* **1981**, *10*, 386-391.
72. Richardson, C. P.; Parr, A. D. *J. Environ. Eng.* **1988**, *114*, 792-809.
73. Hairsine, P.B.; Rose, C.W. *Wat. Resour. Res.* **1992**, *28*, 237-243.

74. Hillel, D. *Environmental Soil Physics*, Academic Press: San Diego, CA, 1998.
75. Nearing, M.A.; Parker, S.C. *Soil. Sci. Soc. Am. J.* **1994**, *58*, 1612-1614.
76. Rivlin, J.; Wallach, R.; Grigorin, G. *J. Contam. Hydrol.* **1997**, *28*, 21-38.
77. Yan, M.; Kahawita, R. *Wat. Res.* **2000**, *34*, 3335-3344.
78. Nearing, M.A; Norton, L.D.; Bulgakov, D.A.; Larionov, G.A.; West, L.T.; Dontsova, K.M. *Wat. Resour. Res.* **1997**, *33*, 865-876.
79. Hairsine, P.B.; Rose, C.W. *Wat. Resour. Res.* **1992**, *28*, 245-250.
80. Lick, W.; Chroneer, Z.; Rapaka, V. *Wat. Air Soil Pollut.* **1997**, *99*, 225-235.

Chapter 10

Assessing Atrazine Input and Removal Processes in the Chesapeake Bay Environment: An Overview

Haydee Salmun[1,2] and Kristin Goetchius[1]

[1]Department of Geography and Environmental Engineering,
The Johns Hopkins University, Baltimore MD 21218
[2]Current address: Department of Geography, Hunter College of the City
University of New York, 695 Park Avenue, New York, NY 10021

This study focuses on the behavior of atrazine, a broad-spectrum herbicide extensively used in the Chesapeake Bay watershed in the cultivation of corn and sorghum. Studies show that atrazine can have detrimental effects on aquatic ecosystems. Reactions between atrazine and reduced sulfur species present in anoxic sediment porewaters may provide a significant sink for this agrochemical. An integral approach involving field data analysis, laboratory studies and modeling is needed to understand the behavior of atrazine and to assess the toxic effects associated with the discharge of atrazine into the environment. As a first step, available data have been compiled to identify important inputs of atrazine and to estimate the resident mass of atrazine in the Chesapeake Bay. This paper presents a simple mass balance for the northern section of the Bay based on these data. This analysis illustrates that in order to assess and predict the long-term trends of atrazine for different loading scenarios, more comprehensive field data and more sophisticated models are needed that better capture the relevant physical and chemical processes.

Introduction

Estuaries are regions that support a wide variety of marine resources, including unique wildlife habitats and recreational opportunities. Because estuaries are dynamic zones of high biological productivity, most commercially important fish species spend a significant part of their life cycles in estuaries. At the same time, many estuaries are subject to nonpoint sources of pollution brought in via runoff. Organic contaminants originating from agricultural activities within the estuaries' watersheds are thus funneled directly into highly sensitive ecosystems. Even though many of these contaminants are herbicides with low mammalian toxicity, their effects at low concentrations on phytoplankton that are responsible for primary production in such areas are not well understood (1). Herbicides have been implicated in the decline of submerged aquatic vegetation (SAV) in ecosystems such as the Chesapeake Bay (2). Furthermore, they often are highly toxic to fish and other aquatic fauna. A comprehensive review of the current knowledge and understanding of the behavior of pesticides in surface waters can be found in Larson et al. (1).

One of the most prominent groups of herbicides is the *chloro-s-triazines*, of which atrazine is the best-known example. Atrazine is a pre- and post-emergence herbicide for the control of annual and perennial grass, as well as for the annual broad-leaved weeds, and is one of the fourteen organic compounds of potential concern in the Chesapeake Bay (3). In addition to atrazine itself, a variety of environmental degradation products may form, such as deethylated, deisopropylated and hydroxyatrazine. These degradates may be more or less toxic than the parent compound, depending on chemical structure. For detailed studies of metabolites of atrazine and other herbicides in surface waters, see Meyer and Thurman (4). Many of the studies reported in the literature have been conducted in the laboratory or in artificial streams. Natural aquatic ecosystems are very complex, and it is difficult to find suitable controls that would aid in the assessment of the effects of atrazine in surface waters. It is the general consensus, however, that atrazine and its metabolites, if present at sufficient levels, could exacerbate the chronically stressed Chesapeake Bay ecosystem.

Atrazine levels are sufficient to inhibit various species of SAV in localized areas susceptible to agricultural runoff (5), so it does pose a threat in those areas. Significant concentrations of atrazine or persistent exposure to lower levels may result in changes in species composition and diversity, with species susceptible to atrazine being replaced by more resistant ones. For example, the less desirable water milfoil grass proved more resistant to atrazine than several other species of Bay grass that are important food sources for waterfowl (6). Atrazine can also affect aquatic fauna since it appears to be a potent environmental endocrine disrupter. A recent study based in the U. K. investigated the effect of atrazine on the reproductive system and the consequent impact on the reproductive behavior of mature male Atlantic salmon (*Salmo salar* L) (7). It concluded that exposure of the mature males to sub-lethal levels of atrazine in the water inhibited their ability to detect and respond to female priming pheromones.

Atrazine is persistent in aquatic environments, as reflected by its high dissolved concentrations in tributaries to the Chesapeake Bay (8). It is estimated that about 1% of the total amount of atrazine applied to the Chesapeake Bay watershed reaches the aquatic environment (9, 10). A survey of studies of surface waters conducted between 1976-1993 found detectable atrazine concentrations in 67% of the analyzed samples (11). There was no indication of a decreasing trend of atrazine in surface water over a continuous span of 14 years during the study period despite a reported decrease in its use. Data on atrazine concentration collected by Hall et al. (12) in 1995-1996 from the top meter of the water column during the high and low periods of atrazine use show that atrazine continues to be detected in all major and many secondary tributaries as well as in the Bay proper. The concentrations reported in the latter study appear to be smaller, although not significantly. The persistence of atrazine in the Bay depends in part on the dynamics of mixing processes, as well as on chemical and biological reactions in the water column and sediment layers. Recent laboratory work suggests that atrazine may undergo transformation in the presence of naturally occurring inorganic sulfur nucleophiles (polysulfides) and that abiotic reactions with these reduced sulfur species may constitute a significant removal mechanism for atrazine in anoxic marine sediment porewaters (13, 14).

Effective management of these chemicals is required to minimize their potential adverse effects on living communities. To develop managerial and practical guidelines concerning their use in the watershed, it is necessary to study their transformation and transport from the application site, which can be effectively accomplished by an approach that uses a combination of field data and process modeling. This chapter provides an overview of the various hydrologic and chemical processes that occur in the Chesapeake Bay that may affect the fate and transport of atrazine. A summary of data on atrazine input to the Bay from different sources is presented, and a simple mass balance argument is formulated based on these data. The failure of the mass balance argument to provide a consistent picture of the fate of atrazine in the Bay demonstrates the need for more detailed data, complemented by a modeling approach that better captures the physics of transport, mixing, and chemical processes.

A complete description of the processes that govern the fate of agrochemicals in the Bay is still beyond our current scientific knowledge. Simplified analyses are thus often used to gain some insight into the trends of chemical concentrations in the Bay. The next section presents a descriptive summary of the relevant physical and chemical processes occurring in the Bay. A compilation of data on atrazine inputs to the Bay is summarized in the following section. These data are used to estimate a resident atrazine mass, which is compared to estimations made from measured field values. We then present a simple mass balance for the Upper Chesapeake Bay region. This is followed by a brief discussion of the use of simple models in combination with field data to describe the behavior of atrazine in Bay waters and surrounding environment. We conclude with a summary of the main points discussed in this overview.

Relevant Physical and Chemical Processes

The Chesapeake Bay is the largest estuary in the United States. It drains a 164,000 square kilometer watershed with an extremely high ratio of land and population to volume of water (Figure 1). The distance between the mouth of the Bay to the mouth of the Susquehanna River is about 320 kilometers. The length of the shoreline is in the range of 7,000 kilometers long (15). The Bay is relatively narrow and shallow, with a mean width of approximately 15 kilometers and a mean depth of 10 meters, although the actual depth ranges from a few meters to about 30 meters in the deep channel. It holds about 18 trillion gallons of water and has a total surface area of about 8,000 square kilometers. The mean hydraulic residence time of the Bay is approximately 90 days (16). The overall circulation in the Bay is characterized by the flow of fresh water toward the ocean in the surface layer and the landward flow of saline water in the deep layer arising from the entrainment of salty water at the head of the estuary, known as the gravitational circulation. The mean velocity associated with this circulation is approximately 0.1 m/s (17). The physical characteristics of the Chesapeake Bay are summarized in Table I.

Table I. Physical Characteristics of the Chesapeake Bay

Watershed area	$164,000 \text{ km}^2$
Distance from Susquehanna River to mouth of Bay	320 km
Mean width	15 km
Mean depth	10 m
Maximum channel depth	30 m
Surface area	$7,800 \text{ km}^2$
Upper Bay Surface Area	$1,480 \text{ km}^2$
Volume	18 trillion gallons
Hydraulic residence time	90 days
Shoreline length	7,000 km

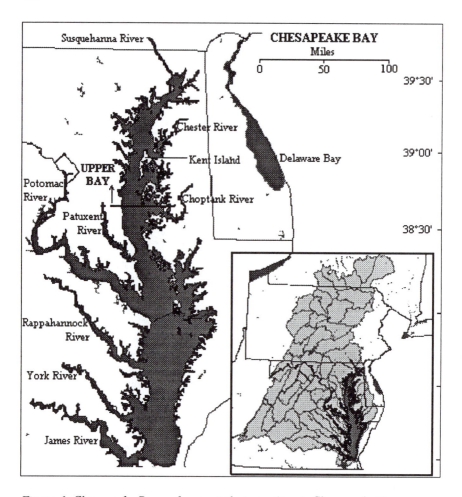

Figure 1. Chesapeake Bay and major tributaries (inset: Chesapeake Bay drainage basin).

The Bay is classified according to circulation and stratification patterns as a partially mixed estuary (*18*). Mixing for the most part is due to turbulent motion caused primarily by tidal action and secondarily by climate conditions.

   To assess the impact agrochemicals could have on the well-being of estuarine ecosystems and to determine the role estuaries may play in effecting contaminant removal, an integral approach that includes the relevant hydrologic and chemical processes must be considered. Agrochemicals such as the chlorinated triazine herbicides tend to undergo abiotic hydrolysis relatively slowly; for example; the half-life for the uncatalyzed hydrolysis of atrazine is 1800 years at pH 6.97 and 25°C (*19*). They are at best slowly degraded by microbes, which are present at low concentrations typical of aquatic environments (*20, 21, 22*). Moreover, they tend to be of low volatility, and sufficiently hydrophilic that they are not efficiently removed from the water column via scavenging by settling particles (*23*). Peak agrochemical loading often occurs in the spring, coinciding with seasonal anoxia during which sulfate (abundant in sea water) may undergo dissimilatory microbial reduction to generate reactive sulfur nucleophiles. Concentrations of polysulfides, which are particularly reactive sulfur nucleophiles, can attain higher values in the sediment porewaters. In order for atrazine that is introduced into fresh waters to react with polysulfides, it must first mix down the water column to reach the sulfidic sediment porewaters. This complicates attempts to evaluate the significance of reactions with sulfur nucleophiles based simply on a comparison of reaction half-lives with overall hydraulic retention times; rates of vertical mixing also play a key role and must be included in modeling approaches.

Atrazine in the Chesapeake Bay

Systematic investigation of pesticides in the Chesapeake Bay watershed began in the 1970's in response to the observed decline in SAV and fish populations during that period. Investigations typically spanned a few years and tested for specific families of herbicides (e.g., chloro-s-triazines and chloroacetanilides) and insecticides (e.g., organophosphates and chlorinated hydrocarbons) in various media. Detection limits and consistency in the methods of data collection improved with time. The data that are summarized below consist of measurements of concentrations of atrazine in the inputs to surface waters at several locations throughout the Bay. We should stress that data are sparse; spatial and temporal distributions have to be estimated and/or extrapolated from measurements in the top 0.5-1.0 m of the surface layer.

Inputs of Atrazine

The three major pathways for atrazine to enter the Chesapeake Bay are surface runoff, groundwater inflow, and wet and dry atmospheric deposition. Processes such as hydraulic flushing, air-water transfer and chemical reactions will then influence the final concentration and distribution of atrazine in the Bay.

Surface Runoff

There are few studies on the watershed scale that estimate the atrazine loading to the Bay due to agricultural activity. Input of atrazine and other herbicides, such as metolachlor, from surface runoff primarily occurs in the months of May through August, as demonstrated in Figure 2. In the following analysis it is assumed that all of the surface runoff enters the Bay via river flow.

Three tributaries, the Susquehanna, the Potomac, and the James Rivers account for approximately 80 to 85 percent of the fresh water flow into the Bay from the northern and western regions. These three tributaries also dominate herbicide input to the Bay (8). Foster and Lippa (8) present the most comprehensive data sets available. They estimated that 2,700 kg of atrazine were loaded into the Bay in the period 1992-1993: 1,700 kg/yr via the Susquehanna, 780 kg/yr via the Potomac, and 220 kg/yr via the James. Similar estimates of atrazine loading were reported by the U. S. Environmental Protection Agency (USEPA) (15, 24) for the Susquehanna and the James Rivers. Godfrey et al. (25) showed that the relative error in the estimation method used by Foster and Lippa (8) was less than 40%. The Chester and Choptank Rivers are the major pathways of atrazine input from the Eastern Shore of the Bay. Based on concentration and flow rate, these rivers account for a combined mass load of approximately 100 kg/yr. The total input of atrazine via surface runoff is thus 2,800 kg/yr.

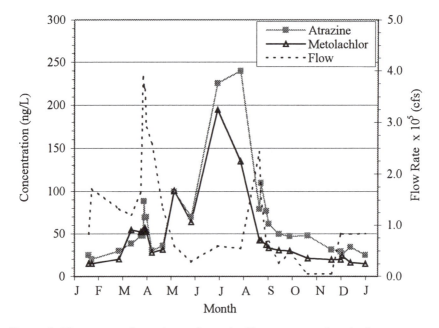

Figure 2. Flow rate and atrazine and metolachlor concentrations in the Susquehanna River in 1994 (26).

Groundwater

To estimate the amount of atrazine input via groundwater inflow, both concentration and groundwater seepage measurements are needed from areas surrounding the estuary. Studies conducted within the Bay's watershed in the 1980s and early 1990s measured concentrations of atrazine in groundwater ranging between 0.3 to 3 μg/L (*27, 28, 29*) throughout the year. To estimate atrazine loading due to groundwater inflow we assume that a representative atrazine concentration of approximately 1 μg/L exists in the groundwater that discharges into the Bay. Although atrazine concentrations vary seasonally depending on time of application and type of bedrock, this estimate is a conservative one and is taken to characterize atrazine concentration in groundwater throughout the year. Fewer studies are available on groundwater seepage rates due to the experimental difficulty associated with such measurement. Reay *et al.* (*30*) reported an average groundwater discharge into the Bay in the southern region of the Eastern Shore of 0.35 L/(m^2 hr), which corresponds to a flow velocity of 0.0084 m/day. They also observed that discharge into the Bay occurs predominantly within a 50-meter distance offshore. Assuming that the estimated shore length is 7,000 km and neglecting the bottom-slope effects over this distance, the bottom area for inflow is estimated as 350 km^2 (50 m x 7,000 km). Consequently, the estimated atrazine groundwater loading to the whole Bay, computed as groundwater flux x discharge concentration x bottom area, is 1,100 kg/yr.

Atmospheric Deposition

Despite the low Henry's law coefficient for atrazine, it has been detected in rainfall in agricultural regions in the U.S. (*31, 32*). Goolsby *et al.* (*32*) estimated atrazine deposition rates that range between 11 and 60 μg/m^2/yr in the northeastern region of the U.S. Wu (*31*) estimated the wet deposition at Rhode River Watershed (MD) as 102 μg/m^2/yr in 1977 and as 9.7 μg/m^2/yr in 1978. USEPA (*15*) estimated the combined wet and dry deposition of atrazine to the Bay as 770 kg/yr. Recent studies by Harman-Fetcho *et al.* (*33*) estimated the wet deposition of atrazine per rain event over a study period from April 17-June 26, 1995. Data from twenty four events were used to estimate that the total wet deposition flux of atrazine in Solomons, MD was 6,400 ng/m^2 over the entire study period, ranging from 5 to 1,300 ng/m^2 per event. If these measurements can be considered characteristic for the entire Bay, they would result in a load of approximately 50 kg of atrazine over the 70 days reported for the study period, which is comparable to the lower bounds reported in the literature (*31, 32*). Here we use 102 μg/m^2/yr as an upper limit for the amount of atrazine introduced to the whole Bay resulting in an atmospheric deposition load of approximately 800 kg/yr, which agrees very well with the estimate by the USEPA.

Total Inputs

The total amount of atrazine entering the whole Bay via the various pathways is 4,700 kg/yr. The estimates are summarized in Table II. The National Oceanic and Atmospheric Administration (NOAA) reports that approximately 481,582 kg (1,061,707 lb) of atrazine is applied to the Chesapeake Bay drainage area annually (*11*). Using this value and assuming that the atrazine loading is 1% of that mass, approximately 4,800 kg/yr of atrazine are washed into the Bay. This value closely agrees with the one presented here.

Atrazine inputs to the Bay derive mainly from: (1) surface runoff, estimated to contribute about 60% of the total load, (2) groundwater flux, contributing about 23% of total atrazine load, and (3) atmospheric input: wet and dry deposition, which constitutes 17% of the atrazine load to the Bay.

Table II. Summary of Atrazine Inputs to the Chesapeake Bay

Source	Entire Bay		Upper Bay	
	kg/yr	%	kg/yr	%
Surface Runoff	2,800	60	1,800	84
Groundwater	1,100	23	200	9
Atmospheric	800	17	150	7
TOTAL	**4,700**	**100**	**2,150**	**100**

Resident Atrazine Mass in the Bay

To date, only sporadic temporal and spatial measurements exist for atrazine concentration in the surface waters of the Chesapeake Bay, and the quality of the data only allows for an order-of-magnitude estimate of the resident atrazine mass. To obtain this estimate, we assume that the bulk of this herbicide resides in the water column and the sediment layer, with negligible amounts in biota and the surface microlayer, a layer of up to 1 mm thick at the air-water interface.

Samples for atrazine concentration in surface waters are typically taken from the top 0.5 m in the Bay. Reported concentrations for the data collected during 1976-1993 range between 0.03-4.3 µg/L, and show no evidence of spatial or temporal trends (*34*). Data histograms (not shown here) yield a log-normal distribution with a mean and variance of 0.7 µg/L and 0.99 µg/L, respectively. Applying this mean concentration to a 6-m thick, well-mixed surface layer, and multiplying by the surface area in Table I, we can compute the average mass of atrazine in the surface layer. The average mass for the Bay is 32,760 kg (0-125,890 kg, 95% confidence interval), of which 6,224 kg (0-23,919 kg, 95% confidence interval) is assumed to be in the northern 110 km of the Bay. For reference, we note that these values are of the same order as those reported by Schottler and Eisenreich (35) for the atrazine inventory in, for example, Lake Michigan of the Great Lakes.

The lack of a standard protocol for sediment sampling introduces additional variability in reported concentrations to the natural variability encountered in the Bay. Kroll and Murphy (34) detected atrazine in 93% of their sediment samples taken in 1978 at a Maryland site, with a maximum measured concentration of 2.15 ppb. Boynton et al. (36) measured atrazine concentrations below 1 ppb in the sediment layers in the eastern and western tributaries of the Bay in 1980. More recently, Eskin et al. (37) reported no detectable atrazine concentrations in 40 stations covering the main stem and tributaries of the Bay. The difference between this work and earlier studies may reflect a removal of atrazine from the sediment layer, or, simply, better accuracy and reliability in the analytical methods. It is reasonable to adopt the most recent results and to consider the atrazine mass resident in the sediment layer insignificant.

A Simple Mass Balance for the Upper Chesapeake Bay

A large fraction of atrazine is introduced to the Bay via surface runoff in the northern section. The largest runoff contribution to the Chesapeake Bay is the Susquehanna River, which discharges directly into the Upper Bay. In addition, the Chester and Choptank Rivers discharge into the northern section of the Bay. A larger amount of data on atrazine concentrations is available for this region resulting from a larger network of stations. Consequently, a first step toward understanding atrazine behavior in surface waters and sediment porewaters of the entire Bay is to understand its behavior in the northern section and we focus on this region in the remaining of the present discussion. The Upper Bay extends approximately 110 km south of the outlet of the Susquehanna River (Figure 1). Its mean depth ranges from 12 m over the first 60 km to 25 m in the remaining 50 km. As for the entire Bay, total input and output can be computed for the Upper Bay to provide a simple budget analysis, and the results compared to resident mass that was calculated from measured concentrations.

The Susquehanna River fluvial input of atrazine in the Upper Bay region is 1,700 kg/yr (8). The Chester and Choptank Rivers discharge approximately 100 kg/yr. Both contributions yield a total surface flux of 1,800 kg/yr into this region. To estimate the amount of atrazine introduced as groundwater inflow, we assume that the loading to the northern region of the Bay via this pathway is proportional to the ratio of its surface area (1.48×10^7 m^2) to that of the whole Bay (7.8×10^7 m^2), which is 0.19. Thus, the input in groundwater to the upper region is 200 kg/yr. In a similar fashion, from estimates for the entire Bay we estimate that the atmospheric deposition to the Upper Bay is 150 kg/yr. In summary, a total of 2,150 kg/yr is loaded to the Upper Bay with surface runoff estimated to contribute about 60% of the total load, groundwater flow contributing about 23% of total atrazine load, and atmospheric input constituting 17% of the atrazine load. These results are summarized in Table II.

Measured atrazine concentration range between 0.03 and 4.3 µg/L, with a mean value of 0.7 µg/L. The total surface area of the Upper Bay is 1.48×10^9 m^2 and if we assume that atrazine is reasonably well distributed over the top 6 m (the surface mixed layer) then the volume over which atrazine concentration is quasi-uniform is approximately 9×10^9 m^3. We note that considering only the first meter of the water column, where data from measurements are reported, does not change this simple mass balance argument significantly. With these assumptions (consistent with the data as reported in the literature) the total mass of atrazine resident in this part of the Bay ranges from 270 kg for the lowest reported concentrations to 38,700 kg for the upper limit. The latter is highly unrealistic. If, on the other hand, we use the mean concentration value of 0.7 µg/L, we obtain a total mass of about 6,000 kg. The outflow from the Upper Bay over a depth of 6 m is approximately 11,000 m^3/s (24). The rate of mass of atrazine removed by this flow is obtained by multiplying the outflow value times the concentration value. The lowest concentration value (0.03 µg/L), if assumed constant throughout the year, results in about 10,400 kg/yr removal of atrazine from this portion of the Bay, or 5,200 kg/yr if that concentration is present only half the year. Yet, we can only account for 2,150 kg of atrazine as input, hence this simple mass balance argument indicates that either a substantial source of atrazine is unaccounted for or, more likely, that concentrations are lower and are not uniform in time and space. Furthermore, atrazine must be leaving this region in the top layer by some diffusion-like mechanism. To achieve steady state, atrazine concentrations in the outflow surface water need to be 0.006 µg/L, but there is no evidence that supports this estimate. More and better data and models are clearly needed to elucidate some of these important issues.

Modeling Atrazine Behavior in the Chesapeake Bay

Recent work suggests that rates of atrazine reaction with reduced sulfur species present in the sediment porewaters may be sufficiently rapid to provide an important removal mechanism in sulfate-reducing environments (13). The most significant reaction of atrazine with reduced sulfur nucleophiles occurs with polysulfides. This reaction can be modeled as a pseudo first order reaction assuming a polysulfides concentration range in the sediment porewaters of 0.02-0.2 mM (38) and a second order rate constant of 5.6 x 10^{-3} M^{-1} s^{-1} (14).

The multi-box model known as MASAS (Modeling Anthropogenic Substances in Aquatic Systems developed by Ulrich, 1991) was used in the hope that simulations with this approach could guide the development of more realistic models to study this potential removal mechanism. We assumed an idealized estuary with some basic aspects of the characteristics of the Chesapeake Bay, such as gravitational circulation and river flow. Three types of chemical reactions were incorporated in the model: photolysis, hydrolysis, and abiotic reaction with reduced sulfur species in the sediment layer. Simulations revealed that, although reduced sulfur species in the sediment have the potential for degrading atrazine, stratification and reduced mixing across the pycnocline minimize the importance of such reactions. Simulated atrazine concentrations in the Bay showed that it tends to accumulate in the upper water layers, even during the weak stratification in fall period. A large discrepancy was observed between simulated and measured atrazine concentrations in surface waters, with measured concentrations being one order of magnitude higher than simulated values. Although MASAS has been used successfully to describe atrazine behavior in Swiss lakes (39, 40), our results confirm the limitations of this type of model for use in large estuaries. These models assume that the rates of vertical mixing are smaller than reaction rates and that both are much smaller than horizontal mixing rates. This assumption is not valid in most estuaries.

In particular, horizontal advection and horizontal diffusion in the Chesapeake Bay are comparable while vertical diffusion is a fast process that acts over short distances, and a model must account for all three. In this environment, atrazine that is discharged to the surface waters could be horizontally distributed over a distance of 1 km over a period of one week, since the time scale of horizontal advection-diffusion processes is 10^4-10^5 s (approximately 3 hours). As atrazine is distributed horizontally, it also mixes vertically down the water column. With the estimates of vertical diffusivity for the Bay that are available in the literature, for a depth of 10-20 m the time scale for vertical diffusion processes is on the order of 15 minutes, and can be as short as 3 minutes. The sulfidic waters are in the sediment porewaters and atrazine needs to be transported to the water-sediment interface in order to encounter and react with reduced sulfur species. The characteristic horizontal and vertical scales that describe the flow in the Bay indicate that it is possible for atrazine to reach the depth of the water-sediment interface before it is horizontally transported out of the system. The subsequent exchange at the water-sediment interface depends on many factors, including half-life of atrazine, the hydraulic residence time of the bottom layer, turbulent processes, and other characteristics of the water column above the sediment layer. Simple box models cannot capture the dynamics necessary to describe these exchanges that ultimately govern the fate of atrazine in the Bay.

The Bay environment includes many fresh, brackish and salt water marshes and tidal flats which cover an area of approximately 1,700 square kilometers. These marshes are present along both the Eastern and Western Shores of the Chesapeake Bay and constitute an important component of the Bay ecosystem, in part because they are highly reactive environments and may play an important role in removing pesticides before they reach the main estuary. If these removal processes result in the detoxification of contaminants, they might be considered for use in treating agricultural runoff.

Salt marshes are shallow systems, with a water column depth that ranges from less than 0.2 m to 0.6 m (41). They can be tidally flooded for long periods, although the degree of inundation depends on the geographic location and season. During periods of flooding salt marshes are characterized as anoxic environments, and high concentrations of reduced sulfur species have been detected (42). Because of their shallowness, anoxic characteristics, and the likely presence of atrazine in salt marshes, there exists a potential for abiotic reactions between atrazine and polysulfides. Atrazine can reach a salt marsh via groundwater flux, atmospheric deposition, tidal flow, and surface runoff. In a shallow water column atrazine reaches the sediment porewaters where it can react with polysulfides, which have been detected in sediments of several salt marshes in the area, including the Great Marsh in Delaware (42). The polysulfides concentration in sediment porewaters in salt marshes of the Bay region is reported to be as high as 0.3 mM (43). Studies have also shown that concentrations of polysulfides in salt marshes vary seasonally (42), with concentrations peaking during periods of anoxia. During spring and summer, salt marsh vegetation injects oxygen into the upper sediment layer (44), which may partially oxidize hydrogen sulfides and bisulfides resulting in the formation of polysulfides (45). Therefore, peaks in polysulfides concentration are likely to coincide with the peaks in atrazine concentration, making spring and summer the most probable seasons to observe the reaction between them. To investigate the fate of atrazine in salt marshes, and the potential role these marshes play in affecting the concentration of atrazine in the Bay proper, modeling studies and more detailed data are needed.

Summary and Conclusion

An overview of the behavior of atrazine, an extensively used corn herbicide, in the Chesapeake Bay environment has been presented. Atrazine has a differential effect on varying species of submerged grasses and could shift species composition in areas where it reaches the Bay or tributaries in large quantities. There is concern that it can become an added stressor to an environment already heavily affected by excess nutrients, rapid changes in land use and population growth. There is also concern about the effects of various atrazine metabolites that may have an increased or reduced toxicity depending on their chemical structure. The Chesapeake environment includes the presence of salt marshes, which may play a role in transforming agrochemicals. Salt marshes are very complex systems and there is very little or no data readily available on atrazine behavior in marshes surrounding the Bay. There is some limited information on atrazine concentrations for the Bay itself and some tributaries. Some of the physical processes that must be included in models of salt marshes are also relevant to the estuary.

From our investigation we conclude that available data on atrazine, although not comprehensive, constitute an adequate basis for modeling studies and allow for preliminary estimates of the fate of atrazine in the Chesapeake Bay. However, a great limitation on these studies resides in the lack of observations throughout the water column. Box models that have successfully been used to study atrazine behavior in lakes cannot capture the basic processes that govern transport in estuaries as complex as the Bay. To look at different loading scenarios and to attempt to predict trends, we need a multi-dimensional convective-diffusive model with chemical processes included. We stress here again that a great limitation on initializing and validating such models resides in the lack of observations of atrazine distribution with depth in the water column. We anticipate that the study of atrazine input and removal mechanisms in a salt marsh will be hampered by a similar lack of detailed data and the appropriate hydrodynamic model to be used with them.

Acknowledgements

The first author gratefully acknowledges the assistance of Y. Farhan with the initial compilation of the data. Discussions with A. Lynn Roberts and Katrice Lippa helped us understand the issues associated with the relevant chemical reactions. We also thank the anonymous reviewers for their comments and suggestions. This research has been partially funded by the US Environmental Protection Agency through STAR Grant R826269-01-0. The contents of this manuscript do not necessarily reflect the official views of the USEPA, and no official endorsement should be inferred.

References

1. Larson, S. J.; Capel, P. D.; Majewski, M. S. *Pesticides in Surface Waters. Distribution, Trends, and Governing Factors*; Pesticides in the Hydrological System; Ann Arbor Press, Inc.: Chelsea, MI, 1997; Vol. 3, pp. 373.

2. Glotfelty, D. E.; Taylor, A. W.; Isensee, A. R.; Jersey, J.; Glenn, S. *J. of Environmental Quality* **1984**, *13*, 115-121.
3. U.S. Environmental Protection Agency, 1991. Chesapeake Bay Toxics of Concern List, *Report prepared by the Chesapeake Bay Program Toxics Subcommittee's Living Resources Subcommittee's Joint Criteria and Standards Workgroup*, Annapolis, Maryland.
4. *Herbicide Metabolites in Surface Water and Groundwater*; Meyer, M. T.; Thurman, E. M., Eds; ACS Symposium Series 630; American Chemical Society: Washington, D. C., 1995; pp. 318.
5. Solomon, K. R.; Baker, D. B.; Richards, R. P.; Dixon, K. R.; Klaine, S. J.; LaPoint, T. W.; Kendall, R. J.; Weisskopf, C. P.; Giddings, J. M.; Gesiy, J. P.; Hall, L. W.; Williams, W. M. *Environ. Toxic. Chem.* **1996,** *15,* 31-46.
6. Jones, T. W.; Winchell, L. *J. Environ. Qual.* **1984**, *13*, 243-247.
7. Moore, A.; Waring, C. P.; *Pesticide Biochem. Physiol.* **1998**, 62, 41-50.
8. Foster, G.; Lippa, K. *J. Agric. Food Chem.* **1996**, *44*, 2447-2454.
9. Wu, T. L. *K. Environ. Qual.* **1980**, *9*, 459-465.
10. Kemp, W. M.; Means, J. C.; Jones, T. W.; Stevenson, J. C. Herbicides in Chesapeake Bay and their effects on submerged aquatic vegetation. In *Chesapeake Bay Program Technical Studies: A Synthesis*. United States Environmental Protection Agency: Washington, D.C., 1982; pp. 502-566.
11. Johnson, W. E.; Plimmer J. R.; Kroll R. B.; Pait A. S. In *Perspectives on Chesapeake Bay, 1994: Advances in Estuarine Sciences*; Nelson, S.; Elliott, P., Eds.); Chesapeake Bay Program; 1994, 105-146.
12. Hall, L. W., Jr.; Anderson, R. D.; Kilian, J.; Tierney, D. P; *Environ. Monitoring and Assessment.* **1999**, *59*, 155-190.
13. Lippa, K. A.; Roberts, A. L. *ACS, Division of Environmental Chemistry, Preprints of Extended Abstracts* **1998**, *38*(2), 128-129. *ACS, National Meeting and Exposition Program*; Washington, D.C., 2000.
14. Lippa, K. A.; Klotz, J. C.; Roberts, A. L. *ACS, National Meeting and Exposition Program*; Washington, D.C., 2000.
15. U.S. Environmental Protection Agency, 1994b. *Chesapeake Bay basinwide toxics reduction strategy reevaluation report*. Chesapeake Bay Program's Toxics Subcomittee Report 117/94.
16. Hagy, J. D.; Boynton, W. R.; Sanford, L. P; *Estuaries*. **2000**, *23*, 328 – 340.
17. Schubel, J. R.; Pritchard, D. W. *Estuaries* **1986,** *9*, 236-249.
18. Pritchard, D. W. *Estuaries*, **1967**, Lauff, G., Ed., Amer. Assoc. Adv. Sci, 37-44.
19. Plust, S. J.; Loehe, J. R.; Feher, F. J.; Benedict, J. H.; Herbrandson, H.F. *J. Org. Chem.*, **1981**, *46*, 3661-3665.
20. Howard, P. R. *Handbook of Environmental Fate and Exposure Data for Organic Chemicals;* Lewis Publishers: Chelsea, MI, 1991; Vol. 3, 1-684.
21. *The Pesticide Manual: A Worldwide Compendium;* Worthing, C. R., Ed.; British Crop Protection Council, Farnham, Surrey, 1991.

22. *Agrochemicals Desk Reference: Environmental Data;* Montgomery, J. H., Ed.; Lewis Publishers: Chelsea, MI, 1993; 1-625.
23. Meakins, N. C.; Bubb, J. M.; Lester J. N. *Marine Poll. Bulletin* **1995,** 30, 812-819.
24. U.S. Environmental Protection Agency, 1994. *Response of the Chesapeake Bay water quality model to loading scenarios.* Chesapeake Bay Program Technical Report Series 101/94.
25. Godfrey, J.T.; Foster, G. F.; Lippa, K. A. *Environ. Sci. Technol.*, **1995,** *29*, 2059-2064.
26. Foster, G. D.; Lippa, K. A.; Miller, C. V.; *Environ. Toxic. Chem.* **2000**, *19*, 992-1001.
27. Mostaghimi, S.; McClellan, P. W.; Cooke, R. A. *Water Sci. Technol.* **1993**, *28*, 279-387.
28. Isensee, A. R.; Nash, R. G.; Helling, C. S. *J. Environ. Qual.* **1994,** *19*, 434-440.
29. Ritter, W. F.; Scarborough, R. W.; Chirnside A. E. *J. Contaminant Hydrology* **1994**, *15*, 73-92.
30. Reay, W. G.; Gallagher, D. L.; Simmons, G. M., Jr. *Water Resources Bulletin* **1992**, *28*, 1121-1134.
31. Wu T. L. *Water, Air, and Soil Pollution* **1981,** *15*, 173-184.
32. Goolsby, D. A.; Thurman, E. M.; Pomes, M. L.; Meyer, M. T.; Battaglin, W. A. *Environ. Sci. Technol.* **1997**, *31*, 1325-1333.
33. Harman-Fetcho, J. A.; McConnell, L. L.; Rice, C. P.; Baker, J. E. *Environ. Sci. Technol.* **2000,** *34,* 1462-1468.
34. Kroll, R. B.; Murphy, D. L. *Pilot monitoring project for 14 pesticides in Maryland surface water*; Maryland Department of the Environment: Baltimore, MD, 1993.
35. Schottler, S. P.; Eisenreich, S. J. *Environ. Sci. Technol.* **1997**, *31*, 2616-2625.
36. Boynton W. R.; Means, J. C.; Stevenson, J. C.; Kemp, W. M.; Twilley, R. *Concentrations of the herbicides atrazine and linuron in agricultural drainage, tributary estuaries and littoral zones of Upper Chesapeake Bay*; University of Maryland, Center for Environmental and Estuarine Studies: Solomons, MD, 1983.
37. Eskin, R.; Rowland, K. H.; Alegre, D.Y. Contaminants in Chesapeake Bay sediments 1984-1991. USEPA: 1996.
38. MacCrehan, W.; Shea, D. In *Geochemical Transformations of Sedimentary Sulfur*; Vairavamurthy, M.; Schoonen, M. A., Eds.; ACS Symposium Series; American Chemical Society: Washington, DC, 1995; Vol. 612, 294-325.
39. Ulrich M. M; Müller, S. R.; Singer, H. P.; Imboden, D. M.; Schwarzenbach, R. P. *Environ. Sci. Technol.* **1994**, *28*, 1674-1685.

40. Müller, S. R.; Berg, M; Ulrich, M. M.; Schwarzenbach, R. P. *Environ. Sci. Technol.* **1997**, *31*, 2104-2113.
41. Leonard, L.; Luther, M. *Limnolog. Oceanogr.* **1995,** *40*, 1474-1484.
42. Luther, G. W. III; Church, T. M.; Scudlark, J. R.; Cosman, M. *Science* **1986**, *232*, 742-749.
43. Boulègue, J.; Lord, C. J.; Church, T. M. *Geochim. Cosmochim. Acta* **1982**, *46*, 453-464.
44. Cutter, G. A.; Velinsky, D. J. *Marine Chemistry* **1988,** *23*, 311-327.
45. Luther, G. W. III; Giblin, A. E.; Varsolona, R. *Limnol. Oceanogr.* **1985**, *30*, 727-736.

Chapter 11

The Urban Atmosphere: An Important Source of Trace Metals to Nearby Waters?

Robert P. Mason, Nicole M. Lawson, and Guey-Rong Sheu

Chesapeake Biological Laboratory, Center for Environmental Science, University of Maryland, P.O. Box 38, Solomons, MD 20688–0038

Atmospheric deposition of trace metals from urban environments on the perimeter of large water bodies is a potential source to these systems. There is little data, however, to support or refute this notion. Here we discuss a study of Baltimore's urban air and its potential impact on the northern Chesapeake Bay and compare our results to those of other studies. Elevated concentrations of metals, especially lead, zinc and mercury were measured at an urban sampling site compared to a rural location. The difference was most marked for lead with the annual depositional flux almost three times higher in the city. Normalized fluxes at a rural site were mostly similar to those measured previously around the Chesapeake Bay and were also similar to those measured recently at other rural sites in Maryland and around the Great Lakes. The results of our study suggest that local atmospheric inputs from urban sources should be included in any evaluation of atmospheric deposition to lacrustrine, coastal or estuarine systems.

Introduction

Atmospheric deposition provides an important fraction of the heavy metal inputs (e.g. lead (Pb), cadmium (Cd), and mercury (Hg)) to many water bodies, even those that are removed from local inputs (*1, 2*). The metals in deposition are derived from both natural and anthropogenic (point and non-point) sources to the atmosphere (*3-5*). Combustion, waste incineration and other industrial sources are important contributors of the more toxic elements, such as Hg, Cd, Pb, As and Se (*6, 7*). In the USA, the EPA has targeted anthropogenic sources of Hg for regulation (*7, 8*) to reduce Hg inputs to the atmosphere, and it is likely that future regulatory policies will focus on the other metals and metalloids that are volatilized to the atmosphere during high temperature combustion (*6*). Each element's emissions to the atmosphere has a dominant anthropogenic source inventory although soils are important sources of the crustal elements Al, Fe and Mn, as well as Cr (*9-11*). Waste incineration is an important source for Pb, Cd and Zn and Cr (*9-11*). Mercury emissions occur predominantly from coal combustion, medical and municipal waste incineration (*7, 12*). Metal production results in the emissions of many trace elements but is particularly important for Cu and Ni (*1*). Petroleum combustion (coal and oil) is an important source of the more volatile metals (*1, 13*). It has been estimated that the anthropogenic load to the atmosphere has increased total worldwide emissions such that estimates of the anthropogenic contribution to the total atmospheric input are: 85% for Cd; 60-70% for Pb, Zn and Ni; 50-60% for Hg and Cu; 40% for Cr and 10% for Mn (*1, 11, 12*). For Al and Fe, natural inputs constitute greater than 90% of the total.

There have been a number of recent studies focusing on understanding the long-range transport and deposition of pollutants to remote environments, including the open ocean (*1, 2, 13, 14*), but there has been less focus on quantifying the local deposition from point and area sources. Urban air usually contains elevated concentrations of trace elements and pollutants, which could enhance depositional fluxes to nearby water bodies. Recent studies have focused on quantifying the impact of pollutants from the Chicago metropolitan area on Lake Michigan (*15, 16*), and from the Baltimore metropolitan area on the Chesapeake Bay (*17-22*). While it has been acknowledged that the concentrations of metals in precipitation are likely higher in urban environments as a result of anthropogenic inputs, there have been few studies of the concentrations and deposition of metals in urban areas. Their potential impact on nearby water bodies needs to be examined as many urban environments are in coastal, lacustrine or estuarine settings. For this reason, we studied the potential atmospheric input of inorganic contaminants to the upper Chesapeake Bay from Baltimore.

The Chesapeake Bay is the largest estuary in the United States and the Baltimore metropolitan area, one of the heavily industrialized areas on the east coast, is located on the western shore of the Chesapeake Bay. The Chesapeake Bay has a large surface-to-volume ratio (mean depth is 7 m) and it is therefore particularly vulnerable to the influence of atmospheric deposition (3). Previous studies have demonstrated that Baltimore's urban atmosphere is an important source of PAHs and PCBs to the Chesapeake Bay (17, 18). Previous studies around the Chesapeake Bay have quantified the concentrations of inorganic and organic contaminants in wet and dry deposition at rural sites in the watershed (3-5, 20, 23). Additionally, a paper discussing the impact of urban-derived gaseous and particulate mercury appears as part of this book (21). In this chapter the focus is on metals in wet deposition.

For estuarine and coastal environments, direct deposition to the water's surface of some pollutants can be as important as runoff from the watershed given that the metals are strongly retained within the terrestrial environment (22, 24). Atmospheric deposition appears to be an important source to the Chesapeake Bay of metals such as Hg, Pb and Cd (3, 5, 20). However, as estimations are based on the measurements of metal deposition at rural sites, they could potentially underestimate the impact of deposition if the urban influence were significant. The present study was aimed at examining the difference in concentration and deposition in an urban and a rural setting to help address this important question.

Sampling Sites and Methods

Three atmospheric sites in the Chesapeake Bay watershed, and in close proximity to the Bay itself, were continuously monitored in 1997/98 as part of a large EPA-funded study of the atmospheric speciation of metals and organic contaminants and of the concentration of these constituents in wet and dry deposition (17, 18, 20, 21). There was an urban site, located on the roof of the Science Center (SC) in downtown Baltimore (Figure 1) and two more rural sites. These sites were the Chesapeake Biological Laboratory (CBL) at the mouth of the Patuxent River (about 60 km from the nearest city) and Stillpond (STP), on the eastern shore of the northern Chesapeake Bay and within 20 km of Baltimore (Figure 1). Previous studies occupied sites at Elms, Wye and Lewes (Figure 1) and a site in western Maryland near Frostburg (FRB).

Samples were collected every nine days for gaseous and particulate contaminants in the atmosphere and rain collections were integrated over the same time period (17, 20, 22). In the first year of study trace metal samples at SC

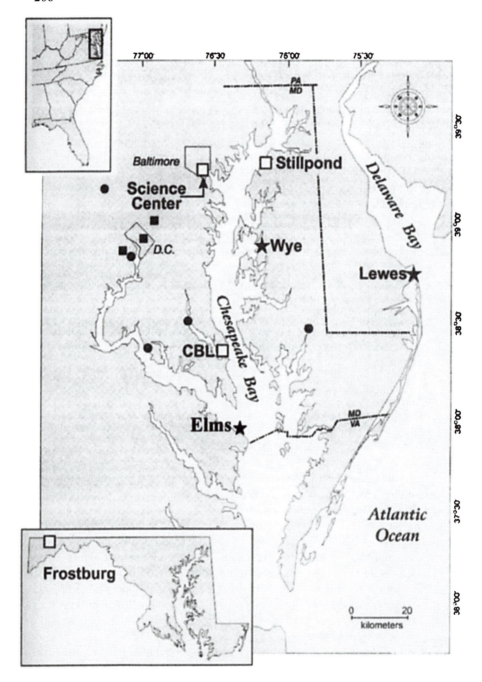

Figure 1. Map of the sampling sites. Locations of our study sites are the Science Center in downtown Baltimore (SC) and the Chesapeake Biological Laboratory (CBL). Other sites discussed in the chapter are Elms, Wye and the site in western Maryland, near Frostburg.

and STP were analyzed by the University of Delaware while CBL analyzed samples from SC and CBL in the second year of study. These data are reported here. Precipitation collections were made using a MIC-B automatic precipitation collector that had been modified to allow the use of glass funnels for Hg collection, a polyethylene funnel for trace metals and another for major ions (*25, 26*). Prior to use, the polyethylene bottles for metals were washed with low metal-containing detergent, then rinsed and soaked in HCl for two days. After rinsing with Q-water, the bottles were stored with dilute HCl until use. All equipment was kept bagged when not in use. Rain funnels were cleaned in a similar fashion and were changed after each sample period. Methods for major ions and Hg are given elsewhere (*20, 26*). The trace metal sampling protocols were essentially equivalent to those outlined in EPA Method 1669 (Sampling Ambient Water for the Determination of Trace Metals at EPA Water Quality Criteria Levels) and associated guidance documents (*e.g.* Guidance on Establishing Trace Metal Clean Rooms in Existing Facilities), as detailed in EPA Standard Methods (*27*), and adapted for precipitation sampling.

All samples were brought back to CBL for processing in a laminar flow hood. Samples were spiked with Optima nitric acid to 1% acid, and stored in the refrigerator until trace metal analysis was performed. Initially metals were analyzed by Graphite Furnace Atomic Absorption using a Perkin-Elmer Instrument. However, most samples were analyzed using a Hewlett-Packard 4500 Inductively Coupled Plasma-Mass Spectrometer (ICP-MS). Samples for metals, which had been acidified for at least a week, were analyzed without further preparation. External calibration curves were used, based on a mixed standard with automatic blank correction. Calibration curve regression coefficients exceeded 0.99 on all occasions. The detection limits for each metal were determined using the standard deviation of blank samples. The detection limits were all less than or equal to the values outlined in the EPA Guidance on the Documentation and Evaluation of Trace Metals Data Collected for Clean Act Compliance Monitoring (*27*). They were 0.002 μg/L for Cd; 0.03 for Pb and ranging from 0.1 to 0.2 μg/L for the other metals. These values are all significantly smaller than the concentrations found in precipitation. Relative standard deviations for duplicate analysis of all metals analyzed by ICP-MS were less than 5%, and spike recoveries of matrix additions were between 95 and 105%.

Results and Discussion

The concentration of the metals in precipitation at the two sites are shown in Figures. 2 and 3. Additional information on the average major ion concentrations are given in Table I. The concentrations of Na and Cl at CBL

208

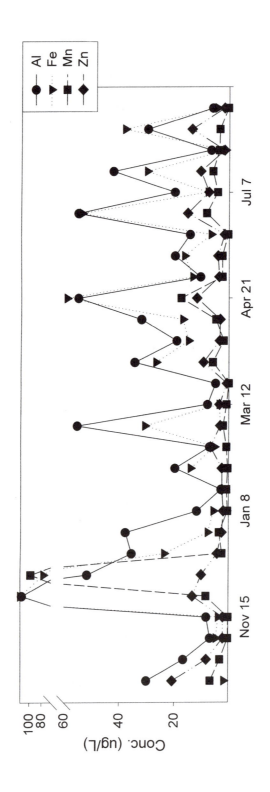

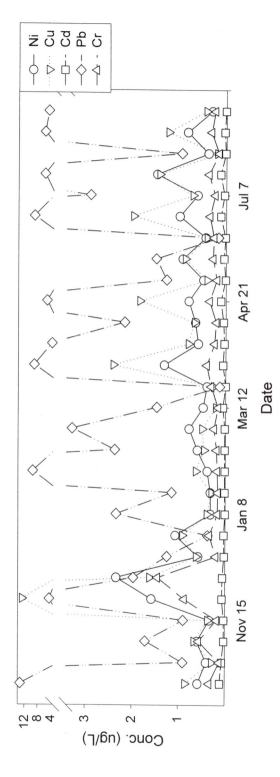

Figure 2: Concentration of metals in precipitation collected at the Science Center.

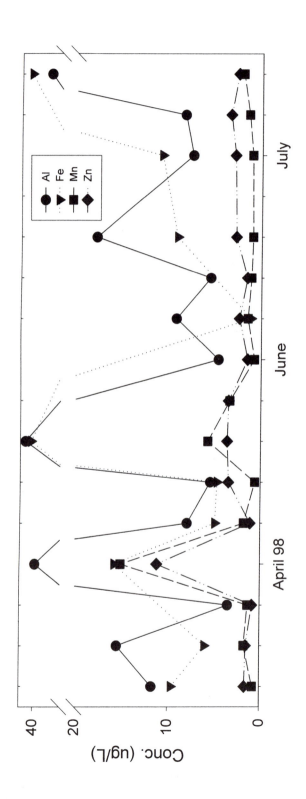

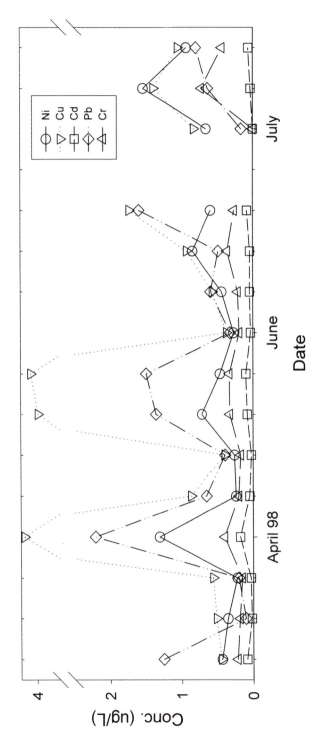

Figure 3: Concentration of metals in precipitation collected at the Chesapeake Biological Laboratory

were higher than SC and this suggests a larger entrainment of ocean-derived air masses at CBL during precipitation events. The pH of the rain and the average concentrations of the other major ions were similar at both sites. Surprisingly, the sulfate concentrations were comparable and this suggests that local coal burning sources do not contribute substantially to the sulfate in precipitation collected at SC. Others have similarly concluded that the sulfate signal in Maryland is derived mostly from regional rather than local sources (9, 28).

The highest metal concentrations were those for the crustal elements Fe and Al, as expected. There should be a strong correlation between Fe and Al at SC if these two metals are derived mostly from crustal sources and this is indeed the case (Figure 4). The correlation is significant and the slope of the correlation line is close to one. The slope is to some degree related to the few high Fe values found. While the relative concentration of Fe to Al is similar to that found in previous studies around the Chesapeake Bay (3, 5, 23) it is somewhat higher than the crustal ratio of 0.6 (29). It is assumed in crustal ratio calculations that the Al in wet deposition is derived from soil particles. However, up to 10% of the Al in rural Maryland particulate is from coal combustion sources (9, 28). This fact should be kept in mind when comparing concentrations at the two sites, and in evaluating relative fluxes.

Table I. Chemical Composition of Major Ions in Rainwater Collected at the Chesapeake Biological Laboratory (CBL) and the Science Center (SC).

Parameter	CBL	SC
Average pH	4.23	4.21
H^+ (μeq/L)	58.9±28.4	61.7±55.2
Ca^{2+} (μeq/L)	7.9±5.6	15.3±19.1
Mg^{2+} (μeq/L)	7.3±4.8	6.1±6.5
K^+ (μeq/L)	1.7±0.5	1.5±1.3
Na^+ (μeq/L)	25.9±20.8	9.4±11.4
NH_4^+ (μeq/L)	26.1±14.7	25.3±24.7
Cl^- (μeq/L)	31.9±16. 9	23.3±13.2
NO_3^- (μeq/L)	31.9±9.3	35.7±28.9
SO_4^{2-} (μeq/L)	69.7±31.7	68.1±64.8

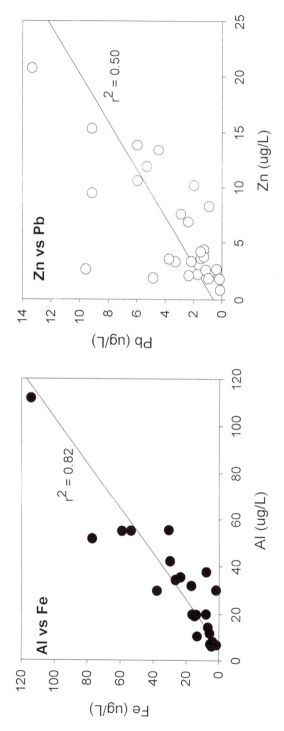

Figure 4: Correlations between metals at the Science Center.

Clearly, while the dominant source of Fe is from soil even in this urban environment, there are a few occasions where the Fe concentration exceeds that expected based on the Al concentration, indicating local impact. For the occasions where Fe was highest there were also elevated concentrations of other trace metals, particularly Pb and Cu, although there is not an entirely consistent pattern as the metals do not always co-vary (Figure 2). For example, there were occasions where the Pb concentration was elevated but that of Fe was not. Although the dataset is smaller at CBL, there is similarly a good correlation between Fe and Al (Figure 5). Again, elevated Fe concentrations were associated with higher concentrations of Pb and Cu. The slope of the Fe:Al line is 0.83, closer to the crustal value but somewhat higher, but the slope is again driven by a few high values. In a previous study, the ratio of the volume weighted mean concentration of Fe/Al at Elms, a site within 20 miles of CBL, was 0.84 for rain collected between 1990 and 1993 (3). More recent results from Stillpond and Lewes, Delaware (Figure 1) also show similar relationships (23). Thus, our results are consistent with these earlier studies.

Manganese concentrations were significantly elevated above that expected from crustal ratios (Mn:Al around 0.01) at both sites. Clearly there are sources of Mn other than soil. Given this, the lack of a strong correlation between Al and Mn is expected. Again, our results are also consistent with those of other studies (3) and are also consistent with measurements in Florida where a Fe:Al ratio similar to that of soil was found for wet deposition but the Mn:Al ratio was substantially elevated (30). The similarity of the ratios and concentrations at both sites in this study suggests that while there are potentially sources of Mn, and to a lesser extent Fe, in the city besides soil, the relative elevation in concentration reflects a regional rather than a local signal.

At SC, a multiple regression analysis of the data suggested that Al and Fe correlated, at the 99% level, strongly with all metals except Zn, Cd, Pb (correlated at the 95% level; Table II). The Mn concentration varies with that of Fe, Al and with Cd and Zn reflecting a dual source signal of both a soil-derived and a pollution-derived component. Chemical mass balance analysis of aerosols collected at various rural sites around the Chesapeake Bay (9, 28) suggested that Mn particles were derived from soil (50%), steel production (about 30%) and regional coal combustion - this fraction reflects the contribution from upwind (out-of-state) combustion emissions (28) and thus the correlation with Cd and Zn, which have mostly an incineration source, is curious. At rural sites, previous studies found that Fe aerosol was derived from soil (70%) and from steel production (30%). Overall, the regional sulfate signal is the most important contributor to aerosol mass (64% at CBL) with the soil component as the other important mass fraction (about 20% at CBL; ref. 9). Steel production-derived particles are a small fraction (<1%) so that the influence of this source for Mn and Fe is small. Clearly, the additional source of Mn from coal combustion contributes to the differences between Fe and Mn at SC. At CBL, Al correlates with most metals but Fe is not strongly related to any metal except Al. The non-soil signal for Mn is stronger at CBL as its concentration tracks that of the other trace metals more closely than it does either Al or Fe (Table III), in agreement with previous studies (9, 28).

Mercury did not correlate with either Fe or Al at SC (Table II). Three of the heavy metals (Pb, Cd and Zn) were well correlated at SC and the relationship for Pb and Zn is shown in Figure 4. The sources of Pb and Zn are dominated by waste incineration while Cd has other sources (vehicle emissions, steel processing and coal) in addition to incineration. Concentrations of Pb ranged from low values of less than 1 μg/L to above 10 μg/L (Figure 2). Such elevated concentrations were not found at CBL where the highest Pb concentration measured was less than 3 μg/L (Figure 3). Zinc concentrations were somewhat elevated at SC compared to CBL - the highest Zn concentration at SC was twice that of CBL. The correlation between Zn and Pb at CBL was not as strong as at SC and the relationship is strongly driven by one high concentration event (Figure 5; Table III). The strong correlation of these three metals in the city likely reflects the proximity of the SC site to waste incineration sources, both municipal and medical, which are present in urban Baltimore. While Cr is released by incineration, it also has other anthropogenic sources (e.g. metal plating), and an important soil component, and thus its concentration does not strongly track that of Pb and Zn at SC.

Table II: Regression Statistics for Comparison of the Concentration of Metals at Science Center.

	Al	Cr	Mn	Ni	Cu	Zn	Cd	Pb	Fe
Hg	0.03	0.04	0.21	-0.02	-0.04	0.64	0.49	0.56	-0.14
Al		0.72	0.66	0.78	0.83	0.55	0.50	0.37	0.93
Cr			0.40	0.73	0.84	0.43	0.33	0.26	0.74
Mn				0.53	0.42	0.68	0.76	0.46	0.65
Ni					0.67	0.51	0.55	0.36	0.72
Cu						0.41	0.27	0.21	0.87
Zn							0.80	0.76	0.52
Cd								0.60	0.43
Pb									0.32

NOTE: Values greater than 0.6 represent a correlation that is significant at the >99% confidence level; values above 0.45 represent a correlation at above the 95% confidence level.

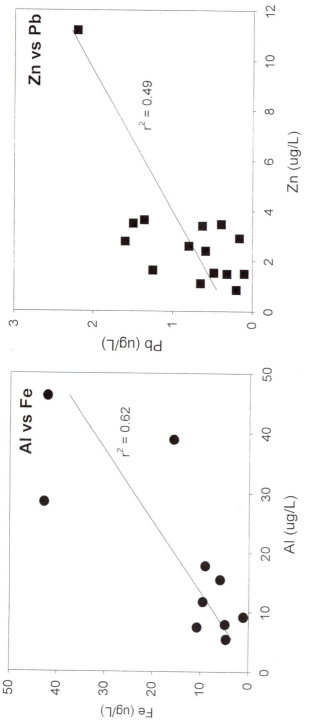

Figure 5: Correlations between metals at the Chesapeake Biological Laboratory.

Table III: Regression Statistics for Comparison of the Concentration of Metals at the Chesapeake Biological Laboratory.

	Al	Cr	Mn	Ni	Cu	Zn	Cd	Pb	Fe
Hg	0.34	-0.17	0.40	0.54	0.41	0.52	0.14	0.22	0.25
Al		0.64	0.73	0.67	0.80	0.64	0.69	0.72	0.79
Cr			0.50	0.65	0.56	0.42	0.69	0.63	0.47
Mn				0.72	0.79	0.93	0.82	0.71	0.31
Ni					0.58	0.74	0.64	0.59	0.53
Cu						0.73	0.80	0.81	0.34
Zn							0.78	0.71	0.25
Cd								0.95	0.26
Pb									0.32

NOTE: Values greater than 0.7 represent a correlation that is significant at the >99% confidence level; values above 0.52 represent a correlation at above the 95% confidence level.

For the other metals, Hg correlates with Zn, Pb and Cd at SC but not at CBL. Similarly, Hg did not correlate with other metals in Florida (*30*). The metals Cu, Ni and Cr are strongly correlated at both sites. Mercury, which is transported globally, also likely has a two component signal from local sources – coal combustion emissions and waste (medical and municipal) incineration while Pb and Cd have their primary sources as waste incineration (*9, 11*). The overall inter-relationships of Pb, Cd and Hg suggest that overall, in the city, Pb and Cd do have different sources. The metals Pb, Cd, Cu and Zn are well correlated at CBL (Table III). This suggests that the correlations found for Hg at SC reflect the urban sources and not the regional picture.

Based on the measurements presented here and those of others one can conclude that the data at CBL reflects a regional signal of metals in wet deposition and not specifically the diluted signal from Baltimore, or another urban center. The correlations at CBL between metals are less significant than those of the urban location and this is expected as this site is likely impacted by a variety of sources with different trace metal signatures, such as ocean-derived precipitation. Storm direction is an important determinant of observed concentration.

To directly compare the two sites, annual deposition fluxes were calculated. For each 9-day collection, the flux was calculated from the measured concentration and the amount of rain i.e.:

$$\text{Flux (mg m}^{-2}) = [\text{conc.(mg/m}^{-3})]*\text{rain depth (m)}$$

and the yearly flux was estimated by summation. For CBL, a full year's data was not collected so the flux was scaled to that of a year. The fluxes estimated are gathered in Table IV and compared to that of other investigations around the

Table IV: Yearly Fluxes for Metals in Wet Deposition in Maryland. Data from Elms and Wye taken from ref. 3; data from Western MD from refs. 25 and 30; data from the Great Lakes region from ref. 31. All fluxes are in mg m^{-2} yr^{-1}.

Metal	Great Lakes 93/94	Elms 90-93	Wye 90-93	STP 95-96	FRB 96-98	CBL 98	SC 98	Ratio
Al	-	6.96	12.7	31.2	31.9	18.5	18.0	0.97
Fe	-	5.49	11.4	17.1	27.2	14.9	12.3	0.83
Mn	1.9-2.4	1.15	1.33	1.5	-	3.04	3.67	1.21
Cr	0.06-0.08	0.035	0.15	0.09	0.78	0.24	0.25	1.04
Ni	0.23-0.29	0.20	0.33	1.12	0.80	0.59	0.52	1.13
Cu	0.57-0.85	0.38	0.27	0.90	0.64	1.49	0.71	0.48
Zn	3.5-5.5	1.56	1.56	6.9	4.41	3.06	4.11	1.34
Cd	0.07-0.09	0.035	0.038	0.058	0.13, 0.12	0.063	0.033	0.52
Pb	0.55-1	0.42	0.47	0.58	0.68, 0.64	0.89	2.52	2.83
Hg	-	-	-	-	0.015	0.014	0.030	2.14

Chesapeake Bay (*3, 23, 26, 31*), and elsewhere (*32*). As expected for an element whose signature is dominated by soil, the fluxes are similar for Al at CBL and SC. Also, this indicates that there is no measurable increase in local dust sources due to the urban nature of the landscape surrounding the SC site; probably because of the location of this site on the roof of a building. The estimated flux of Fe was slightly lower at SC but, given the nature of the estimations, it is probable that fluxes that differ by less than 10% should be considered equal, based on the accuracy of the ICP-MS measurement. For Hg, the error is around 20%. Using this rationale, it is only the fluxes of Fe, Cu, Zn, Cd, Pb and Hg that are different between the two sample locations (Table IV). The fluxes of Pb and Hg are 2-3 times higher at SC, and the Zn flux about 30% higher, which is expected given the known sources (power plants and incinerators) within the downtown Baltimore area.

The substantially lower flux of Cd at SC is curious, especially given its relatively strong correlation with Zn in terms of concentration per collection. The Cd flux at CBL is also about twice that measured in the earlier Chesapeake Bay study (i.e. at Elms and Wye; ref. *3*), is about half that measured at a site in western Maryland over two different study periods (Table IV; refs. *24, 26*) but is similar to that measured at Stillpond in 95/96. Some of these differences may be related to differences in the amount of deposition between years. The fluxes of Al also differ substantially between sites and across years, being the highest in western MD in 96/97 which was an exceptionally wet year (about 50% more wet deposition than on average at the site; ref. *25*). Thus, normalizing the fluxes to Al suggests that while the Cd flux has high temporal and spatial variability across the region this is due, to a large degree, to differences in the amount of rainfall during each of the collection periods.

Comparing the CBL and Elms fluxes it is apparent that the ratio is similar for the crustal elements Al, Fe and Mn, and for Ni (CBL/Elms ~ 2.7; Table IV) but is lower for Pb, Zn and Cd. Thus, by normalizing the concentrations for each period to that of Al, it is apparent that the relative flux for the metals was lower at CBL in 1998 compared to the earlier study (1990-1993) for Cd and Zn, and to a lesser degree for Pb. The relative fluxes for Cu, and especially for Cr, are higher relatively. Overall, while it is difficult to directly compare fluxes across study periods due to differences in rainfall amount and other methodological differences, it is clear that the concentrations of the more toxic trace metals (Pb, Zn and Cd) have decreased in deposition at the rural sites monitored in Maryland while others (Cu, Cr, Ni) have not decreased significantly in the last 10 years. Others have suggested that fluxes of Zn, Cu and Ni were higher in 95/96 compared to the earlier results (*23*). A 5-year study (1995-2000) of Hg at CBL (*20*) did not show a substantial decrease in normalized deposition.

As mentioned by other investigators (*3, 5*) previous estimates of wet deposition to the Chesapeake Bay have not accounted for the potential enhanced deposition associated with the Baltimore urban presence, or that of other nearby urban centers. We can make some predictions on the relative importance of the urban centers based on our measurements of trace metals in deposition at SC and CBL. While the importance of wet compared to dry deposition varies for each metal (80% of the total for Cd, 70% for Zn, 60-65% for Cr, Cu and Ni, 50% for Pb), the overall relative effect of wet and dry deposition will be similar to that of wet only given that particulate scavenging is the dominant source for trace metals in rain (i.e. there is no gas phase to be considered except for Hg; ref. *21*). It should be cautioned, however, that heavy metals on particulate in urban areas are typically associated with coarser material that is rapidly removed from the atmosphere such that the particulate distribution away from sources is shifted toward smaller particles (*16*). These larger particles (>10 μm) contribute substantially to the dry particulate deposition flux (*16*). Thus, the relative dry deposition difference between the urban and rural environment is greater than that for wet deposition, which is derived predominantly from the scavenging of smaller particles. Based on our wet deposition data, the local wet deposition in the vicinity of Baltimore will be increased by a factor of 2.8 for Pb, 1.3 for Zn, and hardly increased for the other trace metals. However, the impact of large particulate deposition may enhance this difference in terms of total deposition. Also, for Hg, given the presence of ionic gaseous Hg species (*21*), the flux will be increased by a factor of 2-3 by both wet and dry deposition.

Mercury and metal wet deposition data from Stillpond (*20, 23*) is comparable to that at CBL and western Maryland and thus we suggest that the urban plume from Baltimore does not strongly impact this site. Thus, we estimate that the urban plume from Baltimore and other urban centers in the upper and lower Bay, will impact about 10% of the Bay's surface. Assuming that the relative increases in metal deposition found for Baltimore are representative of the region, we suggest that the previous estimates of trace metal wet fluxes to the Bay are not underestimated for the metals Cr, Ni, Cu and Cd, or for Al, Fe and Mn. However, the atmospheric flux of Pb is low by about 20%. That of Hg, if based only on rural measurements, would be similarly biased but recent estimates have accounted for the urban influence (*20, 21*). For Zn, the error is less than 10%. To a large degree, these differences are similar to the errors associated with such estimates as the fluxes are influenced by seasonal and annual differences in precipitation amount, wind direction during events and other factors that are not normally adequately addressed by short-term studies. Thus, we conclude that while earlier studies had not adequately quantified the importance of the urban signal in deposition of metals to nearby waters, the errors associated with this are small for most metals. It is likely that the estimates are biased low for Hg and Pb. However, we caution that more studies are required to adequately determine whether our results are valid for other cities on the shores of large water bodies.

Acknowledgements

We thank Joe Steinbacher and Bernie Crimmins and others at CBL for help with sample collection, and for help in bottle preparation and other laboratory activities. This study was funded by the EPA STAR Air Toxics Program, Grant # R825245-01. This is Contribution No. 3462, Chesapeake Biological Laboratory, Center for Environmental Science, University of Maryland.

References

1. Nriagu, J.G.; Pacyna, J.M. *Nature* **1998**, *333*, 134-139.
2. Duce, R.A.; Liss, P.S.; Merrill, J.T.; Atlas, E.L.; Buat-Menard, P.; Jhicks, B.B.; Miller, J.M.; Prospero, J.M.; Arimoto, R.; Church, T.M.; Ellis, W.; Galloway, J.N.; Jickells, T.D.; Knap, A.H.; Reinhart, K.H.; Schneider, B.; Soudine, A.; Tokos, J.J.; Tsunogai, S.; Wollast, R.; Zhou, M. *Global Biogeochem. Cycles* **1991**, *5*, 193-259.
3. Baker, J. E.; Poster, D. L.; Clark, C. A.; Church, T. M.; Scudlark, J. R.; Ondov, J. M.; Dickhut, R. M.; Cutter, G. In *Atmospheric Deposition of Contaminants to the Great Lakes and Coastal Waters*; Baker, J. E., Ed.; SETAC Press: Pensacola, FL, 1997; pp 171-194.
4. Mason, R. P.; Lawson, N. M.; Sullivan, K. A. *Atmos. Environ.* **1997**, *31*, 3531-3540.
5. Scudlark, J. R.; Church, T. M. In *Atmospheric Deposition of Contaminants to the Great Lakes and Coastal Waters*; Baker, J. E., Ed.; SETAC Press: Pensacola, FL, 1997; pp 195-208.
6. DOE. A comprehensive assessment of toxic emissions from coal-fired power plants, phase I results. E&E Research Center, University of North Dakota final report, 1996 , Contract #DE-FC21-93MC30097.
7. USEPA. *Mercury Study Report to Congress*, 1997, EPA-452/R-97-004; Office of Air: Washington, DC.
8. USEPA. *Deposition of Air Pollutants to Great Waters*, Second Report to Congress, 1997, EPA 453-R-97-011, Washington, DC
9. Ondov, J.M.; Quinn, T.L.; Han, M. Maryland Department of Natural Resources, Chesapeake Bay and Watershed Programs Report: CBWP-MANTA-AD-96-1, 1996, pp. 1-111.
10. Gordon, G.E. *Environ. Sci. Technol.* **1988**, *22*, 1132-1142.
11. Nriagu, J.O. *Nature* **1989**, *338*, 47-49.

12. Mason, R. P.; Fitzgerald, W. F.; Morel, F. M. M. *Geochim. Cosmochim. Acta* **1994**, *58*, 3191-3198.

13. Ondov, J.M.; Wexler, A.S. *Environ. Sci. Technol.* **1998**, *32*, 2547-2555.

14. Kim, G.; Alleman, L.Y.; Church, T.M. *Global Biogeochem. Cycles* **1999**, *13*, 1183-1192.

15. Offenberg, J.H.; Baker, J.E. *Environ. Sci. Technol.* **1997**, *31*, 903-914.

16. Holsen, T. M.; Zhu, X.; Khalili, N. R.; Lin, J. J.; Lestari, P.; Lu, C.-S.; Noll, K. E. In *Atmospheric Deposition of Contaminants to the Great Lakes and Coastal Waters*; Baker, J. E., Ed.; SETAC Press: Pensacola, FL, 1997; pp 35-50.

17. Offenberg, J. H.; Baker, J. E. *J. Air & Waste Manage. Assoc.* **1999**, *49*, 959-965.

18. Bamford, H. A.; Offenberg, J. H.; Larsen, R. K.; Ko, F-C.; Baker, J. E. *Environ. Sci. Technol.* **1999**, *33*, 2138-2144.

19. Mason, R.P.; Lawson, N.M.; Sullivan, K.A. *Atmos. Environ.* **1997**, *31*, 3541-3550.

20. Mason, R. P.; Lawson, N. M.; Sheu, G.-R. *Atmos. Environ.* **2000**, *34*, 1691-1701.

21. Sheu, G-R.; Lawson, N.M.; Mason, R.P. In *Speciation and Distribution of Atmospheric Mercury over the Northern Chesapeake Bay*, this volume.

22. Mason, R. P.; Lawson, N. M.; Lawrence, A. L.; Leaner, J. J.; Lee, J. G.; Sheu, G.-R. *Mar. Chem.* **1999**, *65*, 77-96.

23. Kim, G.; Scudlark, J.R.; Church, T.M. *Atmos.Environ.* **2000**, *34*, 3437-3444.

24. Church, T.M.; Scudlark, J.R.; Conko, K.M.; Bricker, O.P.; Rice, K.C. Maryland Department of Natural Resources, Chesapeake Bay and Watershed Programs Report: CBWP-MANTA-AD-98-2, 1998, pp. 1-87.

25. Landis, M.S.; Keeler, G.J. *Environ. Sci. Technol.* **1997**, 31, 2610-2615.

26. Castro, M.S.; Mason, R.P.; Scudlark, J.R.; Church, T.M. Final Report, Maryland Department of Natural Resources Power Plant Research Program, 2000.

27. USEPA *Methods for the Analysis of Water and Wastewaters.* 1997, CD ROM

28. Wu, Z.Y., Han, M.; Lin, Z.C.; Ondov, J.M. *Atmos. Environ.* **1994**, *28*, 1471-1486.

29. Turekian, K.K.; Wedepohl, K.H. *Geol. Soc. Am. Bull.* **1961**, *72*, 175-192.

30. Landing, W.M.; Perry, J.J.; Guentzel, J.L.; Gill, G.A.; Pollman, C.D. *Water Air Soil Poll.* **1995**, *80*, 343-352.

31. Lawson, N.M.; Mason, R.P. *Water Res.*, in press.

32. Sweet, C.W.; Weiss, A.; Vermette, S.J. *Water Air Soil Poll.* **1998**, *103*, 423-439.

Chapter 12

Speciation and Distribution of Atmospheric Mercury over the Northern Chesapeake Bay

Guey-Rong Sheu, Robert P. Mason, and Nicole M. Lawson

Chesapeake Biological Laboratory, Center for Environmental Science, University of Maryland, P.O. Box 38, Solomons, MD 20688–0038

Baltimore's urban air is a potentially important source of mercury (Hg) to the northern Chesapeake Bay. Elevated total atmospheric Hg (THg) concentrations were detected at a sampling site in downtown Baltimore (4.4 ± 2.7 ng/m^3), as compared to a rural site (1.7 ± 0.5 ng/m^3). The urban air was also enriched with reactive gaseous Hg (RGHg) and particulate-bound Hg (Hg-P). The annual dry depositional fluxes of RGHg and Hg-P at sites around the northern Chesapeake Bay have been determined, with the fluxes of RGHg ranging from 7 to 121 µg/m^2 and the fluxes of Hg-P from 1 to 34 µg/m^2. These values were the same magnitude as the wet depositional fluxes of Hg measured at the same sites. Local wind direction influenced the concentration of atmospheric Hg detected at the urban sampling site. When air came from the SE, S and SW directions, the urban sampling site tended to be impacted by the local emission sources, with higher THg and Hg-P concentrations detected.

More than 40 states in the USA have issued advisories regarding the consumption of fish containing elevated levels of Hg, which is mostly as the highly toxic monomethylmercury (MMHg; *1-2*). Recent studies have demonstrated that the atmosphere is an important source of Hg to many surface waters and terrestrial environments (*3-8*), especially in areas remote from the point sources. According to Mason et al.'s estimate, for example, the combination of wet and dry particulate atmospheric depositions currently make up 90% of the total Hg input to the open ocean (*9*). However, typically less than 1% of the total atmospheric Hg deposition is MMHg (6, *10-12*). Simple mass balance calculations have demonstrated that the amount of deposited MMHg is not enough to support the MMHg levels found in aquatic systems and *in situ* production is therefore essential (*9-10*). Although the wet deposition is undoubtedly an important pathway of atmospheric pollutants to surface waters (*6, 13-14*), recent studies also suggest the importance of dry deposition (*13-14*). The intensity of dry deposition is directly related to the concentration and speciation of pollutants in the atmosphere. Thus, monitoring the concentration and speciation of atmospheric Hg and understanding its transformation and transport are imperative to the study of Hg biogeochemistry and the development of controlling and remediation strategies.

Three types of Hg are identified in the atmosphere based on their physical and chemical properties: elemental gaseous Hg (Hg^0), reactive gaseous Hg (RGHg, divalent forms), and particulate-bound Hg (Hg-P). The speciation of RGHg is currently unknown, but it is usually assumed to be gaseous $HgCl_2$ or other mercuric halides, which have a higher surface reactivity and water solubility than gaseous Hg^0. Unlike other metal pollutants that tend to exist in the particulate phase in the atmosphere, Hg exists mainly in the gaseous phase because of its anomalous physicochemical properties. For example, elemental Hg has a relatively high vapor pressure of 0.246 Pa at 25°C (*15*) and a low Henry's law constant of 0.11 M/atm at 20°C (*16*). This low Henry's law constant, which means low water solubility, makes the removal of gaseous Hg^0 by rainwater less efficient compared to the more water-soluble RGHg and to the scavenging of Hg-P. Both RGHg and Hg-P are effectively removed from the atmosphere by wet and dry deposition with an atmospheric residence time from a few minutes to weeks. On the other hand, Hg^0 is more stable in the atmosphere with a residence time between 6 months and 2 years (*17-18*). Consequently, gaseous Hg^0 is the dominant form of Hg in the atmosphere with only a few percent of the total present as RGHg and Hg-P (*19-21*). Nevertheless, air could be enriched with RGHg and Hg-P in the close proximity of point sources.

Although Hg^0 is the dominant form in the atmosphere, RGHg may still be a significant component of the total depositional fluxes. However, due to the poor understanding of its speciation, its usually low concentration in the atmosphere, and the lack of proper sampling techniques, the importance of RGHg to Hg biogeochemistry has been neglected until recently. This situation has been improved with several RGHg sampling techniques being developed in the past few years, including a multi-stage filter pack method (*22*), a refluxing mist chamber method (*23-26*), and a KCl-coated annular denuder method (*27*). An examination of these methods has been published elsewhere (*21*). These sampling techniques greatly enhance our ability to evaluate the significance of RGHg to the local and regional Hg biogeochemical cycles.

Urban air usually contains elevated concentrations of trace elements and pollutants, which could enhance depositional fluxes to nearby great waters, such as the Chicago metropolitan area to Lake Michigan (*28*) and the Baltimore metropolitan area to Chesapeake Bay (*29-30*). The Chesapeake Bay is the largest estuary in the United States and the Baltimore metropolitan area, one of the heavily industrialized areas on the east coast, is located on the western shore of the Chesapeake Bay. The Chesapeake Bay has a large surface-to-volume ratio (mean depth is 7 m), and it is therefore particularly vulnerable to the influence of atmospheric deposition (*13*). Previous studies have demonstrated that Baltimore's urban atmosphere is an important source of PAHs and PCBs to the Chesapeake Bay (*29-30*). A mass balance for Hg in the Chesapeake Bay also showed that atmospheric deposition is the major source of Hg with an annual flux of 1300 moles (*3, 6*). This number was derived mainly from the wet and dry particulate deposition data collected in Maryland. The annual wet depositional flux measured in 1997/98 at a downtown Baltimore site was higher than those measured at rural sites, by a factor of 2-3 (*31*), which demonstrated the importance of urban areas as Hg sources through wet deposition to the adjacent waters. However, dry deposition of atmospheric RGHg was not fully considered in this mass balance calculation and the influence of urban air on the dry depositional flux to the Chesapeake Bay was not discussed. Hence, the primary objectives of this study were to characterize the speciation and distribution of atmospheric Hg in rural and urban sites around the northern Chesapeake Bay and to estimate the associated dry depositional fluxes. The significance of RGHg to the air-surface exchange of Hg was of special concern since its role usually has been neglected in the early mass balance studies. The impact of Baltimore's urban air on the dry depositional flux of Hg to the Chesapeake Bay was also evaluated.

Sampling Sites and Methods

Both intensive and long-term atmospheric Hg sampling was conducted in the city and in the northern Chesapeake Bay downwind of the Baltimore metropolitan area. Air was sampled aboard the EPA *OSV Anderson* (SHIP) over the northern Chesapeake Bay and at four land-based sites (Fig. 1): the first site was on the roof of the Maryland Science Center (SC) at downtown Baltimore next to the Inner Harbor, the second site was 9 km south-southeast of SC at the USCG Station Curtis Creek (CC), the third was 17 km east of SC by Hart-Miller Island (HMI), and the fourth was about 42 km east-northeast of SC at USCG Station Stillpond (STP) on the Eastern Shore of Maryland.

Sampling started in February 1997. Two intensive sampling events were conducted aboard the ship over the bay and at all the land sites in February (18th-26th) and July (22nd-31st) 1997 with 24-hr samples being collected daily. Samples were also collected every 9 days at SC (stopped on 12/21/1998) and at STP (stopped on 03/12/1998). Usually the sampling time was 24-h at all the sites except aboard the ship over the bay, where weather and the operation of the ship restricted the air sampling. A total of 70 and 50 samples were collected at SC and STP, respectively. Sixteen samples were collected at all sites during the two intensive sampling events. Total atmospheric Hg (THg), RGHg, and Hg-P were all measured. Total atmospheric Hg was sampled by pumping air through 3 gold-coated quartz sand traps (*32*) and quantified by cold vapor fluorescence spectrometry (CVAFS; *33*). As the gold traps retained RGHg and Hg-P in addition to Hg^0, since they were not pre-removed before entering the traps, the THg represents the sum of all the atmospheric Hg species. Some of the THg data were less than 1 ng/m^3, which is the lower limit of the background value for this latitude (*18, 34*). This could be due to power failure of the pump or to a gas leak from the sampling tubing and hence these low THg data were excluded from further analysis.

Both RGHg and Hg-P were sampled using the filter pack method and the details of this method has been described elsewhere (*21-22*). Acid-cleaned open face 5-stage Teflon filter holders were used. Two 47 mm 0.45 µm Teflon filters and three 47 mm cation exchange membranes were housed in the Teflon filter holder. The first Teflon filter removed the particles while the second filter acted as a backup (or blank). Similarly, the first ion exchange membrane collected the RGHg sample with the second membrane acting as a backup. The third membrane was to prevent backward diffusion of RGHg from the pump. Incoming air first flowed through the Teflon filters to remove particles prior to the ion exchange membranes that were positioned behind the Teflon filters. Filter samples were first oxidized by 0.2 N BrCl solution for at least 30 minutes, then pre-reduced using hydroxylamine hydrochloride, prior to quantification by $SnCl_2$ reduction-CVAFS (*6*). The analytical detection limits (3 times the S.D. of blank filters divided by the average air volume sampled in 24-hr) for RGHg and Hg-P were 7 pg/m^3 and 9 pg/m^3, respectively.

Local wind directions were recorded every 10 minutes at the SC site by our collaborators in 1997. This information enabled us to explore how the local wind direction influenced the atmospheric Hg measured at sites near potential emission sources, such as the SC sampling site.

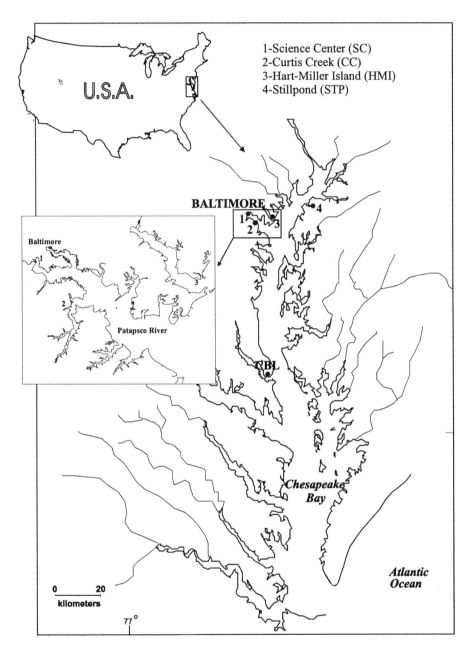

227

1-Science Center (SC)
2-Curtis Creek (CC)
3-Hart-Miller Island (HMI)
4-Stillpond (STP)

U.S.A.

BALTIMORE

Baltimore

Patapsco River

CBL

Chesapeake
Bay

Atlantic
Ocean

0   20
kilometers

77°

Figure 1. Map of the sampling sites.

Results and Discussion

All the atmospheric Hg data collected for this study are summarized in Table I. Atmospheric Hg values reported by other groups are also included for comparison. The atmospheric Hg data measured at CC and SC were similar to those measured at other urban or industrial areas. At a certain site, both RGHg and Hg-P exhibited higher temporal variability compared to THg, as indicated by the ratio between the standard deviation and the mean value. Detailed analyses of our data are presented in the following sections.

Table I. Summary of Our Data and Published Values

Sampling Sites	THg (ng/m³)	RGHg (pg/m³)	Hg-P (pg/m³)	References
CC (intensive)[a]	6.1±1.9	385±495	715±1285	Current study
SC (intensive)[a]	3.6±2.3	120±200	142±392	Current study
HMI (intensive)[a]	2.2±0.7	22±19	44±64	Current study
SHIP (intensive)[a]	2.2±0.8	21±26	25±22	Current study
STP (intensive)[a]	1.7±0.4	21±15	24±29	Current study
SC (all data)[a]	4.4±2.7	89±150	74±197	Current study
STP (all data)[a]	1.7±0.5	24±22	42±50	Current study
Seoul, Korea	1.4~23.2			35
Bordeaux, France	0.7~23.6			36
Guizhou, China	2.7~12.2	150~1460		37
Tennessee	1.9~2.4	50~257		24
Indiana	3.3~4.7	83~156		24
Northern Wisconsin	1.2~1.8		6~63	38
Florida	1.8~3.3		34~51	39
Detroit, Michigan			94	40
Rural Michigan			10~22	40
Toronto, Canada			3~91	41
Urban U.K.			<10~2020	42

a. These data are presented as mean±standard deviation.

Spatial Distribution of Atmospheric Mercury

Since the intensive samplings were conducted at four land sites and also aboard the research vessel both in February and July 1997, this dataset is ideal for the analysis of the spatial distribution of atmospheric Hg around and over the northern Chesapeake Bay. These data are summarized in Figure 2. For the

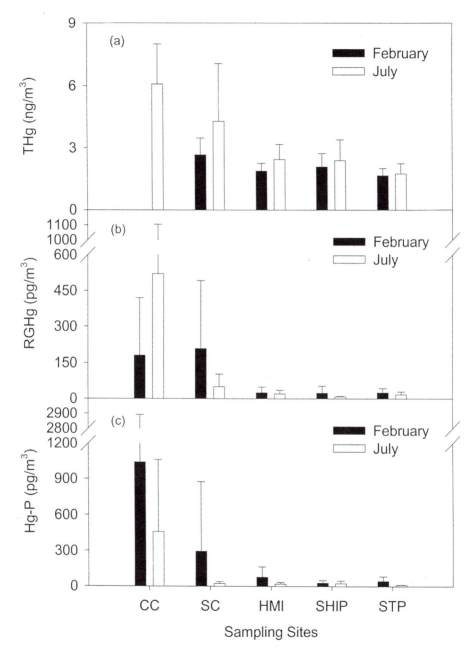

Figure 2. Summary of the intensive sampling data.

THg, a west-to-east concentration gradient was observed, with higher THg measured on the western shore sites, and the gradient was more evident in July (Figure 2a). This could be due to the difference in local wind direction between February and July 1997. The winds were predominantly coming from the east-northeast direction in February, while in July the wind directions were more variable. Among the five sampling sites, CC, HMI and SC are on the western shore of the Chesapeake Bay. Both CC and SC showed elevated THg concentrations, while the HMI measurements were similar to those measured over the Chesapeake Bay (SHIP). The measurements made at STP, which is a rural site on the Eastern Shore and downwind of the Baltimore metropolitan area, were slightly lower than those measured at HMI and SHIP. Similar to the spatial distribution of THg, both RGHg and Hg-P also exhibited a west-to-east gradient over the northern Chesapeake Bay (Figure 2b-c), with higher concentrations measured on the western shore sites.

Expressing the concentrations of RGHg and Hg-P as a relative unit (percent THg) is another way to explore the spatial variability of atmospheric Hg speciation (Table II). On average about 20% of the THg was RGHg and Hg-P at CC. At SC, about 10% of THg was RGHg and Hg-P. On the other hand, at the other three sites, RGHg and Hg-P only made up 3~4% of THg. Therefore, not only were elevated RGHg and Hg-P concentrations measured at CC and SC but also a greater portion of THg measured at the same sites was in fact RGHg and Hg-P. This suggests that both CC and SC were influenced by local Hg emission sources. As the Hg enriched air mass moved across the Chesapeake Bay, both RGHg and Hg-P were quickly depleted, possibly due to the deposition to the surface water.

Table II. Concentrations of RGHg and Hg-P Expressed as %THg

	CC	SC	HMI	SHIP	STP
RGHg	10.8±15.2	5.0±9.8	1.2±1.1	1.6±2.7	1.3±0.9
Hg-P	8.8±14.8	4.5±10.3	2.6±3.2	1.1±0.7	1.5±0.9

Atmospheric organic pollutants, such as PCBs and PAHs, were also sampled by other groups during the same February and July intensive samplings and the data have been published (*29-30*). Like our atmospheric Hg data, the distribution of atmospheric PCBs and PAHs varied spatially, with higher concentrations in the city as compared to the downwind sites east of the city. Therefore, the Baltimore metropolitan area is likely contributing elevated contaminant loading through dry deposition to the Chesapeake Bay.

Estimates of Dry Depositional Fluxes of Atmospheric Mercury

One way of estimating the dry depositional flux of a pollutant is using the atmospheric concentration of the pollutant times its dry deposition velocity (V_d) and this technique was employed for this study. The average concentrations of each Hg species sampled during the intensive studies were used. Published values of V_d of each Hg species have been summarized elsewhere (21). For this study, $V_d = 1$ cm/s is assumed for RGHg and 0.15 cm/s for Hg-P. Also, as Hg^0 is not very water-soluble and the Chesapeake Bay is a source of Hg^0 to the atmosphere (3), its dry depositional flux was not calculated. Wet depositional fluxes that have been estimated and published elsewhere (6, 31) are also included for comparison.

Since the dry depositional fluxes were directly proportional to the atmospheric concentration, it is not surprising to see higher fluxes at CC and SC (Table III), knowing the existence of the west-to-east Hg concentration gradient. It is also clear that the dry depositional fluxes are important Hg sources to the Chesapeake Bay. At the rural site (STP), the combination of the dry RGHg and Hg-P depositional fluxes was more than half of the wet flux. The total dry fluxes of RGHg and Hg-P were more than the wet flux at the city site (SC). Therefore, this simple flux estimate not only demonstrates the importance of dry deposition to the Hg budget, especially the dry deposition of RGHg, but also demonstrates the importance of the urban atmosphere as a significant Hg input to adjacent surface waters.

Table III. Annual Wet and Dry Depositional Fluxes of Atmospheric Hg

Species	CC	SC	HMI	SHIP	STP
RGHg	121	38	7	7	7
Hg-P	34	7	2	1	1
Wet Deposition		30	25		14

Units are $\mu g/m^2$.

Temporal Variability of Atmospheric Mercury

Almost 2-yr of data were collected at SC and over 1-yr of data were collected at STP. This dataset enabled us to analyze the temporal variation of atmospheric Hg over the northern Chesapeake Bay area. These data are summarized in Figure 3. At STP, the annual average values and associated standard deviations of THg (1.69±0.51 vs. 1.69±0.41 ng/m³) and RGHg (24±21 vs. 21±27 pg/m³) were about the same for 1997 and 1998, although the average

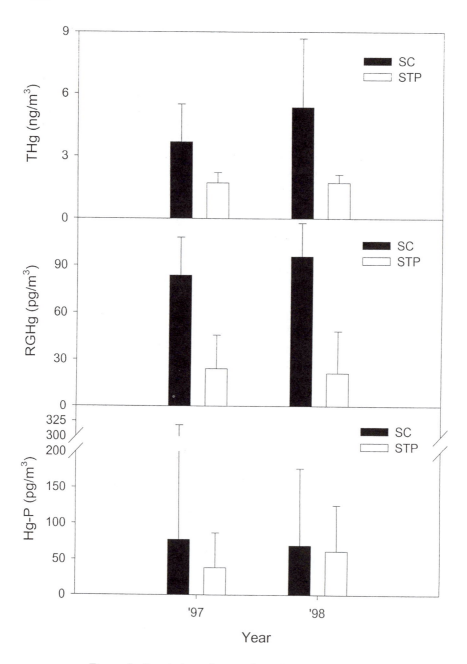

Figure 3. Speciation of atmospheric Hg at SC and STP

Hg-P of 1998 (61±64 pg/m³) was higher than the value of 1997 (39±48 pg/m³). This could be due to the fact that the Hg-P was only sampled in the first 3 months of 1998 and Hg-P was usually higher in the colder seasons. On the other hand, at SC the annual average values and associated standard deviations of THg and RGHg were higher in 1998 (5.3±3.3 ng/m³ and 96±149 pg/m³) than in 1997 (3.7±1.8 ng/m³ and 84±152 pg/m³), while the average Hg-P was slightly lower in 1998 (69±107 pg/m³) than in 1997 (78±241 pg/m³). This result may suggest that the STP is a regional background site and therefore not showing much annual variation in atmospheric Hg.

The annual dry depositional fluxes were re-estimated for SC and STP using the long-term dataset. The new estimates are summarized in Table IV and, again the dry depositional fluxes were of the same magnitude as the wet depositional fluxes at SC and STP, although the new SC estimate was lower than that calculated previously using the intensive data.

Table IV. Refined Estimates of Annual Dry Depositional Fluxes

Species	SC-1997	SC-1998	STP-1997	STP-1998
RGHg	26	30	8	7
Hg-P	4	3	2	3
Wet Deposition	30		14	

Units are μg/m².

The monthly average concentrations and standard deviations of each Hg species are shown in Figures 4-6 (Note different concentration scales in each figure). No consistent monthly THg patterns were observed (Figure 4). In 1997, the monthly average THg showed a uniform pattern at SC through the year. However, in 1998, the average THg was higher in the colder months. Like the THg, no evident temporal patterns of monthly average RGHg were observed at both sites (Figure 5). On the other hand, the distribution of monthly average Hg-P exhibited a seasonal pattern, with higher Hg-P in the fall and winter (Figure 6). This could be due to the higher anthropogenic Hg-P emissions in cold seasons or enhanced Hg^0 deposition onto ambient aerosols at lower temperatures.

Influences of Local Wind Direction on Atmospheric Mercury

Local wind direction was recorded concurrently with the atmospheric Hg sampling at SC in 1997. This helped us to understand how the local wind

234

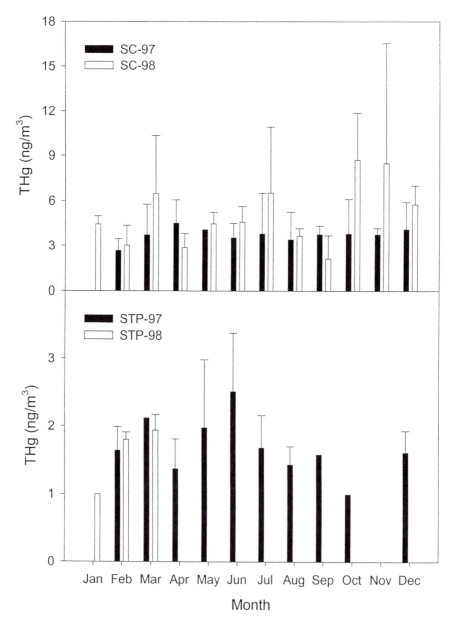

Figure 4. Monthly average concentrations of THg at SC and STP.

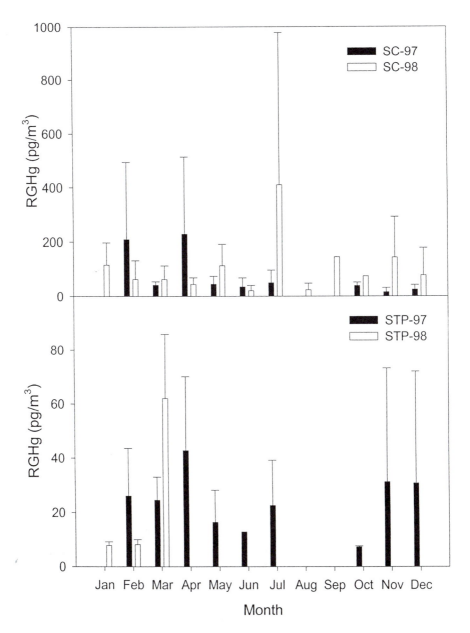

Figure 5. Monthly average concentrations of RGHg at SC and STP.

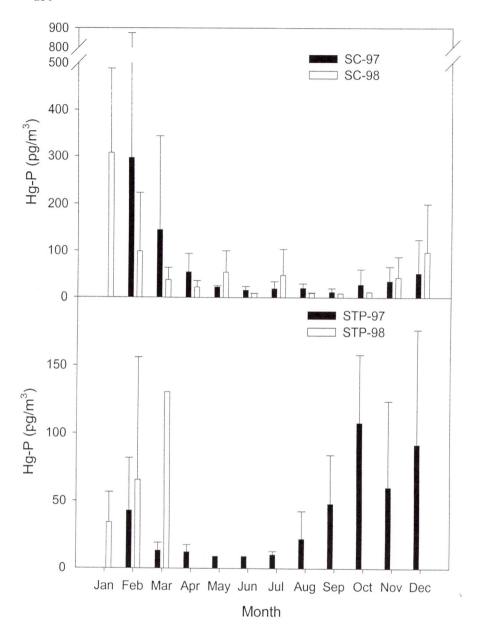

Figure 6. Monthly average concentrations of Hg-P at SC and STP.

direction influenced the concentration and speciation of atmospheric Hg. This dataset was first sorted from the lowest to the highest THg concentrations and a clear trend was observed with higher THg concentrations usually associated with winds coming from the SE, S and SW directions. Coincidentally, several potential local stationary sources of Hg are located on the SE, S and SW of the Baltimore area, including power plants and waste incinerators. A couple of these potential sources are in the close proximity of the CC and SC sampling sites. Therefore, it is reasonable to separate the 1997 dataset to two subsets based on the wind direction, with SE, S and SW directions as a group and the other wind directions as another group (Figure 7 and Table V). In Figure 7, the top and bottom error bars represent the 90th and 10th percentiles of the dataset, respectively. The upper and lower boundaries of the box represent the 75th and 25th percentiles of the dataset. The horizontal line in the box is the median. Mean values and standard deviation are summarized in Table V.

Table V. Influences of Wind Direction on Atmospheric Hg at SC

Wind Direction	THg (ng/m^3)	RGHg (pg/m^3)	Hg-P (pg/m^3)
SE, S and SW	4.2±2.1	50±37	51±82
Other directions	3.2±1.4	109±197	36±41

Since potential Hg sources are located on the SE, S and SW of the SC sampling site, air from these directions usually contained higher THg and Hg-P concentrations as expected (Table V). However, it is surprising to see air from these same directions had lower average RGHg content compared to air from other directions, as the anthropogenic emission is usually assumed to be the most important RGHg source. The same dataset showed that the air from the SE, S and SW always contained measurable RGHg (18~87 pg/m^3) indicating the influence of local emission sources, while air from other directions was sometimes depleted of RGHg (< 7 pg/m^3). The higher average RGHg associated with air from other directions was mainly due to the few high values measured with air from the E and NE directions. For example, 5 out of the 7 high RGHg events (>100 pg/m^3) were associated with wind from N and NE. These rare high RGHg events suggest there are unidentified RGHg sources located on the E and NE of the SC site, or may reflect *in situ* production of RGHg under specific environmental conditions, which are still not well understood.

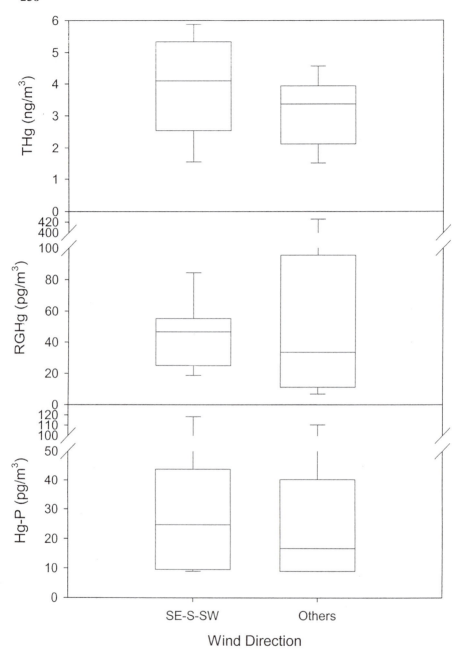

Figure 7. Statistical distribution of atmospheric Hg concentrations based on wind directions.

Conclusions

Our 1997/98 study in the northern Chesapeake Bay region has demonstrated that Baltimore's urban air is an important source of Hg to the Chesapeake Bay. Air sampling was conducted at four land-based sites (CC, HMI, SC and STP) and aboard a research vessel over the surface water (SHIP). The CC, HMI and SC sites are on the western shore of the bay: SC is in downtown Baltimore, CC is in an industrial area SE of the city, and HMI is on the east of the city. Some potential stationary sources of Hg, including power plants and waste incinerators, are located on the SW-S-SE region of the city. The STP site is located in the rural area of the Eastern Shore, which is downwind of the Baltimore metropolitan area.

The intensive sampling data showed higher atmospheric Hg concentrations at CC and SC (Table I). These values are similar to those reported for other urban/industrial areas. Intermediate values of atmospheric Hg were detected at HMI and SHIP, while the lowest values were always found at STP (Table I). As the urban air moved eastward across the Chesapeake Bay, both RGHg and Hg-P were depleted faster than THg, suggesting mechanisms other than mixing and dilution were acting on these two species. Dry deposition to surface water is probably the main mechanism responsible for this quick depletion of RGHg and Hg-P. The composition of atmospheric Hg also changed dramatically as the air moved eastward, with a high fraction of THg as RGHg and Hg-P at CC and SC (Table II). The changing speciation supports the notion of the removal of RGHg and Hg-P by dry deposition during transport.

The annual dry depositional fluxes of RGHg and Hg-P were estimated using the intensive sampling data (Table III). The dry depositional flux was greater than the wet flux at SC and was about half of the wet flux at STP. Therefore, it is clear that the dry deposition is an important pathway for Hg to enter the Chesapeake Bay, with higher fluxes associated with urban air.

Long-term data were collected at SC and STP. The STP data did not show significant differences in atmospheric Hg concentration and speciation between 1997 and 1998 (Figure 3). This suggests that the STP site may represent the regional background signal and thus is not showing much annual variation of atmospheric Hg. No evident seasonal pattern of THg and RGHg were observed at both sites (Figures 4-5); however, a distinct seasonal pattern was observed for Hg-P, with higher ambient Hg-P in colder months (Figure 6). This may be due to the higher anthropogenic emission intensity of Hg-P in cold seasons or enhanced Hg^0 deposition onto ambient aerosols at low temperatures.

Concentrations of atmospheric Hg measured at SC were influenced by local wind directions. With potential Hg sources on the SW-S-SE directions of the site, elevated THg and Hg-P were detected with air originating from these directions (Table V). However, air that came from the same directions contained only about 50% of the RGHg, on average, as measured in air from other directions (Table V). This was due to few high RGHg concentrations measured with air from the E and NE directions. The exact causes responsible for these irregular measurements are not clear, however.

In summary, atmospheric Hg showed great spatial and temporal variability around the northern Chesapeake Bay region, with elevated Hg values usually measured in Baltimore's urban air. This urban plume is an important source of Hg to the Chesapeake Bay through wet and dry deposition. The dry deposition of RGHg is of special concern since it was usually neglected in the early mass balance calculations. High RGHg values were detected occasionally at SC with air coming from the directions where no sources have been identified. This could be due to unknown sources or the *in situ* production in the air under specific environmental conditions, which is still not well understood and needs further investigation.

Acknowledgements

We thank Joe Steinbacher and Bernie Crimmins and others from CBL for help with sample collection. We also thank the captain and crew of the *OSV* Anderson for their help. We also thank Dr. Ondov of University of Maryland for providing the local wind directions. This study was funded by an EPA STAR grant. This is Contribution No. 3464, Chesapeake Biological Laboratory, Center for Environmental Science, University of Maryland.

References

1. U.S. EPA. *National Forum on Mercury in Fish*, EPA-823-R-95-002; Office of Water: Washington, DC, 1995.
2. U.S. EPA. *Mercury Study Report to Congress*, EPA-452/R-97-004; Office of Air: Washington, DC, 1997.
3. Mason, R. P.; Lawson, N. M.; Lawrence, A. L.; Leaner, J. J.; Lee, J. G.; Sheu, G.-R. *Mar. Chem.* **1999**, *65*, 77-96.
4. Mason, R. P.; Sullivan, K. A. *Environ. Sci. Technol.* **1997**, *31*, 942-947.
5. Fitzgerald, W. F.; Engstrom, D. R.; Mason, R. P.; Nater, E. A. *Environ. Sci. Technol.* **1998**, *32*, 1-7.
6. Mason, R. P.; Lawson, N. M.; Sullivan, K. A. *Atmos. Environ.* **1997**, *31*, 3531-3540.
7. Lockhart, W. L.; Wilkinson, P.; Billeck, B. N.; Hunt, R. V.; Wagemann, R.; Brunskill, G. J. *Water, Air, Soil Pollut.* **1995**, *80*, 603-610.

8. Watras, C. J.; Bloom, N. S.; Hudson, R. J. M.; Gherini, S.; Munson, R.; Claas, S. A.; Morrison, K. A.; Hurley, J.; Wiener, J. G.; Fitzgerald, W. F.; Mason, R.; Vandal, G.; Powell, D.; Rada, R.; Rislov, L.; Winfrey, M.; Elder, J.; Krabbenhoft, D.; Andren, A. W.; Babiarz, C.; Porcella, D. B.; Huckabee, J. W. In *Mercury Pollution: Integration and Synthesis*; Watras, C. J., Huckabee, J. W., Eds.; Lewis Publishers: Boca Raton, 1994; pp 153-177.
9. Mason, R. P.; Fitzgerald, W. F.; Morel, F. M. M. *Geochim. Cosmochim. Acta* **1994**, *58*, 3191-3198.
10. Mason, R. P.; Fitzgerald, W. F.; Vandal, G. M. *J. Atmos. Chem.* **1992**, *14*, 489-500.
11. Mason, R. P.; Fitzgerald, W. F. In *Global and Regional Mercury Cycles: Sources, Fluxes and Mass Balance*; Baeyens, W., Ed.; Kluwer Academic Publishers: Netherlands, 1996; pp 249-272.
12. Lamborg, C. H.; Rolfhus, K. R.; Fitzgerald, W. F. *Deep-Sea Res. II* **1999**, *46*, 957-977.
13. Baker, J. E.; Poster, D. L.; Clark, C. A.; Church, T. M.; Scudlark, J. R.; Ondov, J. M.; Dickhut, R. M.; Cutter, G. In *Atmospheric Deposition of Contaminants to the Great Lakes and Coastal Waters*; Baker, J. E., Ed.; SETAC Press: Pensacola, FL, 1997; pp 171-194.
14. Scudlark, J. R.; Church, T. M. In *Atmospheric Deposition of Contaminants to the Great Lakes and Coastal Waters*; Baker, J. E., Ed.; SETAC Press: Pensacola, FL, 1997; pp 195-208.
15. Schroeder, W. H.; Yarwood, G.; Niki, H. *Water, Air, Soil Pollut.* **1991**, *56*, 653-666.
16. Sanemasa, I. *Bull. Chem. Soc. Jpn.* **1975**, *48*, 1795-1798.
17. Shia, R-L; Seigneur, C; Pai, P.; Ko, M.; Sze, N. D. *J. Geophys. Res.* **1999**, *104*, 23747-23760.
18. Slemr, F.; Langer, E. *Nature* **1992**, *355*, 434-437.
19. Lindqvist, O. *Tellus* **1985**, *37B*, 136-159.
20. Ebinghaus, R.; Jennings, S. G.; Schroeder, W. H.; Berg, T.; Donaghy, T.; Guentzel, J.; Kenny, C.; Kock, H. H.; Kvietkus, K.; Landing, W.; Mühleck, T.; Munthe, J.; Prestbo, E. M.; Schneeberger, D.; Slemr, F.; Sommar, J.; Urba, A.; Wallschläger, D.; Xiao, Z. *Atmos. Environ.* **1999**, *33*, 3063-3073.
21. Sheu, G.-R.; Mason, R. P. *Environ. Sci. Technol.*, **2001**, *35*, 1209-1216.
22. Bloom, N. S.; Prestbo, E. M.; VonderGeest, E. Presentation at the 4th International Conference on Mercury as a Global Pollutant, Hamburg, August 4-8, 1996.
23. Stratton, W. J.; Lindberg, S. E. *Water, Air, Soil Pollut.* **1995**, *80*, 1269-1278.
24. Lindberg, S. E.; Stratton, W. J. *Environ. Sci. Technol.* **1998**, *32*, 49-57.

242

25. Stratton, W. J.; Lindberg, S. E.; Perry, C. J. *Environ. Sci. Technol.* **2001**, *35*, 170-177.

26. Lindberg, S. E.; Stratton, W. J.; Pai, P.; Allan, M. A. *Fuel Process. Technol.* **2000**, *65-66*, 143-156.

27. Stevens, R. K.; Schaedlich, F. A.; Schneeberger, D. R.; Prestbo, E.; Lindberg, S.; Keeler, G. Presentation at the 5th International Conference on Mercury as a Global Pollutant, Rio de Janeiro, May 23-28, 1999.

28. Holsen, T. M.; Zhu, X.; Khalili, N. R.; Lin, J. J.; Lestari, P.; Lu, C.-S.; Noll, K. E. In *Atmospheric Deposition of Contaminants to the Great Lakes and Coastal Waters*; Baker, J. E., Ed.; SETAC Press: Pensacola, FL, 1997; pp 35-50.

29. Offenberg, J. H.; Baker, J. E. *J. Air & Waste Manage. Assoc.* **1999**, *49*, 959-965.

30. Bamford, H. A.; Offenberg, J. H.; Larsen, R. K.; Ko, F-C.; Baker, J. E. *Environ. Sci. Technol.* **1999**, *33*, 2138-2144.

31. Mason, R. P.; Lawson, N. M.; Sheu, G.-R. *Atmos. Environ.* **2000**, *34*, 1691-1701.

32. Fitzgerald, W. F.; Gill, G. A. *Anal. Chem.* **1979**, *51*, 1714-1720.

33. Bloom, N.; Fitzgerald, W. F. *Anal. Chim. Acta* **1988**, *208*, 151-161.

34. Slemr, F.; Seiler, W.; Schuster, G. *J. Geophys. Res.* **1981**, *86*, 1159-1166.

35. Kim, K.-H.; Kim, M.-Y. *Atmos. Environ.* **2001**, *35*, 49-59.

36. Pécheyran, C.; Lalère, B.; Donard, O. F. X. *Environ. Sci. Technol.* **2000**, *34*, 27-32.

37. Tan, H.; He, J. L.; Liang, L.; Lazoff, S.; Sommar, J.; Xiao, Z. F.; Lindqvist, O. *Sci. Total Environ.* **2000**, *259*, 223-230.

38. Lamborg, C. H.; Fitzgerald, W. F.; Vandal, G. M.; Rolfhus, K. R. *Water, Air, Soil Pollut.* **1995**, *80*, 189-198.

39. Dvonch, J. T.; Vette, A. F.; Keeler, G. J.; Evans, G.; Stevens, R. *Water, Air, Soil Pollut.* **1995**, *80*, 169-178.

40. Keeler, G.; Glinsorn, G.; Pirrone, N. *Water, Air, Soil Pollut.* **1995**, *80*, 159-168.

41. Lu, J. Y.; Schroeder, W. H.; Berg, T.; Munthe, J.; Schneeberger, D.; Schaedlich, F. *Anal. Chem.* **1998**, *70*, 2403-2408.

42. Lee, D. S.; Garland, J. A.; Fox, A. A. *Atmos. Environ.* **1994**, *28*, 2691-2713.

Chapter 13

Atmospheric and Fluvial Sources of Trace Elements to the Delaware Inland Bays

Thomas M. Church, Joseph R. Scudlark, and Kathryn M. Conko

Graduate College of Marine Studies, University of Delaware,
700 Pilottown Road, Lewes, DE 19958

The primary objective of this study was to compare the fluvial versus atmospheric importance of trace element fluxes to the Delaware Inland Bays. The dissolved fluvial loading of selected trace metals was determined based on seasonal sampling (1992-95) of the major streams. Atmospheric wet metal deposition was gauged based on long-term (1982-present) continuous collections at a nearby site. Atmospheric dry deposition was modeled based on measured aerosol concentrations and assumed bi-model deposition velocities. Atmospheric dry flux appears to be more significant for crustal elements (Al, Fe, and Mn), while wet flux is more important for non-crustal trace elements (Cd, Cr, Cu, Pb and Zn). Overall, atmospheric deposition provides at least 5% (Mn) to 30% (Zn) of the total loading to the Inland Bays, with direct flux to the surface waters comprising about half of the total aeolian input. The Inland Bays watershed serves as a reservoir for atmospherically deposited Al, Cr, Cu and Zn. An examination of basin yields (kg/m^2) for the major tributaries reveals the accelerated weathering of uniquely bog iron deposits in two marshy tributaries, while another exhibits exceptionally large export of anthropogenic metals Cd, Cr, Cu and Pb.

Introduction

Within the past two decades, atmospheric deposition is increasingly recognized as a biogeochemically important and quantitatively significant non-point source of sulfur and nitrogen oxides to surface waters (*1,2,3*). Many trace metals are co-emitted from similar high temperature emission processes. Thus atmospheric mobilization and deposition of other contaminants in acid precipitation may have equally important ecological implications. For example, the atmospheric input of trace elements to mid-Atlantic estuaries and shelf waters has been shown to rival those from other sources (4,5).

Largely as the result of such studies, the 1990 Clean Air Act Amendments contain specific provisions (Section 112m, the so-called "Great Waters" section) which legislatively mandate an examination of atmospheric contaminant deposition to coastal and inland waters. This includes smaller east coast estuaries such as Delaware's Inland Bays (Rehoboth and Indian River) since their combined output to local shelf waters can be equivalent. With this in mind, the primary objective of this study was to accurately assess the relative atmospheric and fluvial loading of trace elements to the Inland Bays.

Study Area

The Delaware Inland Bays are comprised of three interconnected estuaries: Rehoboth Bay, Indian River Bay and Little Assawoman Bay, located on the mid-Atlantic coast of Delaware (Figure 1). Although dwarfed in size by large east coast estuaries such as the Chesapeake and Delaware Bays, the Delaware Inland Bays are more typical of the numerous small, shallow, poorly-flushed systems that are found along the Atlantic and Gulf coasts.

For the purposes of this study we have focused on the two primary bays, Rehoboth and Indian River. Freshwater inputs are derived from a number of lateral sources (total discharge of 8.6 m³/s), 80% of which originates from groundwater. Overall, the Inland Bays encompass a surface area of 8.81×10^7 m², and drain a watershed of 6.81×10^8 m². Thus they possess a large watershed: open water area ratio (8:1) typical of coastal plain estuaries. Land use is primarily agricultural (40%) and forest (38%), and the underlying geology is predominantly quartz sand and silts as part of the Omar coastal plain formation. Seasonally, urban areas comprise only 10% of the watershed, but have a major impact on water quality. The bays are highly eutrophic due to nutrient loading resulting from the transient summer population density, and surrounding agricultural activities. The coal-fired Indian River power plant impacts both the atmospheric as well as fluvial loading

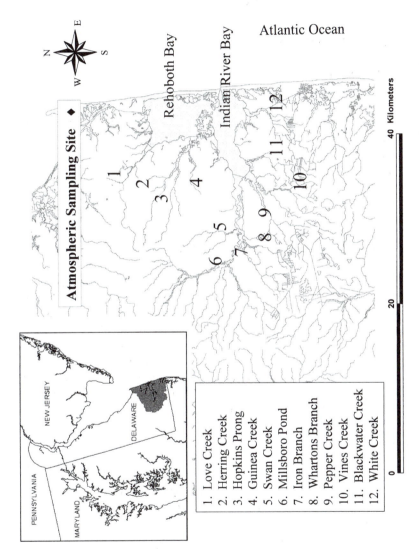

Figure 1. Map of the Inland Bays, showing the stream sampling locations and atmospheric collection site

Atmospheric Sampling Site ◆

Rehoboth Bay

Indian River Bay

Atlantic Ocean

1. Love Creek
2. Herring Creek
3. Hopkins Prong
4. Guinea Creek
5. Swan Creek
6. Millsboro Pond
7. Iron Branch
8. Whartons Branch
9. Pepper Creek
10. Vines Creek
11. Blackwater Creek
12. White Creek

0   20   40   Kilometers

NEW JERSEY

PENNSYLVANIA

MARYLAND

DELAWARE

to the bays in the form of fugative blowoff and runoff from the surrounding storage piles of fly ash.

Methods

Atmospheric Wet and Dry Deposition

Precipitation was sampled at a long-term atmospheric research site near Lewes, DE in a remote area of the 4000 acre Cape Henlopen State Park (Figure 1). Major ions (H^+, Na^+, K^+, Ca^{+2}, Mg^{+2}, Cl^-, SO_4^{-2}, and NO_3^-) have been sampled on an event/daily basis continually since 1977. The trace elements (Al, As, Ba, Cd, Cu, Cr, Fe, Mn, Ni, Pb, Se, Sb, and Zn) have been sampled in parallel since 1982, using rigorous "ultra-clean" trace metal sampling and handling techniques. The precipitation trace elements were all analyzed by graphite furnace atomic absorption spectrophotometry (GFAAS), as verified using EPA standard reference materials as described elsewhere (6,7).

Precipitation amount was determined using a continuously-recording Belfort (gravimetric) rain gauge. Based on a comparison of predicted (rain gauge) with collected precipitation depths, the overall collection efficiency was >95%.

There does not exist a widely-accepted method for the routine determination of dry deposition by direct means. Thus, this study adopted the traditional approach of estimating dry flux based on aerosol concentrations measured at Lewes times modeled dry deposition velocities. Aerosol data collected at the Lewes site during independent studies was utilized for this purpose(8,9). To derive dry flux rates, the average aerosol concentrations were first apportioned into crustal and non-crustal components, based on published soil compositions (10), and using Al as a crustal normalizer. Deposition velocities representative of coarse and accumulation mode particles were then applied to the crustal and non-crustal fractions (0.5 and 0.1 cm/s, respectively) (11).

Stream Water

The major streams discharging to the Rehoboth and Indian River Bays (based on freshwater discharge) were sampled quarterly, under variable flow conditions. Sampling was conducted during all four seasons during each of the three years, starting 4/92 and ending 3/95. During the first year of this study, the 12 streams, indicated in Figure 1 were sampled. Because our preliminary results indicated that the trace metal loading is dominated (>90%) by seven of these streams, during

Years 2 and 3 our sampling concentrated on just those systems. In addition, to ensure that our sampling location for Swan Creek was not tidally influenced, during Years 2 and 3 it was moved a short distance inland, above an impoundment. However, this did not appear to significantly alter the observed high metal concentrations or loading fluxes.

The sampling sites were chosen to be above tidal influence and well-mixed, at locations which facilitated the measurements for stream discharge. This was usually a straight stream segment with a uniform cross-section, such as a culvert, which was surveyed and marked for future reference. At Millsboro Pond, sampling was conducted near the USGS gauging station.

Grab samples of surface stream water were manually obtained in acid-washed 2 liter LDPE bottles and stored refrigerated until processed (within 4 hours of collection). In the laboratory, the samples were filtered through a 0.45 μm Gelman Mini-Capsule® filter, using a peristaltic pump and employing trace metal clean techniques. Aliquots for trace metal analyzes were acidified to 0.4% (v/v) using ultra-pure (doubly quartz re-distilled) HCl, and stored frozen. Analyzes for "dissolved" (<0.45μ) Al, Fe, Mn and Zn were conducted using a Jobin-Yvon Model 70 Plus inductively-coupled plasma atomic emission spectrophotometer. Analyzes for Cd, Cr, Cu and Pb were performed using a Varian Spectra 20 graphite furnace atomic absorption spectrophotometer.

The analytical detection limits (ppb) of the trace elements in both the precipitation and stream samples are (for GFAA, ICP-AES respectively): Al (0.12, 3.0), As (0.005, nd), Cd (0.006, nd), Cr (0.10, nd), Cu (0.12, nd), Fe (0.05, 1.0), Pb (0.12, nd), Se (0.005, nd), Zn (0.14, 0.5), where nd means "not determined".

Steam discharge (m^3/s) was calculated as the product of the measured stream cross-sectional area (m^2) and the instantaneously measured flow (m/s), estimated using a neutrally-buoyant surface drogue. The fluvial loading at the time of each sampling was calculated as the measured concentration times the freshwater discharge. Estimated stream discharges were reconciled by comparison with the measured values at the USGS gauging station on Millsboro Pond.

Data Quality Assurance

A comparison of representative field blanks for precipitation and stream water shows them at less than 10% of the average metal concentration, which also represents an approximate limit of analytical confidence. For both precipitation and stream water, this was achieved in all instances. Thus, the field blanks are only employed diagnostically to verify that background contamination is maintained at acceptable levels.

To establish the accuracy of our analyzes, a number of externally-certified reference solutions were employed. These included EPA Quality Assurance Check

Solutions (WP386, ICP 7 & 19, TMA 989) in addition to the Riverine Water Standard Reference Material (SLRS-2, St. Lawrence River), provided by the National Research Council of Canada. The results of our analytical performance agreed within the standard uncertainty provided by these reference standards for all metals reported.

Results

The atmospheric wet and dry (crustal and non-crustal) fluxes are summarized in Figure 2. This figure is based on the annual average over the long term (1982-1995) for wet deposition and periodic dry deposition measurements from 1982-1993. While inter-annual variations in wet flux (and presumably dry flux) on the order of 50% or less are observed, there are no apparent long-term trends. The notable exception is Pb, which has decreased 6-fold in the past decades due to the phasing out of alkyl-Pb gasoline additives (12). Thus, for computing contemporary atmospheric Pb loading to the Inland Bays, the long-term average wet and dry fluxes have been scaled downward (by a factor of 3) to accurately reflect current concentrations and fluxes for the study period.

As shown in Figure 2, with the exception of the normalizer Al, and for only Fe, at least half of atmospheric metal fluxes are derived from non-crustal, presumably anthropogenic sources. For Fe, dry flux provides a contribution equal or greater than wet flux, due to the fact that soil particles tend to be quite large and are subject to gravitational settling near their source. However, this depends on uncertainties in assumed gravitational dry deposition velocities for micron size particles (13) which greatly depend on wind speed and relative humidity (14). Conversely, for most other non-crustal elements enriched as high temperature combustion condensates in sub-micron aerosols, precipitation provides the most efficient atmospheric scavenging mechanism. Consequently, for such elements, wet flux furnishes at least half the input.

The average annual fluvial trace metal loadings to the Inland Bays were computed as the sum of the annual measured loadings from each stream, pro-rated to account for input from the streams not sampled. The scaling factors utilized were 1.27 for 1991 (12 streams sampled) and 1.34 for 1992 and 1993 (7 streams sampled). Thus, the calculation inherently assumes that the metal concentrations in the streams not sampled are equal to the mean of the sampled streams. In the absence of major point sources, this is probably a reasonable approximation, given the similar land use patterns within the watershed.

There was considerable spatial variability (12-67%) in the trace metal fluxes observed in each of the streams over the last three years of sampling. Indian River,

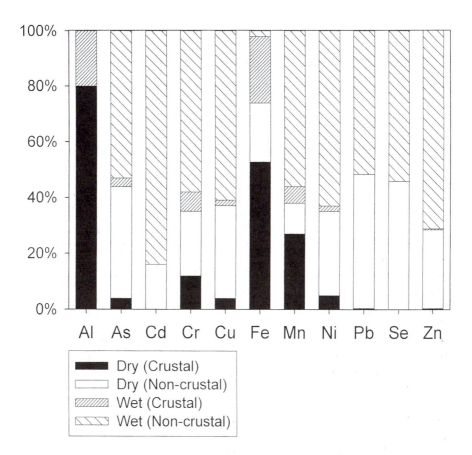

Figure 2. The crustal/non-crustal apportionment of atmospheric wet and dry trace metal fluxes observed at Lewes, DE

which has the greatest freshwater discharge and largest drainage basin, generally exhibits the highest metal fluxes. There is a noticeable increase in the concentrations of several metals (eg: Fe) between the (northern) streams which discharge into Rehoboth Bay and the (southern) streams that discharge into Indian River Bay (Figure 3).

There is also a fairly high degree of inter-annual variability in the fluvial metal loading, which cannot be accounted for simply by differences in precipitation and freshwater discharge. In fact, the least year-to-year variability is associated with those elements derived primarily from the weathering of crustal sources (eg., Al, Fe, Mn), while the greatest is noted for those elements usually ascribed to anthropogenic sources. Only Pb, Cd and Mn exhibit a consistent (decreasing) trend over the three years of sampling. For Pb, the decreasing atmospheric flux suggests the phasing out of leaded gasoline additives, although the precipitation signal has not decreased as dramatically (15). Fluxes of Zn, Al, Cr and Cu were all greatest during 1993, although there is no obvious explanation. Such temporal variability underscores the importance of sampling over multiple years to obtain a representative assessment of riverine loading

The most notable of this behavior is Vines Creek, which exhibited consistently low pH values (4.07 - 6.04) and disproportionately high metal loadings relative to its water discharge. It was noted that the streambed sediments in Vines Creek had a distinctive orange-red Fe-oxyhydroxide hue. In fact, the sampling site is nearby a relict forge which had been used to produce bog iron.

A simple mass balance can be constructed based on the total (wet + dry) atmospheric inputs and fluvial outputs of the Inland Bays watershed (6.81×10^8 m^2). Assuming the atmosphere provides the only significant trace metal source to the drainage basin, the difference between the input and output represents the maximum net watershed accumulation. This comparison (Table 1) reflects varying behavior among the metals. On an annual basis, atmospheric Pb inputs appear to be balanced by stream export. In contrast, Al, Cr, Cu and Zn inputs are retained in the watershed to varying degrees, presumably sequestered in the soil. Thus, for these elements, the watershed serves as a reservoir, which may be periodically remobilized by acidic precipitation events, or eventually when saturated. The remaining elements (Fe, Mn and Cd) appear to exhibit net loss from the watershed, suggesting that they posses other internal sources. In fact, since only the *dissolved* fluvial export was determined, it may be underestimated for particle-reactive elements (e.g. Fe). Thus, for these elements the net watershed loss should be viewed as a *minimum* value. It is well-known that Mn possess a strong vegetative source (16), and its primary soil/sediment phase (MnO_2) is highly redox sensitive and thus readily solubilized in groundwaters. A major source of Cd could be from the application of PO_4 fertilizers, suggesting that agricultural runoff could be an important source to the Inland Bays.

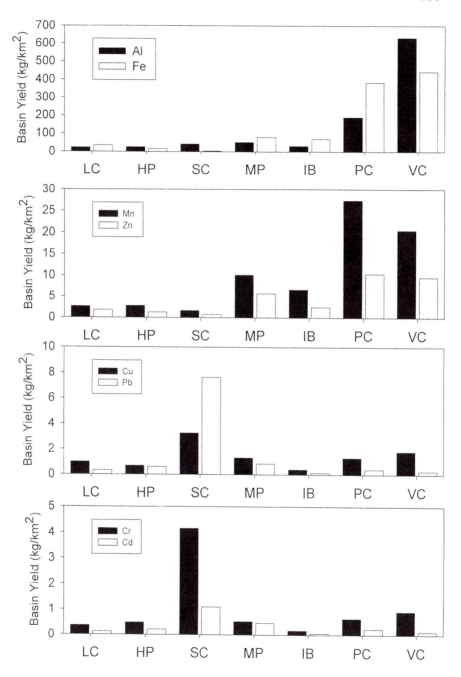

Figure 3. A comparison of trace metal basin *yields (kg/m2) for the major watersheds which drain the Inland Bays: Love Creek (LC), Hopkins Prong (HP), Swan Creek (SC), Millsboro Pond (MP), Iron Branch (IB), Pepper Creek (PC) and Vines Creek (VC)*

Table 1. Mass Balance of Trace Metals to the Inland Bays Watershed

	Atmos. Input	Stream Export	Percent watershed
	-----[kg/yr] -----		accumulation(+) or loss (-)
Al	57,258	47,150	+18%
Cd	84	143	-70%
Cr	470	422	+10%
Cu	708	673	+5%
Fe	25,640	52,340	-104%
Mn	1348	4650	-245%
Pb	681	680	0
Zn	4338	2385	+4

The relative contribution of atmospheric trace element deposition to the Inland Bays can be evaluated by comparing the fluvial and atmospheric fluxes. The fluvial loading encompasses the net input from all upstream point and non-point sources, including atmospheric input, weathering of soils, groundwater and municipal or industrial discharges. Not included in the stream loading term are inputs from downstream sources, including direct ground water seepage to the bays from the unconfined aquifer. Atmospheric loading includes both direct input to surface waters, as well as indirect flux transmitted through watersheds, which is ultimately reflected as a component of the fluvial loading. Direct atmospheric deposition to the Inland Bays is calculated as the product of the total aeolian flux ($\mu g/m^2/yr$) and the combined surface area of Rehoboth and Indian River Bays ($8.3 \times 10^7 m^2$).

One way to examine the impact of differing land uses on the transmission of trace elements is by comparing the annual basin yields of each sub-basin, i.e., the stream export per unit area. This comparison (Figure 3) reveals that the basin yields fall into two categories. The first includes those elements (Al, Fe, Mn and Zn) whose behavior are geochemically dominated, which exhibit exceptionally large yields from Pepper creek and Vines Creek. As previously discussed, the metal export from these systems is governed by the weathering of their unique underlying soil deposits. The second grouping, containing elements that are anthropogenically derived (Cu, Pb, Cr and Cd), exhibit the most prominent yields from Swan Creek. The Swan Creek watershed is typical of the Inland Bays in that it drains primarily agricultural (37%) and forested (58%) terrain, and there are no known major point sources which would otherwise contribute the observed high basin yields. Thus, the source of the anomalously enriched trace element basin yields remains unresolved.

For computing direct atmospheric loading *via* watershed throughput, we apply a transmission factor of 10% to all metals, which is based on the previous studies of Lindberg and Turner (*16*). This is probably a conservative estimate since the cited study applies to a forested ecosystem having organic-rich soils. Such soils would be expected to retain metals more efficiently than those of the present study. These have agricultural/suburban land uses with poorly-buffered, sandy soils typical of the Inland Bays watershed.

The comparison of metal fluxes (Figure 4) reveals that atmospheric input provides at least between 5% (Mn) and 27% (Zn) of the annual trace metal loading to the Inland Bays. Approximately 2/3 of the atmospheric input is directly to the bay surface waters, with the remainder representing the transmission of atmospheric inputs to the watershed. While the relative atmospheric input is significant, given the apparent absence of major trace metal point sources in the Inland Bays, it is not as large as might be anticipated. Compared with nearby estuaries, the relative atmospheric loading for the Inland Bays is greater than that reported for Delaware Bay (*17*), but somewhat less than that estimated for Chesapeake Bay (*5*).

Discussion

The greatest sources of error associated with gauging the atmospheric flux of trace elements to the Inland Bays are (1) extrapolating measurements made at a single atmospheric site to the entire watershed, (2) accurately quantifying dry deposition, and (3) estimating the indirect atmospheric loading *via* watershed transmission.

1. With regard to the spatial representation of atmospheric flux measurements, the wet and dry deposition of nitrogen and trace elements in the mid-Atlantic (*1,18*) do not indicate large regional spatial gradients. Thus, within the relatively small Inland Bays area, extrapolating from a single site to derive basin-wide loading probably provides a reasonable estimation.
2. The aerosol concentrations used to estimate dry deposition exhibit remarkably good agreement, despite being based on differing sampling and analysis methodologies applied over a 6-year period. This confirms both their accuracy and regional consistency in the absence of major local sources. However, the fluxes are based on a range of deposition velocities reported for such purposes, and have an uncertainty of at least a factor of 2 (19, 20). Corroborating evidence for these estimates is provided by direct bulk (wet + dry) deposition measurements (*5*) conducted for selected elements (As, Pb and Zn). These data indicate that at least for the elements examined, the total atmospheric deposition (measured wet + calculated dry) is in fairly good agreement with total deposition based on measured bulk (MB) collections.

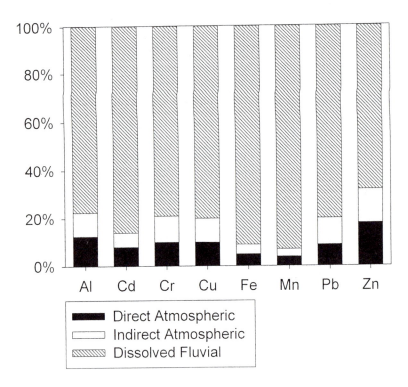

Figure 4. A comparison of relative fluvial and atmospheric trace metal inputs to the Inland Bays. The atmospheric input is apportioned into that which is direct to the water surface, and that indirect to the land surface, assuming a conservatively low (10%) transmission factor.

3. While the direct deposition to surface waters can be gauged with a fair degree of confidence, the primary uncertainty associated with resolving atmospheric inputs to coastal waters is in gauging the indirect loading as it relates to watershed transmission/retention (*21*). The watershed transmission of atmospherically deposited trace metals is fundamentally related to the chemical properties of the metal of interest as well as the geomorphology, stream pH, land use, and basic hydrological characteristics which are unique to each watershed (*16*). The limited published data indicate that the degree of retention in forested watersheds can be quite small (2-34%). Thus, due to the large watershed:open water area ratio of the Inland Bays (8:1), the indirect atmospheric loading may be appreciable compared with the direct input to the Bay surface (Figure 4).

    Although the assessment of atmospheric and fluvial inputs are on an annual basis, their relative importance may vary on shorter time scales. Both atmospheric and fluvial inputs exhibit seasonal cycles which are not synchronous. Atmospheric inputs are greatest during summer, reflecting seasonal emission trends and meteorological transport patterns. Conversely, the largest fluvial loadings are generally associated with the spring freshet (late winter/early spring). In contrast, during late summer/early fall, typical base flow conditions reflect high rates of evapotranspiration in the drainage basin. Thus, during summer periods of high biological production, when maximum atmospheric loading are coupled with minimum stream inputs, aeolian fluxes would provide an even greater contribution.

On shorter time scales, both riverine and atmospheric inputs can be highly episodic and are innately coupled, i.e., large precipitation events typically result in elevated stream transport. It has been shown (*4*) that the most prominent 10% of the precipitation events account for one third of the annual N wet deposition at Lewes. Similarly, for dry deposition, periods of high turbulence can lead to episodes of intensified deposition. Due to the differences between deposition velocity as a function of wind speed, the dry deposition associated with a wind speed of 20 m/s for 10 minutes a day would be equivalent to the deposition associated with an entire week at an average wind speed of 5 m/s (*18*). Such episodic behavior may have major implications in terms of ecosystem response, which in many instances may be more important than the cumulative loading.

However, even the dissolved freshwater metal inputs are geochemically scavenged in the mixing zone of each sub-estuary (*5,22*), so that in this regard, the fluvial inputs represent **maximum** input to the open waters of the Inland Bays. In contrast, atmospheric deposition represents direct, predominantly dissolved (i.e., readily-assimilated) inputs to surface photic zones.

Subsequent studies of two primitive Piedmont watershed in western Maryland (*23,24*) has shown 10% watershed transmission to be a lower limit for most trace elements reported here. For all elements other than Fe and Pb, the so-called transmission factor (import divided by export) were significantly greater than 0.1 (10%). Thus either both watersheds, though quite separate and different in land use, are saturated for many metals, or their soil chemistries are quite similar, or there are other non-point sources in our watershed not defined. This is consistent with the relatively small (positive) accumulation in the Delaware Inland Bays watershed (Table 1). However, this assumes only atmospheric sources, which is unlikely in this agrarian and residential watershed.

Conclusions

A comparison of atmospheric *versus* fluvial loading of selected dissolved trace elements to Delaware's Inland Bays indicates that on an annual basis, aeolian input provides as much as a third of the total input for Zn and at least 5% for the other metals. On shorter time scales, the relative atmospheric and fluvial loadings can vary significantly. Approximately half of the atmospheric input is directly to the surface waters of the bays, a majority of which is derived from wet deposition. The transmission of atmospheric inputs through the watershed varies among elements and for each sub-basin. Overall, atmospheric flux appears to provide quantitatively and qualitatively important inputs for a number of trace metals of ecological concern.

Acknowledgments

This study was made possible by a grant from the Delaware State Resources Program. (Project No. 04, 1995)

References

1.   Scudlark, J.R. ; Church, T.M. *Estuaries* **1993,** 16, 747-759.
2.   Fisher, D.; Oppenheimer, M. *Ambio* **1991**, 20, 102-108.
3.   Scudlark, J.R.; Church, T.M. Report to Center for Delaware Inland Bays, Delaware Department of Natural Resources and Environmental Control, 1999.

4.   Scudlark, J.R.; Conko, K.M.; Church, T.M. *Atmos.Environ.* **1994**, 28, 1487-1498.

5.   Scudlark, J. R.; Church, T.M. In *Atmospheric Deposition of Contaminants to the Great Lakes and Coastal Waters*, Baker, J.E., Ed.; SETAC, Pensacola, FL, 1997, Chap. 10, pp 195-208.

6.   Tramontano, J.M.; Scudlark, J.R.; Church, T.M. *Environ. Sci. Technol.* **1987**, 21, 749-753.

7.   Scudlark, J.R.; Church, T.M.; Conko, K.M.; Moore, S.M. USDA Forest Service Report NC-150, National Atmospheric Deposition Program Symposium Proceedings, Philadelphia, PA., 1992, pp 57-71.

8.   Wolff, G.T.; Kelley, N.A.; Ferman, M.A.; Ruthosky, M.S.; Stroup, D.P.; Korsog. P.E. *J. Air Poll. Control Assoc.* **1986**, 36, 585-591.

9.   Gordon, G.E.; Han, M.; Mizohata, A; Joseph, J. Report No. CBRM-TR-94-1, Maryland Dept. of Natural Resources, 1994, 102 pp.

10.   Turekian, K.K.; Wedepohl, K.H. *Geol. Soc. Am. Bull.* **1961**, 72, 175-192.

11.   Slinn, S.A.; Slinn, W.G.N. *Atmos. Environ.* **1980**, 14, 1013-1016.

12.   Church, T.M.; Scudlark, J.R. USDA Forest Service Report NC-150, Symposium Proceedings, Philadelphia, PA, 1992, pp 45-56.

13.   Sievering, H. *Atmos. Environ.* **1984**, 128, 2271-2272.

14.   Slinn, W.G.N. 1983. In *Air-Sea Exchange of Gases and Particles*, Liss,P; Slinn, W.G.N., Eds.; D. Reidel, Hingham, MA, 1983, pp 299-405.

15.   Shen, G.T. Ph.D. Thesis, MIT, Cambridge, MA., WHOI-86-37, 1986.

16.   Lindberg, S.E., Turner, R.R. *Water Air Soil Poll.* **1988**, 39, 123-156.

17.   Church, T.M.; Tramontano, J.M.; Scudlark, J.R.; Murray, S.L. In *Ecology and Restoration of the Delaware River Basin*, Majumdar, S.K.; Miller, E.W.; Sage, L.E., Eds.; Philadelphia Academy of Sciences, 1988.

18.   Wu, Z.Y.; Han, M.; Lin, Z.C.; Ondov, J.M. *Atmos.Environ.* **1994**, 28, 1471-1486.

19.   Arimoto, R.; Duce, R.A. *J. Geophys.Res.* **1986**, 91, 2787-2792.

20.   Gatz, D. F.; Chu, L.C. In *Toxic Metals in the Atmosphere,* John Wiley and Sons, New York, 1986, pp 355-391.

21.   Valigura, R.A.; Baker, J.E.; Scudlark, J.R.; McConnell, L.L. In *Perspectives on Chesapeake Bay, 1994: Advances in Estuarine Science*, Nelson, S.; Elliott, P, Eds.; Chesapeake Research Consortium Pub. No. 147, 1994.

22.   Chester, R.; Murphy, K.J.T.; Lin, F.J.; Berry, A.S.; Bradshawand, G.A.; Corcoran, P.A. *Mar. Chem.* **1993**, 42, 107-126.

23.   Church, T.M.; Scudlark, J.R.; Conko, K.M. Chesapeake Bay Research and Monitoring Division Report, Maryland Department of Natural Resources, 1998, 87 pp.

24.   Castro, M.; Scudlark, J.R.; Mason, R.P.; Church, T.M. Maryland Power Plant Research Program Report PPAD-AD-1, 2000, 80 pp.

Environmental Impacts
and Monitoring

Chapter 14

Development and Application of Equilibrium Partitioning Sediment Guidelines in the Assessment of Sediment PAH Contamination

Robert J. Ozretich[1], D. R. Young[1], and D. B. Chadwick[2]

[1]NHEERL, Western Ecology Division, U.S. Environmental Protection Agency, Newport, OR 97365
[2]Environmental Science Division, U.S Navy SPAWAR Systems Center, San Diego, CA 92152

The U.S. Environmental Protection Agency used insights and methods from its water quality criteria program to develop ESGs. The discovery that freely-dissolved contaminants were the toxic form led to equilibrium partitioning being chosen to model the distribution of contaminants in sediments. Empirical and mechanistic models encompassing narcosis theory, critical body burdens, and chemical additivity became the framework of a defensible ESG for PAH mixtures. An assessment of Sinclair Inlet, WA sediment is made utilizing bulk sediment and interstitial water concentrations.

Introduction

The Clean Water Act of 1972 (CWA) provided a framework for establishing numerical chemical criteria for both water and sediment with the goal of protecting 95% of the aquatic species that could reside in these matrices. The U.S. Environmental Protection Agency (EPA) identified chemicals that were associated with degraded habitats. This resulted in a consent decree agreement with industries listing 129 chemicals as priority pollutants. This list consisted of individual metals, unsubstituted polynuclear aromatic hydrocarbons (PAHs), pesticides and PCBs. By the early 1980s numerical Water Quality Criteria (WQC) were published for these priority pollutants (*1*). Adherence to WQC has resulted in great positive changes in the uses and communities that many water bodies now support.

While the water column contaminant concentrations in the 1980s began to meet and exceed applicable standards, many benthic communities continued to be severely stressed. The most sensitive species were often missing and much of their biomass consisted of opportunistic species (*2-4*). Recognition of the direct and indirect ecological and human health risks posed by contaminated sediment prompted the EPA to embark on the development of defensible chemical-specific concentration limits in sediments (*5*).

This chapter outlines the development of sediment guidelines for nonionic organic chemicals by EPA. Additionally, an assessment of sediment from an industrial waterway is made using the guidelines as they are currently promulgated for mixtures of PAHs (*6*).

Lessons Learned from WQC Development

To establish chemical specific criteria EPA developed the following general data needs and procedures that were first used in creating WQC (*7*): 1) Determine the LC50 and EC50s of organisms from several fresh and salt water genera for each chemical. 2) Rank the LC50s of each chemical for all genera, choosing a concentration that was lower than those of 95% of the tested species, calling it the final acute value (FAV) for that chemical. 3) Compute a mean acute-to-chronic (ACR) ratio (LC50/EC50). 4) Compute a final chronic value (FCV) by dividing the FAV by the ACR. The FCV then became the WQC for individual chemicals with units of mass per water volume. For sediment quality criteria (SQC) development, infaunal and epibenthic species were tested and the basic computational procedures for arriving at individual chemical criteria were followed.

In developing WQC databases the following quantitative structure activity relationship (QSAR) was reported for nonpolar organic compounds: the effects on an organism as expressed as an LC50 was linear (Figure 1) with the compound's octanol-water partition coefficient (K_{ow}) (*8-9*). It was hypothesized that the mode of action of this group of chemicals was through a reversible interaction with a target tissue that brings about a narcotic-like affect. At the LC50 the concentration in the target tissue is thought to be species specific and is referred to as the critical body burden. The concentration of the chemical present at the critical body burden is defined to be a toxic unit (TU). With a fractional TU present, fewer than 50%

262

of the exposed animals would die. It was found that fractional TUs, of different chemicals with the same, Type I, mode of narcotic action as PAHs, were additive (10). Although the TU term was introduced to express the concept of additivity it was not incorporated into WQC.

The finding from WQC research that was potentially most difficult to address in sediment was that the freely-dissolved concentration of a toxicant was the best predictor of toxicity, not its total water concentration. This resulted in water criteria including adjustment factors related to water hardness and pH (11).

Figure 1. Log LC50 of an organism versus log Kow of chemicals including a

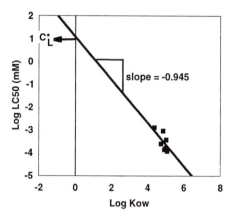

universal slope and its species-dependent y-intercept. C^*_L is the critical body burden in the lipid fraction of an organism. (Reproduced from reference 6)

These factors were a recognition of the equilibrium that exists between dissolved metals and various inorganic ligands. WQC for metal and organic priority pollutants do not contain similar adjustments based on the presence of organic ligands, which are at low concentrations in most surface waters.

However, sediment porewater can contain high concentrations of dissolved organic matter (DOM) spanning the size spectra from simple sugars to complex macromolecules. DOM is quantified by its carbon content as dissolved organic carbon (DOC). Sediments are dominated by heterogeneous solid phases of varying grain sizes and organic carbon loadings. The reversible association of nonionic organic compounds with the organic matter in bulk sediment is illustrated in Figure 2 and is represented by equilibrium partition coefficients, K_{oc} and K_{DOC}. These carbon-normalized coefficients are the ratios of the organic phase concentration to the freely-dissolved concentration; for compound Z ([μg Z/kg OC] divided by [μg Z/L]), with final units of L/kg OC.

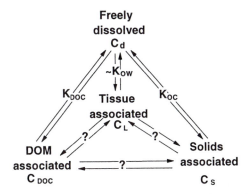

Figure 2. Distribution of organic pollutant within sediment in the presence of an infaunal organism. The ?s represent uncertain partitioning between phases that would require direct contact.

The equilibrium distribution of an organic pollutant among the three non-living, phases of Figure 2 is depicted for the dissolved components in the following:

$$C_d = C_{DOC} \; / \; m_{DOC}K_{DOC} \; (12) \tag{1}$$

where C_d is the freely-dissolved (µg/L water) concentration, C_{DOC} is the concentration associated with DOC (µg/kg OC), m_{DOC} is the concentration of dissolved organic carbon, (kg OC/L water), and K_{DOC} is the equilibrium partition coefficient (L/kg OC) for the organic pollutant between water and DOC. The solid and dissolved components of Figure 2 are depicted in the following:

$$C_{SOC} = C_S / f_{oc} = C_d K_{oc} \tag{2}$$

where C_{SOC} is the carbon-normalized sediment concentration (µg/kg OC), C_S is the sediment concentration (µg/kg-dry solids) and f_{oc} is the organic carbon fraction of the sediment (kg OC/kg-dry solids). Each compound's K_{oc} need not be measured as it could be computed:

$$\log_{10}K_{oc} = 0.983\log_{10}K_{ow} \; (12) \tag{3}$$

where K_{ow} is the readily-available octanol/water partition coefficient. In an equilibrated sediment the solid phase contains the bulk of organic contaminants because it contains the majority of organic carbon (f_{oc} of 0.005 to 0.05 vs ms of 10^{-5} to 10^{-4}) and $K_{oc} / K_{DOC} \cong 10^2$ to 10^3 (13). This physical/chemical model was used by EPA as the basis for using equilibrium partitioning (EqP) theory in ultimately establishing sediment quality guidelines (14).

A consequence of the equilibrium assumption is that the chemical activity of a pollutant is equivalent in all phases and the concentration in any phase can be computed from any other phase if the partition coefficient, pollutant and organic carbon concentrations of the phase are known. Since WQC for nonpolar organic chemicals were based on toxicity tests of essentially freely-dissolved chemicals (low DOC water), the sediment concentration of a chemical equilibrated with the WQC concentration could be considered the sediment quality criterion computed by substituting WQC for C_d in eq 2.

To test the applicability of this theoretical framework in sediment, and to verify predicted exposures, bioassay(15) and pore water sampling procedures (16,17) were developed and analytical methods were adapted to quantify the freely-dissolved concentrations (13,18). In laboratory exposures, sediment organic carbon in combination with compound K_{ow} were found to be good predictors of freely-dissolved chemical and organism response (19,20).

Given the success of the biological and analytical protocols in determining the LC50s of sediment-dwelling organisms and validating the robustness of the EqP theory, EPA felt ready, essentially using eq 2, to publish carbon-normalized sediment quality criteria for individual PAHs, acenaphthene, phenanthrene and fluoranthene (21-23).

Reconciling Laboratory and Field Studies

While the sediment criteria were being developed field surveys of sediment contamination found acutely toxic samples with concentrations of individual listed PAHs that would suggest no toxicity (24,25). The dominant contaminants in many of these sediments were PAHs. Considering chemicals individually in the development of WQC made sense for the majority of priority pollutants because in their uses as solvents, or chemical process feed stocks, their spills would result in single compound problems. Of the 16 priority pollutant PAHs (all of which are nonalkylated "parent" compounds), only naphthalene is a significant industrial chemical whereas the others, such as benz(a)pyrene (a carcinogen), occur as mixtures at low concentrations in crude oils and in high concentrations in petroleum-based preservatives (creosote) and combustion byproducts (26,27). Alkyl-substituted parent PAHs co-occur with the listed PAHs but are found at relatively higher concentrations in oil and at lower concentrations in the other materials. The listed parent PAHs are in Table I (1-16).

Recognition of the low probability of a single priority pollutant PAH (except naphthalene) ever reaching high enough concentrations to cause problems prompted EPA to withdraw the individual PAH criteria and rethink their approach.

Table I. EPA list of 34 PAHs (6) considered in ESG assessment of sediment. Results from 23 compounds in sediment and IW are shown.

Compound	$\log K_{ow}$	Station 2B ESGTU_{FCV} Sediment[a]	IW[b]
1. naphthalene (NAP)	3.36	.0058	.0005
2. acenaphthylene	3.22	.0096	.0018
3. acenaphthene	4.01	.0034	.0027
4. fluorene (FL)	4.21	.0062	.0084
5. phenanthrene (PHEN)	4.57	.0383	.0006
6. anthracene (ANT)	4.53	.0262	.0004
7. fluoranthene (FLU)	5.08	.0769	.0077
8. pyrene (PY)	4.92	.0737	.0093
9. benzo(a)anthracene (BaA)	5.67	.0323	.0048
10. chrysene (CHR)	5.71	.0649	.0067
11. benzo(b)fluoranthene	6.27	.0330	.0272
12. benzo(k)fluoranthene	6.29	.0364	.0175
13. benzo(a)pyrene	6.11	.0282	.0122
14. indeno(124cd)pyrene	6.72	.0175	.0146
15. dibenz(ah)anthracene	6.71	ND[c]	ND
16. benzo(ghi)perylene	6.51	.0144	.0125
17. benzo(e)pyrene	6.14	.0254	.0134
18. perylene	6.14	.0091	.0023
C1[d]-NAP (2 compound group)	3.84,3.86	.0053,.0026	.0001,ND
C2-C4 NAP (3 groups)	4.3,4.8,5.3	.0066,.0006,NA[e]	.013,ND,NA
C1-C3 FL (3 groups)	4.7,5.2,5.7	NA,NA,NA	NA,NA,NA
C1-C4 PHEN/ANT (4 groups)	5.0,5.5	.0023,NA	ND,NA
	5.9,6.3	NA,NA	NA,NA
C1 FLU/PY (1 group)	5.3	NA	NA
C1-C4 BaA/CHR (4 groups)	6.1,6.4	NA,NA	NA,NA
	6.9,7.4	NA,NA	NA,NA
	$\sum\text{ESGTU}_{FCV,23}$	0.52	0.16
	$\sum\text{ESGTU}_{FCV,TOT}$	2.1[f]	0.66

[a] computed using eq 7 ($f_{oc} = 0.055$)

[b] computed using eq 8

[c] ND, not detected

[d] C#, represents the level of alkylation of the parent PAH, e.g. C2-NAP would represent all isomers of naphthalene containing 2 methyl groups or one ethyl group

[e] NA, not analyzed

[f] not acceptable for the protection of benthic organisms

The biological significance of the mixtures only became fully understood when viewed through a combination of narcosis theory, chemical additivity, structure activity relationships, and EqP theory. As they came from the studies supporting WQC, these seemingly disparate concepts were incorporated into the ΣPAH model (28). This model predicted acute toxicity of sediment containing PAHs by the following: concentrations of 13 freely-dissolved PAHs (Table I, compounds 1-13) were computed from bulk sediment concentrations using EqP theory (rearrangement of eq 2); QSAR, LC50 vs Kow, for *Rhepoxinius abronius* (an infaunal, estuarine amphipod) was used to compute the LC50 of the compounds, and their fractional TUs were found by dividing the freely- dissolved concentration by the compound's LC50; the fractional TUs were summed and the predicted toxic response was computed from laboratory-based exposures. Although the empirical ΣPAH model's utility was limited to the prediction of acute toxicity of sediments to one bioassay organism, its success in integrating these concepts and predicting toxicity of field sediments (29) supported a more universal and mechanistic approach to developing water and sediment criteria.

Target Lipid Narcosis Model

Di Toro et al. (30) evaluated the LC50 vs K_{ow} slopes for 156 chemicals and 33 species and found that a single slope could represent all the QSARs (Figure 1). By arguing that an organism's lipid and LC50 concentration were equal when exposed to compounds with a K_{ow} of 1.0 they concluded that octanol and lipid were essentially equal and that lipid is the tissue within an organism that is affected directly by narcotic chemicals. They presented a target lipid model that explained the equivalent slopes of the species' QSARs as a manifestation of the essential equality of the linear free energy relationship of octanol and lipid, and that the varying y-axis intercepts (34 µmol/g octanol to 286 µmol/g octanol) for the different species was a manifestation of their varying sensitivities to Type I narcotic compounds. They incorporated these two concepts in the following equation to compute the species-specific critical body burden of baseline narcotic chemicals:

$$\log(C_L^*) = \log(LC50) + 0.945 \log(K_{ow}) \qquad (4)$$

where C_L^* is the critical body burden, µmol/g octanol(lipid); LC50 is species-specific, mM; -0.945 is the universal QSAR slope for baseline chemicals . They found that if the chemical was a ketone, PAH, or halogenated, the critical body burden would be reduced by a factor of ~0.55 in recognition of the greater potencies of these chemical classes. Equation 4 was shown to be valid with published measurements of the body burdens from 5 species that were exposed at their LC50 concentrations.

DiToro et al. (*30*) used the results of eq 4 to rank the species and determine a critical body burden that was lower than 95% of the tested species. Following other WQC protocols (*7*), such as applying an acute to chronic ratio, they were able to provide a uniform, physiological basis for computing WQC for Type I narcotic chemicals. The resulting final chronic tissue value (FC$_t$V) for the baseline narcotic chemicals was 6.94 µmol/g octanol(lipid). The freely-dissolved FCV (WQC) in equilibrium with this tissue would be compound-K$_{ow}$ and chemical class-specific as shown in the following rearrangement of eq 4:

$$\log FCV = \log 6.94 - 0.945 \log K_{ow} + \Delta c \tag{5}$$

where FCV has units of mM; 6.94 is the FC$_t$V (µmol/g octanol (lipid)) and Δc is the chemical class correction. Di Toro and McGrath (*31*) computed sediment quality guideline concentrations (C$_{SQG}$) for individual PAHs by combining equations 2 (expressed on a molar-basis), 3 and 5 to yield the following:

$$\log C_{SQG} = 0.038 \log K_{ow} + \log 6.94 - 0.2628 \tag{6}$$

where C$_{SQG}$ has units of µmol/g OC and 0.2628 is the logarithm of the PAH chemical class correction.

Derivation of EPA's EqP Sediment Guidelines

The premise of the WQC-approach to criteria development was that a body-burden of the FC$_t$V concentration of toxicant could be present and not adversely affect short or long-term survival of 95% of benthic organisms. Using a PAH-only database and employing the target lipid narcosis model (*30*) EPA derived a PAH FC$_t$V of 2.24 µmol/g lipid (*6*). This compared well with 3.78 µmol/g lipid which would result from applying the recommended PAH class correction (0.546) to the baseline FC$_t$V (30). EPA derived individual PAH sediment guideline concentrations (C$_{SQG}$) using eq 6 with 2.24 as the FC$_t$V and no class correction. Accounting for the contribution of co-occurring compounds would be through additivity but the problem faced by EPA was when to stop adding?

Due to the high costs of quantifying organic chemical pollutants in environmental samples the analyte list for many years included only the priority pollutant PAHs. As more federal agencies got involved in monitoring, and standard solutions became available with more analytes, the target list of published studies grew. As more PAHs were looked for they were found. To date, the most geographically extensive and compound-inclusive studies of sediment contamination is EPA's Environmental Monitoring and Assessment Program

(EMAP). The analyte list of these studies (Table 1) includes 18 individual unsubstituted PAHs and 16 groups with alkyl substitutions (34 PAHs). EPA designated this list of 34 PAHs as "total PAH" and based their ESG for PAH mixtures on the following expression of the additivity of Type I narcotic chemicals:

$$\text{sediment based: } \sum ESGTU_{FCV,TOT} = \sum_{i=1}^{34} \frac{C_{di}}{C_{FCVi} * MW_i} \quad (7)$$

or

$$\text{interstitial water (IW) based: } \sum ESGTU_{FCV,TOT} = \sum_{i=1}^{34} \frac{C_{SOCi}}{C_{SQGi} * MW_i} \quad (8)$$

where $\sum ESGTU_{FCV,TOT}$ is the equilibrium partitioning sediment guideline toxic units from the total, 34 PAHs; C_{SOCi} is the measured carbon-normalized sediment concentration of the i-th PAH; MW_i is the molecular weight needed to equalize units of the ratio. Recognizing that C_d is limited by compound solubility, the maximum C_{SOC} allowed in eq 7 equals $S*K_{oc}$, and the maximum C_d equals S, where S is solubility. The assessment of a sediment is:

acceptable for protection of benthic organisms if: $\sum ESGTU_{FCV,TOT} \leq 1.0$
or
not acceptable for protection of benthic organisms if: $\sum ESGTU_{FCV,TOT} > 1.0$

To provide factors to "correct" under-quantified sediments for unmeasured but likely to-be-present compounds, EPA used two extensive EMAP data sets (Louisian and Carolinian Provinces) to compute the ratios of $\sum ESGTU_{FCV,TOT}$ to $\sum ESGTU_{FCV,13}$ and $\sum ESGTU_{FCV,23}$. The 13 and 23 individual compounds comprise commonly reported data sets. The median and 95% certainty factors correcting 13 and 23 compounds to the 34 "total" are 2.75 and 11.5, and 1.64 and 4.14, respectively.

Over the 15 years that EPA spent on finalizing a contaminated sediment strategy, the numeric pass/fail criteria approach was supplanted by ESGs that were to be used as a complement to existing tools in assessing the extent of sediment contamination and the identification of chemicals causing toxicity (32-34).

Assessment of an Industrial Waterway

In conjunction with a study to determine benthic fluxes (35), sediment was sampled in Sinclair Inlet and the Puget Sound Naval Ship Yard in Bremerton, WA

(Figure 3). Surficial sediment (0-2 cm) was collected at ten stations and to -0.5 m at one station. Aliquots of IW separated from sediment (*16*) were processed directly (total concentrations) or passed through untreated C-18 cartridges (DOM-bound concentrations). The freely-dissolved concentrations were determined as the difference between the total and bound concentrations (*16,18*). De-watered sediment was sonicated in acetonitrile. Extracts were cleaned on treated C-18 columns (*36*), and quantified by mass spectrometer in the single ion monitoring mode Twenty-three PAHs and 22 PCB congeners were quantified. The PAHs consisted of 18 unsubstituted PAHs (see Table I) and 5 alkylated PAHs (1 and 2 methyl-, 2,6 dimethyl- and 2,3,5 trimethyl-naphthalene, and 1 methyl-phenanthrene). Prior to extraction all samples were amended with 6 deuterated PAH and 4 PCB compounds in a 10:1 ratio.

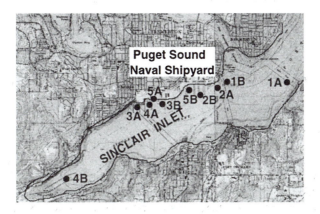

Figure 3. Station locations in Sinclair Inlet, western Puget Sound. Naval Shipyard is in the city of Bremerton, WA.

Results

At five of the ten stations both sediment and IW concentrations were determined (Figure 3, B-stations). Freely-dissolved IW concentrations of PAHs computed from bulk sediment concentrations, assuming equilibrium partitioning (eqs 2 and 3), exceeded measured IW concentrations by factors of 20 ± 4 and 4 ± 0.6 (mean$\pm$SE) for low molecular weight and high molecular weight parent PAHs (Table I compounds 1-6 and 7-13). For PCBs the computed to measured ratios were 0.5 ± 0.1. Comparisons of computed and measured freely-dissolved concentrations of fluoranthene found within depth interval composites (six cores) at Station 5B are shown in Figure 4. Recovery of PCBs and deuterated PAHs spiked at 5 and 50 ng/L averaged 66% and 70%, respectively.

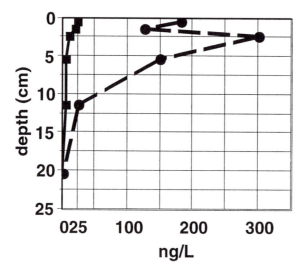

Figure 4. Measured (■) and calculated (●) freely-dissolved fluoranthene at Sinclair Inlet, WA, station 5B.

Using equations 7 and 8, $ESGTU_{FCV}$ values were computed for the 23 individual PAHs in sediment and IW from station 2B and are found in Table I. The $\sum ESGTU_{FCV,23}$ and $\sum ESGTU_{FCV,TOT}$ values for all stations are found in Table II. Conversion of the sediment concentration-based assessment ($\sum ESGTU_{FCV,23}$) to "Total PAH" by applying the 4.14 factor would designate seven of the ten stations as not acceptable the for protection of benthic organisms. None of the five assessments based on measured, freely-dissolved concentrations would fit this category.

Discussion

If only sediment concentrations were used in this assessment, the naval shipyard area of Sinclair Inlet could be considered a "Site of Concern" (*37*) as the proposed national ESG for PAH mixtures was exceeded at a majority of sites among the piers of that facility. However, when the assessments were made on the basis of IW determinations of freely-dissolved concentrations, none among the pier sediment exceeded the ESG. The magnitude of $\sum ESGTU_{FCV,TOT}$ values would not have been sufficient to elicit significant mortality in short-term sediment bioassays which were not done. The consequences of actual values in the 1-3 range might be manifest in a change in the species composition of the sediments which was also not done but would be recommended in a further assessment of these sites (*37*).

Table II. ESG assessment of stations based on
sediment and interstitial water concentrations
of 23 compounds.

Station	$\sum ESGTU_{FCV,23}$ Sed	$\sum ESGTU_{FCV,23}$ IW	$\sum ESGTU_{FCV,TOT}$ Sed	$\sum ESGTU_{FCV,TOT}$ IW
1A	0.22		0.91	
2A	0.22		0.91	
3A	0.30		1.24[a]	
4A	0.30		1.24[a]	
5A	0.37		1.53[a]	
1B	0.27	0.08	1.12[a]	0.33
2B	0.52	0.16	2.11[a]	0.66
3B	0.42	0.14	1.74[a]	0.58
4B	0.05	0.03	0.21	0.12
5B	0.26	0.10	1.08[a]	0.41

[a] not acceptable for the protection of benthic organisms

The large differences between computed and measured freely-dissolved concentrations of PAHs suggest that the assumption of equilibrium was violated for most of these compounds, but not PCBs. Given the greater uncertainties of K_{ow} s used to compute C_d for PCBs, a ratio of 0.5 is indistinguishable from 1.0, whereas the PAH ratios are significantly greater.

What could be the reason for these chemical class differences in bedded sediment? It is likely that in an active port such as this, with large diesel engine-driven vessels, the cause of apparent non-equilibrium in the sediment is the presence of large quantities of soot. These sediments are dominated by pyrogenic (*26,27*) PAHs (maximum of 2100 ug/kg-dry, chrysene, 5A). Trapped within the solid soot matrix are high concentrations of PAHs that were created during combustion and, through condensation, became part of the solid. This isolated mass of PAHs is accessible to vigorous solvent extraction, but not to solvation by water driven by diffusion. Therefore, a large fraction of PAHs from soots are not accessible to the IW for equilibration and are not bioavailable. The low concentrations of PCBs at these sites (maximum concentration 49 ug/kg-dry, PCB-101, 5A) are not from sampling spilled pure Aroclor oil but are from individual PCB congeners associated with sedimented particles equilibrated through the gas or aqueous-phase. An alternate interpretation of these observations is that the system is indeed in equilibrium because there are two K_{oc} s in eq 2 when soot is present, the one of relict organic matter that is approximately equal to K_{ow}, and a second that is represented by super-sorbent soot (*38*). The problem with this interpretation is that the super-sorbent should affect the PCB congeners in a similar way which isn't observed.

Whatever the cause, the bioavailability of PAHs in this study is not well represented by the sediment concentrations, but is by the IW and EPA has set up procedures to evaluate sites of concern for true bioavailability (*14*).

Disclaimer: This information has been funded wholly (or in part) by the U.S. EPA. It has been subjected to the Agency's peer and administrative review, and it has been approved for publication as an EPA document.

References

1. United States Environmental Protection Agency. *Quality Criteria for Water 1986;* EPA 440/5-86-001; U.S. EPA Office of Water Regulation and Standards: Washington, DC, 1987.
2. Swartz, R. C.; DeBen, W. A.; Sercu, K. A.; Lamberson, J. O.*Mar. Pollut. Bull.* **1982,** *13,* 359-364.
3. Swartz, R. C.; Cole, F. A.; Lamberson, J. O.; Ferraro, S. P.; Schults, D. W.; DeBen, W. A.; Lee II, H.; Ozretich, R. J. *Environ. Toxicol. Chem.* **1994,** *13,*949-962.
4. United States Environmental Protection Agency. *The incidence and severity of sediment contamination in surface waters of the United States.* EPA 823-R-97-006; U.S. EPA Office of Science and Technology: Washington, DC, 1997; Vol. 1.
5. Chapman, G. A. In *Fate and Effects of Sediment-Bound Chemicals in Aquatic Systems*; Dickson, K. L.; Maki, A. W.; Brungs, W. A., Eds.; Pergamon Press: Elmsford, NY, 1987; pp 355-377.
6. United States Environmental Protection Agency. *Methods for the derivation of site-specific equilibrium partitioning sediment guidelines (ESGs) for the protection of benthic organisms: PAH mixtures;* Draft Report; U.S. EPA Office of Science and Technology: Washington, D. C. 2000.
7. Stephan, C. E.; Mount, D. I.; Hansen, D. J.; Gentile, J. H.; Chapman, G. A.; Brungs, W. A. *Guidelines for deriving numerical national water quality criteria for the protection of aquatic organisms and their uses;* PB85-227049; National Technical Information Service: Springfield, VA, 1985.
8. Konemann, H. *Toxicology.* **1981,** *19,* 209-221.
9. Veith, G. D.; Call, D. J.; Brooke, L. T. *Can. J. Fish. Aquat. Sci.* **1983,***40,* 743-748.
10. Hermens, J. L. M. In *Handbook of Environmental Chemistry. and Processes;* Hutzinger O., Ed.; Springer Verlag: Berlin, 1989; Vol.2, pp 111-162.
11. United States Environmental Protection Agency. *Federal Register.* **1995,** 60(86), 22231-22237.

12. Di Toro, D. M., et al. *Environ Toxicol Chem.* **1991,** *10,* 1541-1583.
13. Ozretich, R. J.; Smith, L. M.; Roberts, F. A. *Environ. Toxicol. Chem.* **1995,** *14,* 1261-1272.
14. United States Environmental Protection Agency. *Technical basis for the derivation of equilibrium- partitioning sediment guidelines (ESGs) for the protection of benthic species: Nonionic organics contaminants.* EPA-822-R-00-001; U.S. EPA Office of Science and Technology: Washington, DC. 2000.

15. *Annual Book of ASTM Standards*; Standard E 1367-92; American Society for Testing and Materials: Philadelphia, PA, 1993; Vol. 11.04, pp 1138-1163.
16. Ozretich, R. J.; Schults, D. W. *Chemosphere.* **1998,** *36,* 603-615.
17. Ankley, G. T.; Schubauer-Berigan, M. K. *Arch. Environ. Contam. Toxicol.* **1994,** *27,* 507-512.
18. Landrum, P. F.; Nihart, S. R.; Eadie, B. J.; Gardner, W. S. *Environ Sci Technol.* **1984,** *18,* 187-192.
19. Swartz, R. C.; Kemp, P. F.; Schults, D. W.; Lamberson, J. O. *Environ. Toxicol. Chem.* **1988,** *7,* 1013-1020.
20. Swartz, R. C.; Schults, D. W.; DeWitt, T. H.; Ditsworth, G. R.; Lamberson, J. O. *Environ. Toxicol. Chem.* **1990,** *9,* 1071-1080.
21. United States Environmental Protection Agency. *Sediment quality criteria for the protection of benthic organisms: Acenaphthene*; EPA 822-R-93-013; U.S. EPA Office of Water: Washington, DC, 1993.
22. United States Environmental Protection Agency. *Sediment quality criteria for the protection of benthic organisms: Phenanthrene;* EPA 822-R-93-014; U.S. EPA Office of Water: Washington, DC, 1993.
23. United States Environmental Protection Agency. *Sediment quality criteria for the protection of benthic organisms: Fluoranthene;* EPA 822-R-93-012; U.S. EPA Office of Water: Washington, DC, 1993.
24. Swartz, R. C.; Kemp, P. F.; Schults, D. W.; Ditsworth, G. R.; Ozretich, R. J. *Environ. Toxicol. Chem.* **1989,** *8,* 215-222.
25. Tay, K-L.; Doe, K. G.; Wade, S. J.; Vaughn, D. A.; Berrigan, R. E.; Moore, M. J. *Environ. Toxicol. Chem.* **1992,** *11,* 1567-1581.
26. Hites, R. A.; LaFlamme, R. E.; Windsor Jr., J. G. *Geochimica et Cosmochimica Acta.* **1980,** *44,* 873-878.
27. Boehm, P. D.; Farrington, J. W. *Environ Sci Technol.* **1984,** *18,* 840-845.
28. Swartz, R. C.; Schults, D. W.; Ozretich, R. J.; Lamberson, J. O.; Cole, F. A.; DeWitt, T. H.; Redmond, M. S.; Ferraro, S. P. *Environ Toxicol Chem.* **1995,** *14,* 1977-1987.
29. Ozretich, R. J.; S. P. Ferraro; J. O. Lamberson; F. A. Cole. *Environ Toxicol Chem.* **2000,** *19,* 2378-2389.

274

30. DiToro, D. M.; McGrath, J. A.; Hansen, D. J. *Environ. Toxicol. Chem.* **2000,** *19,* 1951-1970.
31. DiToro, D. M.; McGrath, J. A. *Environ. Toxicol. Chem.* **2000,** *19,* 1971-1982.
32. Chapman, P. M. *Ecotoxicol.* **1996,** *5,* 327-339.
33. Swartz, R. C. *Environ. Toxicol. Chem.* **1999,** *18,* 780-787.
34. Ho, K. T.; McKinney, R. A.; Kuhn, A.; Pelletier, M. C.; Burgess, R. M. *Environ. Toxicol. Chem.* **1997,** *31,* 203-209.
35. Chadwick, D. B.; Lieberman, S. H.; Reimers, C. E.; Young, D. *An evaluation of contaminant flux rates from sediments of Sinclair Inlet, WA, using a benthic flux sampling device;* Technical Document 2434; Naval Command, Control and Ocean Surveillance Center, RDT&E Division: San Diego, CA, 1993.
36. Ozretich, R. J.; Schroeder, W. P. *Anal. Chem.* **1986,** *58,* 2041-2048.
37. United States Environmental Protection Agency. *Methods for the derivation of site-specific equilibrium partitioning sediment guidelines (ESGs) for the protection of benthic organisms: nonionic;* EPA 822-R-00-002; U.S. EPA Office of Water: Washington, DC. 2000.
38. Gustafsson, O.; Haghseta, F.; Chan, C.; MacFarlane, J.; Gschwend, P. M. *Environ. Sci. Tech.* **1997,** *31,* 203-209.

Chapter 15

Small Volume Sampling and GC/MS Analysis for PAH Concentrations in Water above Contaminated Sediments

David R. Young[1], R. J. Ozretich[1], and D. B. Chadwick[2]

[1]NHEERL, Western Ecology Division, U.S. Environmental Protection Agency, Newport, OR 97365
[2]Environmental Science Division, U.S Navy SPAWAR Systems Center, San Diego, CA 92152

This paper describes a sampling and analysis method for measuring trace concentrations of polynuclear aromatic hydrocarbons (PAHs) in small volume (0.5 L) water samples collected above sediment contaminated by these toxic organic compounds. The method was used in conjunction with a Benthic Flux Sampling Device at the Puget Sound Naval Shipyard (Sinclair Inlet, WA). Concentrations near or below 1 ng/L were measured for 24 PAHs. Water concentrations were corrected for recovery and procedural blank (generally <0.5 ng/L). Precision of individual PAH concentrations was high; the median standard error obtained for triplicate samples was 0.1 ng/L. Similarly, the median procedural blank value was <0.1 ng/L for the 24 PAHs. This method compares well with large volume methods reported for measuring PAH compounds in water samples.

Introduction

Polynuclear aromatic hydrocarbons (PAH) and their alkyl homologs are among the more toxic components or byproducts of petroleum and other fossil fuels (*1*). The use of such materials as a major energy source since the Industrial Revolution has distributed PAH compounds throughout the global environment (*2*). Information regarding the concentrations of these contaminants in surface waters exposed to different types and magnitudes of PAH sources, including contaminated sediments, is necessary for understanding the processes which affect PAH transport and fate. The ability to reliably identify and quantify trace amounts of PAHs in small samples greatly enhances the acquisition of such information. Here we report a technique for measuring low (sub-nanogram/liter) concentrations of PAHs in 0.5 liter (L) samples of whole (unfiltered) water, collected from contaminated or control environments.

Background

As part of an effort to reduce contamination around anchorages, the United States Navy developed an instrument for the *in situ* measurement of contaminant release rates from in-place sediments (*3*). The instrument, termed the Benthic Flux Sampling Device (BFSD), consists of a chamber and associated support structure (Figure 1) that can be lowered from a small craft to the bottom and released, isolating a volume (40 L) of water in contact with the sediment. At preprogrammed times, generally over a period of two to three days, a microprocessor-based control system is used to control collection of samples from the trapped volume inside the chamber. The criteria that the BFSD be small enough to be easily transported, and capable of collecting up to six water samples, required the development of a technique to measure environmental concentrations of PAH compounds in a 0.5 liter sample. Such a capability was developed by the U.S. Environmental Protection Agency (EPA) Office of Research and Development, and tested at the Puget Sound Naval Shipyard (PSNS).

Study Site

This study was conducted in Sinclair Inlet, one of several bays located in western-central Puget Sound basin in Washington State (Figure 3 of Ozretich et al., this volume). The inlet is bordered by two population centers, Bremerton on the north and Port Orchard on the south. The Bremerton shoreline is dominated by the piers and shipyard facilities of the PSNS, while the Port Orchard shoreline is characterized by several small marinas and low density development.

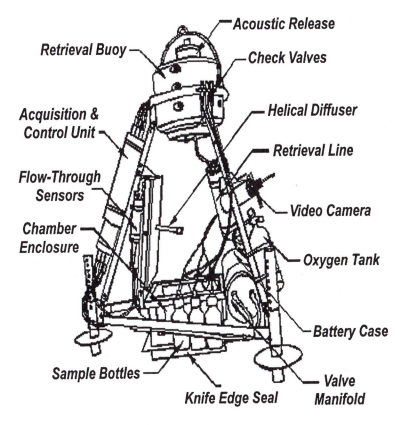

Figure 1. *The Benthic Flux Sampling Device*

Sample Collection

The BFSD is an instrument for *in situ* measurement of constituent flux rates from sediment. A flux out of, or into, the sediment is measured by isolating a volume of water above the sediment, drawing off samples from this volume over time, and analyzing these samples for increases or decreases in water concentration. The BFSD consists of an open-bottomed chamber mounted in a tripod-shaped framework with associated sampling gear, sensors, control system, power supply, and deployment/retrieval equipment (Figure 1). The entire device is approximately

1.2 by 1.2 m wide. The lower part of the framework contains the chamber, sampling valves, sampling bottles, and batteries. The chamber is box-shaped, approximately 40 cm square by 25 cm tall and is constructed of polycarbonate. The top of the chamber is hinged at one edge so it may be left open during deployment to minimize sediment disturbance. The bottom of the chamber forms a knife edge with a flange 5 cm above the base, providing a seal between the box and the sediment.

Samples are drawn off through a 4 mm Teflon tube connected to a manifold of valves into air-filled 500 ml borosilicate glass sample bottles (precleaned by soap and water, distilled/deionized water rinse, followed by kiln-firing at 340 °C for eight hours). Sampling is initiated by the control system that opens the valves at preprogrammed intervals. A bottle then is filled by hydrostatic pressure while venting through a check valve mounted at the top of the frame. Sensors for monitoring conditions within the chamber, including temperature, salinity, pH, and dissolved oxygen, are mounted in the chamber lid. A small pump maintains circulation in the flow-through system to the sensors, and also is used to mix the chamber volume via a helical diffuser. The oxygen system is used to maintain oxic conditions within the chamber by diffusing oxygen at a constant rate through a coil of thin-walled Teflon tubing.

During July 1991, a time series of water samples was collected within PSNS at sites 3B and 5B for analysis of 24 PAH compounds and 22 specific congeners of polychlorinated biphenyls (PCB). After each deployment, blank ferrules were fitted in place of the sampling lines, and the samples were brought ashore for initial processing. Triplicate time-zero water samples also were collected from outside the BFSD (approx. 1 m above the bottom) in precleaned 500 ml borosilicate glass bottles using the Teflon pumping system aboard the survey boat. These samples were processed in a similar manner to that used for the time-series samples. Bulk sediment also was collected at the ten sites within Sinclair Inlet, using a modified Van Veen grab. Samples of the surficial sediment (0 - 5 cm) were transferred into precleaned 500 ml wide-mouth glass jars using a precleaned stainless steel scoop. Upon retrieval the sediment samples were refrigerated, and the water samples were placed on ice and transported within a few hours to a nearby EPA laboratory for immediate processing.

Sample Processing and Analysis

Within a few hours of collection, surrogate compounds were added to each 0.5 liter water sample. The recoveries of these unique compounds were expected to mimic the wide chromatographic elution range of the compounds we were prepared to quantify. For the polynuclear aromatic hydrocarbons, perdeuterated forms of naphthalene, acenaphthene, fluoranthene, chrysene, benzo[b]fluoranthene, and perylene were added at 50 ng/L. Four surrogate PCB congeners, 30, 65, 143 and 166 were added at 5 ng/L. The samples were gently extracted on a shaker table (in the dark, at 80-100 revolutions per minute for 10-18 hours) using 12 ml of 10 percent (volume/volume) isooctane in hexane (pesticide grade). Conceptually this liquid-liquid extraction was similar to the "slow-stirring" method of equilibrating spiked octanol with distilled water for the determination of octanol/water partition coefficients (4,5). Following extraction, the aqueous layer was aspirated from beneath the organic layer which was transferred to a smaller container and reduced in volume to between 0.05 and 0.1 ml with purified nitrogen gas (6). Perdeuterated phenanthrene was added to this final reduced volume and used to determine the recovery of the surrogate compounds.

Aliquots (3 µl) of the extracts were injected into an electronic pressure controlled chromatograph (HP 5890) with mass selective detector (HP 5972A), (Hewlett Packard Co., Palo Alto, California, U.S.A.). The capillary chromatographic column was 60 m long with an internal diameter of 0.25 mm and contained the 0.25µm thick stationary phase DB-5 (J & W Scientific Co., Rancho Cordova, California, U.S.A.). The characteristic ions of PAH compounds are the molecular ion and secondary ions that generally are about one-tenth the intensity of the molecular ion. These ions of the 24 PAH compounds in Table 1 (in retention time order) were monitored (dwell times no less than 60 milliseconds per ion) and quantitative results were obtained by utilizing the relative response factors (RRFs) of the analytes to the perdeuterated surrogate compounds in the standards. It was assumed that the analytes sustained losses similar to their quantitative surrogates. This quantitation procedure provided a direct correction for the procedural losses of PCB and PAH surrogates. The average recoveries for the four PCB and six PAH surrogates from the 22 water samples in this study were 72±2% and 75±3%, respectively (mean ± SE). Analyte identity was verified by closeness of chromatograph retention time ($\pm$ 0.01 minute) and ratio of characteristic ion areas relative to an authentic compound in the standard solutions. For certain samples and PAH compounds, the molecular ion occurred at levels near the estimated detection limit (DL) for this procedure (Table 1), and the secondary ion was below the limit. This precluded second-ion confirmation of the PAH identifications, yielding possible upper-limit concentrations.

One solution of a 6-level series of standard solutions containing the analytes in Table 1 was injected following every 6 samples. The ratio of analyte concentration to surrogate internal standard concentration (5 ppb or 50 ppb) in these standard solutions ranged from 25:1 to 0.025 :1. The average coefficient of variation for the RRFs of the 46 compounds was 17 ± 1 % indicating that the instrument response was quite linear over this 3-orders of magnitude calibration range. All sample peaks were manually integrated.

Table 1. PAH concentrations (ng/L) in unfiltered ambient near-bottom water samples (0.5 L) collected July 21, 1991 from site 5B (Figure 3 of Ozretich et al, this volume).

Compound	Net Conc. (n=3) Ave. ± S.E			Proc. Blank [a] Ave.	Est. DL [b]
Naphthalene	<7.8		-	<7.8	7.8
2-Methylnaphthalene	2.5	±	0.7	1.6	1.6
1-Methylnaphthalene	<1.7		-	<1.7	1.7
Biphenyl	0.4	±	0.1	0.4	0.19
2,6-Dimethylnaphthalene	<7.0		-	<7.0	7.0
2,3,5-Trimethylnaphthalene	1.4	±	0.3	0.4	0.07
Acenaphthylene	0.4	±	0.1	<0.1	0.04
Acenaphthene	2.7	±	0.4	0.5[c]	0.26
Fluorene	1.9	±	0.3	0.2	0.07
Phenanthrene	2.7	±	0.5	0.4	0.07
Anthracene	0.5	±	0.1	<0.1	0.04
1-Methylphenanthrene	0.7	±	0.1	<0.1	0.04
Fluoranthene	5.5	±	0.9	0.1	0.04
Pyrene	2.9	±	0.5	0.1[c]	0.04
Benzo[a]anthracene	0.5[c]	±	0.1	<0.1	0.03
Chrysene	0.7	±	0.1	<0.1	0.03
Benzo[b]fluoranthene	0.6[c]	±	0.1	<0.1	0.03
Benzo[k]fluoranthene	0.4[c]	±	0.0	<0.1	0.03
Benzo[e]pyrene	0.5[c]	±	0.1	<0.1	0.03
Benzo[a]pyrene	0.3[c]	±	0.0	<0.1	0.03
Perylene	0.1[c]	±	0.0	<0.1	0.03
Indeno[1,2,3-cd]pyrene	0.3[c]	±	0.0	<0.1	0.03
Dibenz[a,h]anthracene	<0.1		-	<0.1	0.03
Benzo[g,h,i]perylene	0.3	±	0.0	<0.1	0.03

[a] Average of 2 procedural blank values used to obtain net concentrations
[b] Estimated detection limit (see text)
[c] Values lack GC/MS second-ion confirmation

Generally, the surrogate to analyte area ratios (A_{is}/A_a) of compounds with retention times shorter than benzo[a]anthracene were within the calibration range for all the non-blank samples. The strongly linear instrument response (low CV of RRFs), coupled with the low instrument noise and high analyte sensitivity (≥ 300 area units/pg of injected perdeuterated fluoranthene), allowed us to report concentrations for compounds with areas between one eighth to one half of those calculated from the average A_{is}/A_a of the lowest standard level. This was the case for the majority of compounds in the blank samples, and for those compounds in the water samples with retention times of benzo[a]anthracene and longer (Table 1).

Estimated DL values for the 24 PAH analytes were obtained as follows. Six (empty bottle) procedural blank samples were analyzed in a parallel study, yielding concentration values for an equivalent 0.5 L water sample. The SE values were multiplied by the one-sided critical value of the Student's t-distribution (t.01[5]) statistic to obtain an estimate of the upper 99 percent confidence limit for the actual concentration. This upper limit value was accepted as the estimated detection limit of the method (Table 1). However, we elected not to quantify concentrations below 0.1 ng/L (procedural blank values <0.1ng/L were not used to obtain net concentrations).

The sediment samples were extracted by sonication with acetonitrile and cleaned using C-18 solid-phase sorbent. Total organic carbon (TOC) also was determined in an aliquot of each sample. Details of analysis are provided elsewhere (Ozretich et al., this volume).

Results

With the occasional exception of PCB congener 110, no apparent signals of other congeners were observed; consequently, there will be no further discussion of PCBs. Average values ($\pm$ 1 SE) for blank-corrected (net) concentrations of the 24 target PAH determined in ambient near-bottom water triplicate samples from site 5B, and for the duplicate procedural (bottle) blanks, are listed in Table 1. Concentrations of TOC and the major PAH compounds analyzed in surficial sediment from the ten Sinclair Inlet sites are listed in Table 2.

Discussion

Concentration sums of the "lower molecular weight PAHs" (Table 2, naphthalene to anthracene) obtained for the surficial sediment samples collected at sites 3B and 5B were 1040 and 920 µg/kg dry wt., respectively. Corresponding sums of the "higher molecular weight PAHs" (fluoranthene to benzo[g,h,i]perylene) were 9600 and 5800 µg/kg dry wt. TOC concentrations were

Table 2. Concentrations of TOC (percent dry wt.) and PAHs (µg/g organic carbon) in surficial sediment (0 - 5 cm) collected from ten sites within Sinclair Inlet of Puget Sound, WA (July 1991). (Site locations shown in Figure 3 of Ozretich et al., this volume).

Chemical	1A	1B	2A	2B	3A	3B	4A	4B	5A	5B
							Stations			
TOC (%)	0.3	1.1	1.6	3.7	3.1	2.2	3.7	5.6	5.1	2.4
Naphthalene	3.6	0.5	1.4	2.2	6.8	1.5	2.9	0.1	1.1	1.1
2-Methylnaphthalene	3.0	1.6	1.0	2.4	3.9	1.9	1.3	0.4	1.3	1.5
1-Methylnaphthalene	2.0	0.9	0.2	1.2	2.1	1.2	0.9	0.3	0.6	1.2
Biphenyl	ND	0.3	0.2	0.6	0.7	0.4	0.1	ND	0.3	0.5
2,6-Dimethylnaphthalene	1.7	1.4	1.3	3.4	2.4	2.6	1.6	3.0	2.4	5.6
2,3,5-Trimethylmnaphth.	ND	0.4	0.1	0.3	0.8	0.4	ND	0.1	0.2	0.5
Acenaphthylene	2.5	1.9	2.4	4.3	2.8	3.1	2.6	0.3	2.8	2.0
Acenaphthene	ND	1.4	0.7	1.7	1.9	1.5	0.6	0.1	0.9	0.9
Fluorene	ND	1.4	1.5	3.3	1.9	2.4	1.4	0.2	1.9	1.3
Phenanthrene	3.4	13	6.1	23	6.9	17	7.0	1.5	8.4	8.6
Anthracene	1.9	5.1	5.4	16	4.5	8.3	5.4	0.7	6.9	5.5
1-Methylphenanthrene	1.2	0.9	0.4	1.6	0.5	1.4	0.6	ND	0.7	0.9
Fluoranthene	21	38	16	54	26	41	22	4.9	35	26
Pyrene	34	43	20	51	28	55	27	5.9	38	32
Benz[a]anthracene	14	12	12	27	15	21	19	1.8	21	14
Chrysene	14	15	20	55	21	35	34	2.0	42	20
Benzo[b]fluoranthene	10	15	16	32	19	30	24	3.1	31	18
Benzo[k]fluoranthene	8.8	18	15	36	20	29	21	2.9	26	17
Benzo[e]pyrene	10	11	14	24	17	24	18	2.5	22	14
Benzo[a]pyrene	16	13	17	27	19	24	20	1.8	25	14
Perylene	3.2	4.3	4.8	8.8	6.6	7.8	6.3	0.9	6.8	5.8
Indeno[1,2,3-cd]pyrene	10	6.7	12	19	17	19	16	2.2	18	12
Dibenzo[ah]anthracene	1.0	0.6	1.1	2.8	2.5	3.1	2.7	0.1	3.3	1.0
Benzo[g,h,i]perylene	12	8.5	9.8	16	14	16	13	2.8	13	10

similar at sites 3B and 5B (2.2 and 2.4 percent, respectively). These concentrations indicate a relatively high degree of surficial sediment contamination by PAH compounds at the two PSNS sites where the BFSD was deployed. The time-series data for PAH concentrations within the BFSD chamber (corrected for dilution via outside water that replaced sampled water) gave linear increases with time for several PAHs during each deployment (Figure 2). A linear regression model was used to calculate average flux values for these compounds. The highest fluxes were obtained at Site 3B: acenaphthene (2800 ng/m²/day), fluorene (2200 ng/m²/day), and trimethylnaphthalene (430 ng/m²/day). Substantially lower fluxes were measured at Site 5B (only the first four sample concentrations in the time series were used because the chamber went anoxic from previously oxic conditions). The highest fluxes obtained at Site 5B were for fluoranthene (570 ng/m²/day) and pyrene (530 ng/m²/day).

The reasons for the different sediment-to-water fluxes for a given PAH compound at the two sites are unclear. However, the relatively low variability of the individual water concentrations around their linear regression line, observed for the time series illustrated in Figure 2, indicates the level of precision that can be obtained for PAH compounds at ng/L and sub-ng/L levels using this sampling and analysis procedure. This precision also is illustrated by the low standard errors presented in Table 1. The median SE value for the duplicate procedural blank concentrations is <0.1 ng/L, and the median SE obtained for the Site 5B triplicate water sample concentrations is 0.1 ng/L.

Despite the high levels of surficial sediment contamination by PAH compounds in Sinclair Inlet found in this study, the sampling and analysis technique reported here revealed relatively low PAH concentrations in the unfiltered water samples collected from above these sediments. Reported thresholds of toxicity to aquatic organisms typically exceed the water concentrations measured here by at least three orders-of-magnitude (7).

The specificity and sensitivity of this technique for measuring PAH compounds in small volume water samples appear to compare well with those for other techniques reported in the literature. A summary of key characteristics of various "PAH in water" methods published over the last two decades is presented in Table 3. This comparison shows that half of these marine waters PAH studies have relied upon relatively large (> 4 L) sample volumes, ranging up to 2000 L. The advantage of a large volume strategy is the minimization of contamination during the sampling process, and an increase in the capability of the method to identify individual PAH compounds and quantify them, often in more than one phase. All the studies intended to measure PAH concentrations in both the particulate and the dissolved (plus colloidal) phases relied upon large volumes (35-2000 L). Of the 20 comparison methods listed, only four using ≤ 4 L indicated detection limit values below about 0.1 ng/L, and none indicated sampling volumes as low as the 0.5 L utilized in this study. (It is noted that many of the "Estimated

284

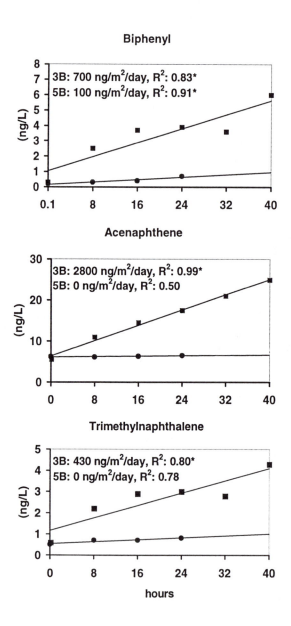

Figure 2. Flux rates computed for stations 3B (■) and 5B (●) (Figure 3 of Ozretich et al., this volume). * indicates data sets with slopes significantly different from zero (p<0.05).

Table 3. Comparison of procedural characteristics for reported methods of measuring PAHs in marine water samples.

Reference Number	Survey Period	Water Vol. (L)	Phases[a]	Isolation Method[b]	Quantif. Method[c]	Detect. Limit (ng/L)[d]
This Study	1991	0.5	W	LLE	GC - MS	0.03-8
8	1980	1270	P&D	F - SPE	CG - FID	0.05
9	1985 - 87	42-55	D	LLE	GC - FID	0.1
10	1986	140	P & D	F - SPE	HPLC - Fl	<0.01
11	1988	4	W	LLE	GC - MS	0.04
12	1988	220 -440	D	SPE	GC - MS	0.01
13	1988	1500-2200	P & D	F - SPE	GC - MS	0.01
14	1989 - 92	25-50	P	F	GC - MS	0.1
15	1989 - 93	2	W	LLE	GC - MS	0.1
16	1990	4	W	LLE	GC - MS	0.2-8
17	1990-91	11	W	LLE	GC - MS	0.2-2
18	1992 - 94	22	W	LLE	HPLC - Fl	0.002
19	1993	10-50	D	F - SPE	GC - MS	0.01
20	1993	230-1400	P & D	F - SPE	GC - MS	0.005
21	1993 - 95	2	W	LLE	HPLC - Fl	1
22	1994	35	P & D	F - SPE	GC - MS	0.003
23	1994 - 95	0.6	W	LLE	C18 - Fl	0.03 - 0.17
24	1996	40	P & D	F - SPE	GC - MS	0.01 - 0.1
25	1996	1	W	LLE	GC - MS	2 - 15
26	1997	250	P & D	F - SPE	GC - MS	0.01
27	1998	1	W	F - SPE	GC - FID	1

[a] **W**: whole (unfiltered) water sample; **P**: particulate phase; **D**: dissolved + colloidal phase.
[b] **LLE**: liquid - liquid extraction; **F - SPE**: filtration - solid phase extraction; **F**: filtration;
[c] **GC-MS**: gas chromatography - mass spectrometry; **GC - FID**: gas chromatography - flame ionization detection; **HPLC - Fl**: high perfomance liquid chromatograph - fluorescence spectrophotometry; **C18 - Fl**: C18 column chromatography - fluorescence spectrophotometry.
[d] Detection limits estimated from values (often for a limited number of PAHs) presented in the cited paper.

Detection Limit (ng/L)" values listed in Table 3 are based on varying types of information obtained from the papers cited, and are approximate only). Thus, the small volume sampling and GC/MS detection system described here for identifying (with independent signal confirmation) and quantifying concentrations of PAH compounds in unfiltered water samples appears to be a useful technique when logistical limitations preclude the sampling and processing of large volumes of water from contaminated or control sites.

Acknowledgments

We thank Kathy Sercu and Fred Roberts (U.S. EPA, Newport, OR) and Scott Echols, John Frazier, and Carolyn Poindexter (AscI, Newport, OR) for assistance in this research. This study was funded in part by the U.S. Naval Facilities Engineering Command (Silverdale, WA), and in part by the U.S. EPA Office of Research and Development Environmental Research Laboratory (Newport, OR). Mention of trade names or commercial products does not constitute endorsement or recommendation for use.

Literature Cited

1. Payne, J. F.; Kiceniuk, J.; Fancey, L. L.;Williams, U. What is a safe level of polycyclic aromatic hydrocarbons for fish: subchronic toxicity study on winter flounder (*P. americanus*). *Can. J. of Fish Aq. Sci.* **1988,** 45, 1983-1993.
2. Cripps, G. C. The extent of hydrocarbon contamination in the marine environment from a research station in the Antarctic. *Mar. Poll. Bul.* **1992,** 25, 288-292.
3. Chadwick, D. B.; Lieberman, S. H.; Reimers, C. E.; Young, D. R. An evaluation of contaminant flux rates from sediments of Sinclair Inlet, WA, using a benthic device. *Technical Document 2434,* Naval Command, Control and Ocean Surveillance Center, San Diego, CA 92152-5001 1993, 95 pp.
4. Brooke, D. N.; Dobbs, A. J.; Williams, N. Octanol:water partition coefficients (p): measurement, estimation, and interpretation, particularly for chemicals with P>10^5. *Ecotox. Environ. Saf.* **1986,** 11, 251-260.
5. deBruijn, J. F.; Busser; Seine, W.; Hermens, J. Determination of octanol/water partition coefficients for hydrophobic organic chemicals with the "slow-stirring" method.. *Environ. Tox. and Chem.* **1989,** 8, 499-512.
6. Ozretich, R. J.; Smith, L. M.; Roberts, F. A. Reversed-phase separation of estuarine interstitial water fractions and the consequences of C-18 retention of organic matter. *Environ. Tox. and Chem.* **1995,** 14, 1261-1272.
7. Knutzen, J. Effects on marine organisms from polycyclic aromatic

hydrocarbons (PAH) and other constituents of waste water from aluminum smelters with examples from Norway. *The Sci. of Tot. Environ.* **1995,** *163,* 107-122.

8. de Lappe, B. W.; Risebrough, R. W.; Walker II, W. A large-volume sampling assembly for the determination of synthetic organic and petroleum compounds in the dissolved and particulate phases of seawater. *Can. J. Fish. Aquat. Sci.* **1983,** *40* (Suppl. 2), 322-336.

9. Gómez-Belinchón, J. I.; Grimalt, J. O.; Albaigés, J. Intercomparison study of liquid-liquid extraction and adsorption on polyurethane and amberlite XAD-2 for the analysis of hydrocarbons, polychlorobiphenyls, and fatty acids dissolved in seawater. *Environ. Sci. Technol.* **1988,** *22,* 677-685.

10. Murray, A. P.; Richardson, B. J.; Gibbs, C. F. Bioconcentration factors for petroleum hydrocarbons, PAHs, LABs and biogenic hydrocarbons in the blue mussel. *Mar. Poll. Bul.* **1991,** *22.12,* 595-603.

11. Bidleman, T. F.; Castleberry, A. A.; Foreman, W. T.; Zaranski, M. T.; Well, D. W. Petroleum hydrocarbons in the surface water of two estuaries in the southeastern United States. *Est. Coast. Shelf Sci.* **1990,** *30,* 91-109.

12. Ehrhardt, M. G.; Burns, K. A. Petroleum-derived dissolved organic compounds concentrated from inshore waters in Bermuda. *J. Exp. Mar. Biol. Ecol.* **1990,** *138,* 35-47.

13. Broman, D.; Näf, C.; Rolff, C.; Zebühr, Y. Occurence and dynamics of polychlorinated dibenzo-*p*-dioxins and dibenzofurans and polycyclic aromatic hydrocarbons in the mixed surface layer of remote coastal and offshore waters of the Baltic. *Environ. Sci. Technol.* **1991,** *25.11,* 1850-1864.

14. Green, G.; Skerratt, J. H.; Leeming, R.; Nichols, P. D. Hydrocarbon and coprostanol levels in seawater, sea-ice algae and sediments near Davis Station in Eastern Antarctica: a regional survey and preliminary results for a field fuel spill experiment. *Mar. Poll. Bul.* **1992,** *25.9-12,* 293-302.

15. Bícego, M. C.; Weber, R. R. ; Ito, R. G. Aromatic hydrocarbons on surface waters of Admiralty Bay, King George Island, Antarctica. *Mar. Poll. Bul.* **1996,** *32.7,* 549-553.

16. Kucklick, J. R.; Bidleman, T. F. Organic contaminants in Winyah Bay, South Carolina I: Pesticides and polycyclic aromatic hydrocarbons in subsurface and microlayer waters. *Mar. Environ. Res.* **1994,** *37,* 63-78.

17. Law, R. J.; Whinnett, J. A. The determination of polycylic aromatic hydrocarbons in seawater from the *Fluxmanche* transect (Dover Strait). *Oceanolog. Acta* **1993,** *16,* 593-597.

18. Witt, G. Polycyclic aromatic hydrocarbons in water and sediment of the Baltic Sea. *Mar. Poll. Bul.* **1995,** *31.4-12,* 237-248.

19. Dachs, J.; Bayona, J. M. Large volume preconcentration of dissolved hydrocarbons and polychlorinated biphenyls from seawater. Intercomparison between C_{18} discs and XAD-2 column. *Chemosph.* **1997**, *35.8*, 1669-1679.
20. Schultz-Bull, D. E.; Petrick, G.; Bruhn, R.; Duinker, J. C. Chlorobiphenyls (PCB) and PAHs in water masses of the northern North Atlantic. *Mar. Chem.* **1998**, *61*, 101-114.
21. Law, R. J.; Dawes, V. J.; Woodhead, R. J.; Matthiessen, P. Polycyclic aromatic hydrocarbons (PAH) in seawater around England and Wales. *Mar. Poll. Bul.* **1997**, *34.5*, 306-322.
22. Gustafson, K. E.; Dickhut, R. M. Distribution of polycyclic aromatic hydrocarbons in southern Chesapeake Bay surface water: evaluation of three methods for determining freely dissolved water concentrations. *Environ. Tox Chem.* **1997**, *16.3*, 452-461.
23. Kira, S.; Sakano, M.; Nogami, Y. Measurement of a time-weighted average concentration of polycyclic aromatic hydrocarbons in aquatic environment using solid phase extraction cartridges and a portable pump. *Bull. Environ. Contam. Toxicol.* **1997**, *58*, 879-884.
24. Axelman, J.; Næs, K.; Näf, C.; Broman, D. Accumulation of polycyclic aromatic hydrocarbons in semipermeable membrane devices and caged mussels (*Mytilus edulis* L.) in relation to water column phase distribution. *Environ. Tox. Chem.* **1999**, *18.11*, 2454-2461.
25. Reddy, C. M.; Quinn, J. G. GS-MS analysis of total petroleum hydrocarbons and polycyclic aromatic hydrocarbons in seawater samples after the *North Cape* oil spill. *Mar. Poll. Bul.* **1999**, *38.2*, 126-135.
26. Utvik, T. I. R.; Durell, G. S.; Johnsen, S. Determining produced water originating polycyclic aromatic hydrocarbons in North Sea waters: comparison of sampling techniques. *Mar. Poll. Bul.* **1999**, *38.11*, 977-989.
27. Zhou, J. L.; Hong, H; Zhang, Z.; Maskaoui, K.; Chen, W. Multi-phase distribution of organic micropollutants in Xiamen Harbor, China. *Water Res.* **2000**, *34.7*, 2132-2150.

Chapter 16

Highly Sensitive Assay for Anticholinesterase Compounds Using 96 Well Plate Format

Nirankar N. Mishra[1], Joel A. Pedersen[2,3], and Kim R. Rogers[1,*]

[1]U.S. Environmental Protection Agency, National Exposure
Research Laboratory, Las Vegas, NV 89119
[2]U.S. Environmental Protection Agency, Region 9,
San Francisco, CA 94105–3901
[3]Environmental Science and Engineering Program,
University of California, Los Angeles, CA 90095

The rapid and sensitive detection of organophosphorus insecticides using a 96 well plate format is reported. Several features of this assay make it attractive for development as a laboratory-based or field screening assay. Acetylcholinesterase (AChE) was stabilized in a gelatin film. The remarkable properties of the dry immobilized AChE preparation include its stability to prolonged storage at room temperature as well as its stability to short term elevated temperatures ($60°C$). The enzyme could be maintained in dry gel form for 365 days at room temperature without substantial loss of activity. The absorbance assay used to measure enzyme activity was evaluated using several solvent systems including water, phosphate buffer, hexane, methanol and ethanol. The microwell assay includes a procedure to oxidize less potent P=S organophosphorus compounds to their more inhibitory oxon forms. The use of this assay to analyze field samples contaminated with mixtures of organophosphorus insecticides is also reported.

One means of reducing uncertainties in human exposure assessment is to better characterize concentrations of hazardous compounds present in the immediate environment of receptor populations. A significant limitation to this approach, however, is that sampling and laboratory analysis of contaminated environmental and biological samples can be slow and expensive; thus, limiting the number of samples that can be analyzed within time and budget constraints. Faster, simpler, and more cost-effective field screening methods can increase the amount of information available concerning the location, source and concentration of pollutants present in the environment (1). Among the compounds of interest for human exposure assessment are pesticides (2), particularly insecticides from the organophosphorus (OP) and carbamate classes which are widely used in agricultural and residential settings. Due to the relatively high toxicity of some of these compounds, a significant number of poisonings occurs each year (3). In addition, these compounds pose a hazard not only to the primary user, but in some cases to emergency healthcare workers as well (4).

The toxic effects of OP and carbamate insecticides are mediated primarily through disruption of cholinergic neurotransmission by inhibition of AChE (5). In addition to their acute toxicity due to inhibition of AChE, these compounds have also been implicated with long term neurological problems such delayed neurotoxic effect (6). OP and carbamate insecticides vary considerably in their overall toxic behavior due to many factors, including absorption, metabolism and interaction of parent compounds and transformation products with various target and non-target proteins. Metabolic activation is particularly important for the phosphorothioates, a subclass of OP compounds characterized by sulfur (P=S) attached to the central phosphorous atom. For this class of insecticides, the parent compound typically shows little anticholinesterate activity whereas the oxidative transformation product (the oxon, P=O) is often highly potent (5).

Although a variety of chromatographic methods have been applied for detection of OP and carbamate insecticides, these techniques are typically expensive and time-consuming. A wide variety of bioanalytical and biosensor methods based on AChE inhibition have also been reported over the past decade (7). Although these methods show considerable promise, they are not well suited for screening large numbers of environmental samples. These assays are relatively simple and particularly sensitive to specific compounds or their transformation products, however, a number of problems hinder their widespread adoption and use. These challenges include such issues as high throughput formats, long-term stabilization of AChE, requirement for organic extraction solvents, oxidative activation of parent compounds and the ability to derive useful information for samples containing mixtures of OP and carbamate compounds. Although significant progress has been reported in such areas as stabilization of immobilized enzymes (8), oxidative activation of parent compounds (9), high throughput formats (*10*), and the use of environmental matrices (*11*), the integration of these concepts and the demonstration of a simple assay format relevant to environmental samples is still of considerable importance to progress in this area. We present here a simple, sensitive, versatile and inexpensive assay format for detection of OP and carbamate insecticides. In addition, we suggest the use of paraoxon equivalents for screening environmental samples contaminated with mixtures of these insecticides.

Experimental Methods

Materials Acetylcholinesterase from an electric eel, (EC 3.1.1.7, 1000U/mg), acetylthiocholine chloride, 5,5'-dithiobis (2-nitrobenzoic acid), D (+) trehalose dihydrate, D (+) glucose, pyridostigmine bromide, and neostigmine bromide were from Sigma Chemical/Aldrich. Aldicarb, carbaryl, carbofuran, chlorpyrifos, chlorpyrifos-oxon, dichlorvos, methomyl, malathion, malaoxon, naled, paraoxon, parathion, trichlorfon, azinphos-methyl, diclofenthion, dimethoate, dimethoate-oxon, terbufos, phosmet, and fenthion were obtained from Chem Service Corp. All others chemicals used were reagent grade. Deionized water (DI) was used for the preparation of all solutions.

Enzyme Immobilization and OP and Carbamate Assay Protocol AChE was dissolved in a solution containing 5% D-(+)-trehalose dihydrate, 5% D-(+)-glucose, 0.1% of gelatin, 1% sodium chloride and 0.002% sodium azide (TGG) and distributed into individual wells of the microtiter plate. The enzyme was dried under a stream of air for 24 hrs at 25°C, after which it was ready for use in the OP inhibition assay. Immediately prior to the assay, the AChE was dissolved in phosphate buffered saline (PBS) solution containing 10 mM sodium phosphate, 100 mM NaCl, pH 7.4. AChE activity was measured using the Ellman method (*12*). The reaction medium contained 75 µL of 1 mM acetylthiocholine chloride, 75 µL of 1 mM 5,5'-dithiobis (2-nitrobenzoic acid), 25 µL of AChE (2.86 ng) and 25 µL of DI water at pH 7.4. For the inhibition assay without prior oxidation of the inhibitors, 25 µL of AChE was incubated for 20 min with 25 µL of the inhibitor at concentrations ranging from 100 µM - 1 pM. For the assay variation that employed oxidation of pesticides containing P=S groups, the inhibitors (25 µL) were first incubated with 5 µL of 0.001% Br_2 solution for 20 min followed by the addition of ethanol to a final concentration of 5% prior to the incubation with AChE.

Assay Procedure for Samples Surface runoff samples collected from four agricultural fields were extracted and concentrated by EPA method 3510C. Solvent extracts were stored at 4°C prior to analysis by gas chromatography and use in the assay. Sample extracts were analyzed on a Hewlett-Packard 6890 Series capillary gas chromatograph equipped with a flame photometric detector following USEPA Method 1657. Prior to use in the assay, samples were solvent exchanged into hexane, taken to dryness under a stream of N_2 then dissolved in the same volume of DI water. Dilutions of the samples were then analyzed using the bromine oxidation variation of the AChE inhibition assay.

Data Reduction and Analysis For inhibition profiles, the means of triplicate data points (at individual inhibitor concentrations) were plotted. IC_{50} values were determined using either a log-logit or four parameter fit. (Correlation coefficients were determined from the best fit of these data.) Error bars representing standard deviations (SD) are only presented in selected inhibition profiles. Paraoxon equivalence (%), PE, was determined using the following relationship:

$$PE = (IC_{50}\ paraoxon/IC_{50}\ compared\ compound)\ x\ 100$$

Results and Discussion

Enzyme Stability Drawbacks for the use of enzyme-based assays in environmental monitoring include storage requirements (e.g., typically below 4°C) and limited shelf life. We report that the use of trehalose, glucose, gelatin, sodium chloride and sodium azide (TGG) for dried AChE preparations dramatically stabilizes activity of this enzyme (Table I).

Table I. Stabilization of Acetylcholinesterase

Storage Media	Storage Temp	Storage Time	Activity (%)
DI Water	25°C	10 min	100
DI Water	25°C	3 hr	0
PBS	25°C	10 min	100
PBS	25°C	3 hr	0
TGG (soln)	25°C	3 days	75
TGG (dried)	25°C	15days	100
TGG (dried)	25°C	60 days	100
TGG (dried)	25°C	365 days	100
TGG (dried)	60°C	10 min	100

The enzyme activity rapidly degraded in water or PBS solution. Although TGG allowed the maintenance of activity in solution for several days, AChE dried in this mixture remained active for extended periods of time (i.e., 1 year) at 25°C or for short periods at 60°C. Although the enzyme immobilization method and stabilization results described here are similar to those previously reported by Nguyen et al. (8), the demonstration of preservation of AChE activity over extended storage periods at 25°C and at high temperature are unique to our study.

Another hindrance in the use of enzymes in environmental assays is that many insecticides display limited solubility in aqueous buffers. Consequently, we explored the use of various organic solvents with the AChE assay (Table II). We observed that aqueous systems and nonpolar solvents such as hexane do not effect enzyme activity whereas the use of polar organic solvents such as methanol and ethanol significantly degrade the enzyme activity.

Table II. Effect of Organic Solvents on AChE Activity

Solvents	Activity[1] (% control)
DI Water	100
PBS	100
Hexane	100
Ethanol	12
Methanol	10
5% Ethanol[2]	98
5% Methanol[2]	100
TGG	100

[1] Activity after 20 min incubation

[2] Aqueous solution

A 5% aqueous solution of these alcohols, however, could be used to increase the solubility of pesticides without significantly decreasing enzyme activity. The effect of organic solvents on the activity of enzymes such as tyrosinase and AChE has been well characterized (13, 14). In general, nonpolar organic solvents such as hexane do not inhibit tyrosinase, but polar organic solvents such as alcohols can cause significant inhibition. Further, Mionette et al. (15) observed effects of organic solvents on acetylcholinesterase assays similar to those we report here.

Inhibition Profiles Inhibition curves were determined for a number of anticholinesterase compounds over micromolar (μM) to picomolar (pM) concentration ranges. Compounds analyzed using this assay included the strongly inhibitory carbamates physostigmine, pyridostigmine and neostigmine (Figure 1) and moderately inhibitory carbamates such as aldicarb, carbofuran, carbaryl and methomyl (Figure 2). The inhibition curves for physostigmine and pyridostigmine showed a typical sigmoidal shape. AChE activity in the presence of neostigmine, however, did not drop below 20% even at the highest concentrations. The relative order of inhibitory potency was neostigmine › physostigmine › pyridostigmine. The rank order of these inhibitors for AChE is the same as previously reported (*17*). Curves for the moderately inhibitory carbamates again showed the characteristic sigmoidal shape with the following inhibitory potencies carbofuran ≈ carbaryl › aldicarb › methomyl (Figure 2). Again, the inhibition profiles were similar in shape and position (except for methomyl) to previous reports using colorimetric or biosensor (*16*) assays.

Inhibition profiles were determined for phosphorothioate OP insecticides such as parathion, malathion, and diazinon (Figure 3). Because these compounds were only weakly inhibitory, the measured concentration range extended from 0.1 nM to 100 μM. The relative order of potency was malathion › diazinon › parathion. The commercially available oxidative transformation products of parathion and malathion (i.e., paraoxon and malaoxon) as well as dichlorvos, were also measured using this assay (Figure 4). The oxidative transformation products were significantly more potent AChE inhibitors than the parent compounds and showed inhibitory profiles comparable to dichlorvos. The cholinesterase inhibition assay yielded similar IC_{50} values for each of these compounds. Indeed, these compounds are typically reported to have inhibition constants within an order of magnitude of each other (*16, 17*).

IC_{50} Values Because of its stability, strong inhibitory effect and commercial availability, paraoxon has often been used as a reference compound for cholinesterase inhibition (*7*). As a result of the widespread use of paraoxon as a reference inhibitor, we elected to compare the relative potency of compounds assayed to this inhibitor in the form of paraoxon equivalence. IC_{50} values and % inhibition relative to paraoxon (paraoxon equivalence) were determined for a wide range of cholinesterase inhibitors (Table III). The extent to which the data fit a four parameter curve fit or log-logit fit are also included as correlation coefficients. The compounds are placed in ascending order of calculated IC_{50} values.

A wide range of bioanalytical assays based on cholinesterase inhibition have been reported over the past decade. IC_{10} values (molar concentration yielding 10% inhibition of the control activity and typically considered as the detection limit) that have been reported for paraoxon using AChE -based inhibition assays vary over the

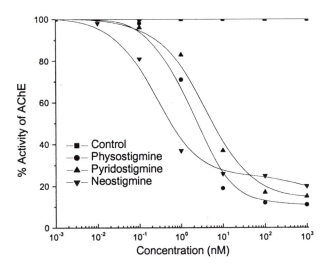

Figure 1. Inhibition profiles for strongly inhibitory carbamates. Data points are means, n = 3.

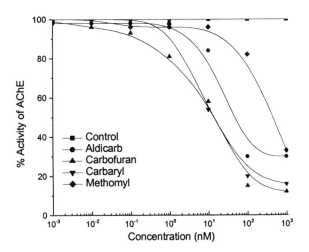

Figure 2. Inhibition profiles for moderately inhibitory carbamates. Data points are means, n = 3.

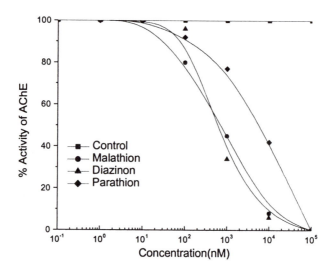

Figure 3. Inhibition profiles for phosphorothioate insecticides. Data points are means, n = 3.

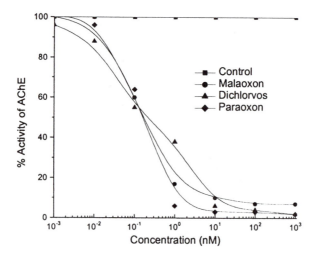

Figure 4. Inhibition profiles for oxon-containing organophosphorus insecticides. Data points are means, n = 3.

following range: 90 nM (*20*), 10 nM (*16, 21*), 0.3 nM (*22*), and 0.1 nM (*11*). The IC_{50} value for paraoxon (0.15 nM) reported here is similar to the lowest IC_{10} values reported using biosensors (*7*), flow injection analysis (*23*), or soluble enzyme assays (*16*).

The rank order for these compounds is similar but not identical to those previously reported in the literature. For example, chlorpyrifos-oxon has been reported to be a more potent inhibitor of cholinesterase than paraoxon (*18, 19*) and methomyl has been reported to be a more potent inhibitor than aldicarb (*16*). A number of factors may account for differences that we observe. The magnitude of difference between paraoxon and chlorpyrifos-oxon IC_{50} values has been shown to be dependent on both the species and tissue from which the cholinesterase is obtained. In addition, assay factors such as contact time (for the enzyme and inhibitor) as well as the removal of excess inhibitor prior to the addition of substrate are critical factors which differ among reported studies. Nevertheless, taken in context of the screening application we propose for this assay, differences in the rank order for various insecticides between our assay and various reports would likely not be a critical issue.

Oxidation of Phosphorothioates The oxidative transformation products of many organophosphorus insecticides are significantly more potent inhibitors of AChE than their parent compounds. In order to make this assay more sensitive with respect to the potential use for screening environmental samples, the phosphorothioate OP insecticides (P=S) were converted to their oxon (P=O) derivatives. We found pretreatment of phosphorothioates by exposure to bromine resulted in relatively rapid and efficient conversion to the oxonate and did not inhibit enzyme activity. More specifically, exposure of samples to 0.001% Br_2 followed by 5% ethanol (to inactivate the bromine) significantly increased the inhibitory potential of parent phosphorothioate OP compounds with no measurable damage to enzyme activity.

Inhibition profiles for chlorpyrifos (parent compound), chlorpyrifos-oxon (commercial standard) and bromine oxidized chlorpyrifos are shown in Figure 5. Although bromine oxidation of the parent compound did not appear to be complete, it facilitated lowering the IC_{50} value for chlorpyrifos into the nM range. The bromine oxidation step for chlorpyrifos yielded an IC_{50} value approximately eight times higher than for the commercially available chlorpyrifos-oxon, but 270 times lower than for the parent compound. The inhibition curves for chlorpyrifos-oxon and bromine oxidized chlorpyrifos display the typical sigmoidal shape above 0.1 nM. Below this concentration, however, they appear biphasic in that AChE is inhibited (to some extent) even at low inhibitor concentrations. Standard deviation error bars were included in this figure to better clarify the anomalies of these curves.

The inhibition profiles for dimethoate (parent compound), dimethoate-oxon and the bromine oxidized parent compound are shown in Figure 6. In this case, the inhibition profiles for the commercially available dimethoate-oxon and bromine oxidized parent compound were similar, suggesting that the bromine oxidation was nearly complete. It is interesting to note that dimethoate-oxon did not show the sigmoidal shape characteristic of most compounds but rather showed an almost log-linear profile.

Table III. Inhibitory Characteristics of Selected Anticholinesterases

Compound	IC_{50} (nM)	Paraoxon Equivalence (%)	Correlation Coefficien
Naled	0.019	789	0.800
Malaoxon	0.093	161	0.989
Paraoxon	0.15	100	0.999
Neostigmine	0.25	60	0.999
Dichlorvos	0.26	58	0.965
Physostigmine	1.8	8	0.998
Chlorpyrifos-oxon	3.2	4.7	0.995
Pyridostigmine	3.9	4	0.999
Carbaryl	8.7	1.7	0.993
Carbofuran	9.7	1.5	0.968
Aldicarb	19.2	0.8	0.991
Methomyl	162	0.1	0.974
Dimethoate-oxon	455	0.02	0.978
Malathion	458	0.03	0.981
Parathion	689	0.02	0.873
Diazinon	760	0.02	0.969
Fenthion	6,240	ND	0.999
Chlorpyrifos	6,740	ND	0.998
Trichlorfon	7,780	ND	0.999
Terbufos	7,860	ND	0.986
Phosmet	8,120	ND	0.987
Azinphos-methyl	9,230	ND	0.979
Diclofenthion	9,890	ND	0.816
Dimethoate	11,900	ND	0.969

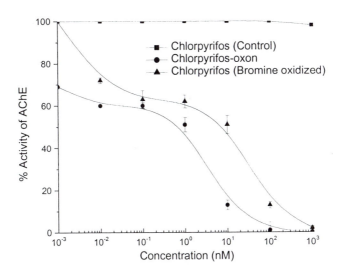

Figure 5. Inhibition profiles for chlorpyrifos and its oxidative transformation products. Data points are means, n = 3.

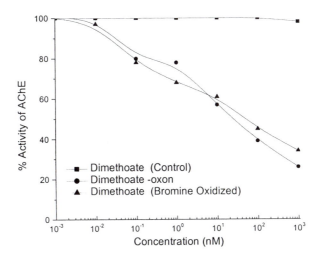

Figure 6. Inhibition profiles for dimethoate and its oxidative transformation products. Data points are means, n = 3.

IC$_{50}$ Values for Oxidized OP Compounds Table IV shows IC$_{50}$ values and paraoxon equivalence values for selected phosphorothioate OP insecticides that were assayed with and without prior bromine oxidation. In addition, where commercially available, the oxon derivatives are also compared. The bromine oxidation protocol significantly increased the sensitivity of this assay to many of these insecticides. In about half of the cases the IC$_{50}$ values were decreased between 20 and 300 times, however, several of the compounds showed only

Table IV. Oxidation and Assay of Selected Phosphorothioate Insecticides

Compound	IC$_{50}$ (nM)	Paraoxon Equivalence (%)
Parathion	689	0.02
Parathion*	11.9	1.3
Paraoxon	0.15	100
Chlorpyrifos	6,740	ND
Chlorpyrifos*	25.1	0.6
Chlorpyrifos-oxon	3.2	5
Malathion	458	0.03
Malathion*	22.1	0.7
Malaoxon	0.093	161
Dimethoate	11,900	ND
Dimethoate*	685	0.02
Dimethoate-oxon	455	0.02
Diazinon	760	0.02
Diazinon*	140	0.10
Terbufos	7,860	ND
Terbufos*	2,286	ND
Fenthion	6,240	ND
Fenthion*	2,300	ND
Trichlorfon	7,780	ND
Trichlorfon*	119	ND
Phosmet	8,120	ND
Phosmet*	8,120	ND
Azinphos-methyl	9,230	ND
Azinphos-methyl*	1,120	ND
Diclofenthion	9,890	ND
Diclofenthion*	1,890	ND

* Parent compound assayed using the bromine oxidation protocol

moderate or no change due to oxidation (e.g., fenthion, phosmet). In the cases of parathion, chlorpyrifos, malathion and dimethoate, the use of commercially available oxon derivatives allowed the estimation of the relative efficiency of the bromine oxidation step.

These results show that differences in the inhibitory behavior for phosphorothioate insecticides and their oxon derivatives vary over a broad range. More specifically, the IC_{50} for malathion and its oxidative transformation product differ by almost 5000 times, whereas in the case of dimethoate, there is only a 20 fold increase in sensitivity between the parent compound and oxon derivative. Consequently, these results indicate that the efficiency of the oxidation was not the sole factor influencing the magnitude of the observed increase in sensitivity for the bromine oxidation step. The kinetics and products of bromine oxidation of phosphorothioate insecticides warrants further evaluation.

Mixtures of OP Compounds In many cases of environmental contamination with OP pesticides where a rapid screening assay would be of value, samples may be expected to contain multiple pesticides. These compounds may be present as parent compounds as well as various oxidative transformation products. Data from Table IV suggest that for mixtures of OP and carbamate insecticides, the most potent AChE inhibitors would dominate the results of the inhibition assay.

In an ideal circumstance, the inhibition curves for the compounds of interest would be identically shaped and placed along the abscissa (log concentration scale) at various positions depending on their relative IC_{50} values. Although the inhibition profiles for different compounds are often similar in shape, we have observed that a number of these curves do not show characteristic sigmoidal shape or similar slopes at their IC_{50} concentrations. Consequently, we investigated the inhibition profile for a mixture of five commonly used insecticides.

A mixture of equimolar concentrations of chlorpyrifos, diazinon, dichlorvos, malathion and parathion (which sum to the plotted concentrations) was analyzed by this assay using the bromine oxidation protocol. Shown in Figure 7 are the inhibition profiles for paraoxon and the previously described mixture. Over a limited concentration range (i.e., 0.1 nM and above), the inhibition profile of the mixture is similar in shape to the paraoxon curve with paraoxon showing greater inhibition than the mixture. At concentrations below 0.1 nM, the curve deviates from its typical shape and the mixture appears to inhibit to a greater extent than paraoxon. This feature in the inhibition profile (i.e., the plateau in the activity at about 80% of maximum at low compound concentrations) was also observed with chlorpyrifos-oxon (see Figure 5).

Environmental Samples Dichloromethane extracts from agricultural runoff samples which were contaminated with mixtures of up to six OP insecticides were analyzed using the assay. Surface runoff water was extracted and contaminants concentrated using EPA Method 3510C. After solvent exchange into hexane, extracts were evaporated, dissolved in DI water, and inhibition profiles obtained using the bromine oxidation protocol after serial dilution (Figure 8). Sample 1110N was the most potent inhibitor, yielding 50% inhibition at a 300-fold dilution followed by sample 1030N at a 20-fold dilution. Samples 930N and onion field behaved nearly the same as each other, yielding 50% inhibition at a 3-fold dilution and sample 945N showed 50% inhibition when undiluted.

Analysis of the dichloromethane extracts by GC-FPD yielded the concentrations of chlorpyrifos, diazinon, dichlorvos, dimethoate, malathion and parathion (methyl) shown in Table V. Each of the samples were contaminated with at least two compounds and two of the samples (1110N and onion field) were contaminated with five compounds. One might anticipate that the most potent inhibitors would contribute the majority of the observed response. In the case of these samples the rank order for IC_{50} values taken from Tables III and IV for the bromine oxidized compounds is dichlorvos › parathion (ethyl) › malathion › chlorpyrifos › diazinon › dimethoate. Viewed from this perspective, for samples 1110N, 1030N and onion field, the dichlorvos concentrations would be expected to determine the observed inhibition. In the case of sample 1110N the observed assay response underestimated the dichlorvos concentration by a factor of five. In the cases of samples 1030N and onion field, the assay underestimated the inhibition expected from dichlorvos by a factor of 100. For sample 930N, chlorpyrifos would be expected to dominate the inhibition of AChE. The observed assay response, in this case, underestimated the contribution from chlorpyrifos by a factor of three. For sample 945N, again chlorpyrifos would be expected to determine the inhibition and, in this case, the assay underestimated the response from chlorpyrifos present in the sample by a factor of five. In summary, the AChE assay reported here underestimated (to various extents based on inhibition profiles in laboratory buffers) the concentrations of OP compounds present in environmental samples that were expected to result in AChE inhibition. Factors that may have contributed to this result include incomplete recovery of the compounds (into the aqueous phase) from the solvent extract, the association of OP insecticides with other organic molecules in the extracts preventing inhibition, or the "bromine demand" exerted by other compounds in the sample matrix reducing oxidation efficiency. The role of sample matrix effects on assay response requires further evaluation. Nevertheless, as a screening tool the AChE inhibition assay (rapidly and inexpensively) identified anticholinesterase activity in all of the extracts from contaminated environmental runoff samples.

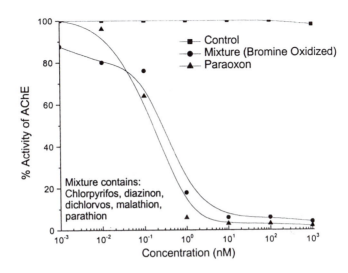

Figure 7. Inhibition profiles for paraoxon and organophosphorus insecticide mixture. Data points are means, n = 3

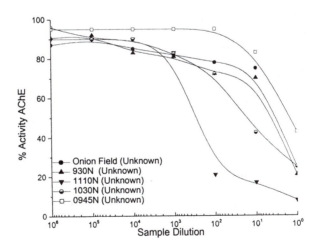

Figure 8. Inhibition profiles for dilutions of environmental samples contaminated with anticholinesterase compounds. Data points are means, n = 3.

Table V. GC-FPD Analysis of Runoff Water Extracts

Sample	Chlorpyrifos (nM)	Diazinon (nM)	Dichlorvos (nM)	Dimethoate (nM)	Malathion (nM)	Parathion (nM)
O.Field	142	605	443	0.86	224	0
930N	262	434	0	2.70	0	0
1110N	185	430	416	3.20	303	0
1030N	0	0	489	0.56	206	0
0945N	137	0	0	0	0	24

Conclusions

This assay is not intended to yield the identities or exact concentrations of the carbamate and OP insecticides present, but rather the relative anticholinesterase activity. More specifically, this screening assay is intended to flag samples which, from a potential risk perspective, warrant further examination for specific carbamate and organophosphorus insecticide contamination. In this respect, the assay performed exceptionally well using standards in laboratory buffers. Although the assay underestimated the inhibition expected from laboratory standards, the AChE-based assay detected anticholinesterase activity present in all of the environmental samples.

Due to its high sensitivity and simplicity in the detection of carbamate and organophosphorus insecticides such as those commonly used in agricultural and residential settings, this assay is a promising candidate for further development as a laboratory or field assay for screening of environmental samples related to human exposure assessment.

Notice: The U.S. Environmental Protection Agency (EPA), through its Office of Research and Development (ORD), funded this research and approved this manuscript for publication. Mention of trade names or commercial products does not constitute endorsement or recommendation of these products by the EPA. NNM is currently a National Research Council Fellow. Environmental samples were collected and analyzed under a cooperative agreement with I.H. (Mel) Suffet at the University of California Los Angeles.

References

1. Rogers, K.R.; Williams, L.R. *Trends Anal. Chem.* **1995**, *14*, 289-294.
2. U.S. Environmental Protection Agency. *Science policy on the use of data on cholinesterase inhibition.* OPP, U.S. EPA: Washington, DC, 1998.
3. National Institute for Occupational Safety and Health. *Worker Health Chartbook*, Publication 200-127, National Institute for Occupational Safety and Health: Cincinnati, OH, 2000.
4. Nosocomial Poisoning Associated with Emergency Department Treatment of Organophosphate Toxicity - Georgia, 2000. *Morbidity Mortality Weekly Rep.* **2001**, *49*, 1156-1158.
5. O'Brien, R. *Toxic Phosphorus Esters*, Academic Press: New York, 1960.
6. Abou-Donia, M.B.; Wilmarth, K.R.; Jensen, K.F.; Oehme, F.W.; Kurt, T.L. *J. Toxicol. Environ. Health* **1996**, *48*, 35-56.
7. Noguer, T.; Leca, B.; Marty, J-L. In *Biosensors for Environmental Monitoring*, Bilitewski, U.; Turner, A.P.F. Eds.; Harwood Academic Publishers: Canada, 2000.
8. Nguyen, V.K.; Ehret-Sabatier, L.; Goeldner, M.; Boudier, C.; Jamet, G.; Warter, J.M.; Poidron, P. *Enzyme Microbial Technol.* **1996**, *329*, 297-304.
9. Kim, Y.A.; Lee, A.S.; Park, Y.C.; Lee, Y.T. *Environ. Res. Sec. A* **2000**, *84*, 303-309.
10. Lui, J.; Tan, M.; Liang, C.; Yin, K.B. *Anal. Chim. Acta* **1996**, *329*, 297-304.
11. Marty, J-L.; Mionetto, N.; Lacorte, S.; Barcelo, D. *Anal. Chim. Acta* **1995**, *311*, 265-271.
12. Ellman, G.L.; Courtney, K.D.; Andres, V., Jr.; Featherstone, R.M. *Biochem. Pharmacol.* **1961**, *7*, 88-95.
13. Wang, J.; Dempsey, E.; Eremenko, A.; Smyth, M.R. *Anal. Chim. Acta* **1993**, *279*, 203-208.
14. Adeyoju, O.; Iwuoha, E.I.; Smyth, M.R. *Talanta* **1994**, *41*, 1603-1608.
15. Mionetto, N.; Marty, J.-L.; Karube, I. *Biosens. Bioelectr.* **1994**, *9*, 463-470.
16. Fernando, J.C.; Rogers, K.R.; Anis, N.A.; Valdes, J.J.; Thompson, R.G.; Eldefrawi, A.T.; Eldefrawi, M.E. *J. Agric. Food Chem.* **1993**, *41*, 511-516.
17. Main, A.R. In *Biology of Cholinergic Function*, Goldberg, A. M.; Hanin, I., Eds.; Raven Press: New York, 1976.
18. Carr, R.L.; Chambers, J.E. *Toxicol. Appl. Pharmacol.* **1996**, *139*, 365-373.
19. Amitai, G.; Moorad, D.; Adani, R.; Doctor, B.P. *Biochem. Pharmacol.* **1998**, *56*, 293-299.
20. Kumaran, S.; Tran-Minh, C. *Anal. Biochem.* **1992**, *200*, 187-194.
21. Marty, J.L.; Sode, K.; Karube, I. *Electroanal.* **1992**, *4*, 249-252.
22. Skládal, P. *Anal. Chim. Acta* **1992**, *269*, 281-287.
23. Gunther, A.; Bilitewski, U. *Anal. Chim. Acta* **1995**, *300*, 117-125.

Chapter 17

Temporal and Spatial Variation in Monitoring of Ambient Urban Air Pollutants

Margaret L. Phillips, Nurtan A. Esmen, Daping Wang, and Thomas A. Hall

Department of Occupational and Environmental Health, University of Oklahoma Health Sciences Center, Oklahoma City, OK 73104

Air pollution data from monitoring stations are widely used as a surrogate for human exposure. Automated continuous monitors provide greater temporal resolution than cumulative samplers, but cost and practical constraints generally limit their use to small numbers of sites in any urban area. Spatial and temporal variation of air pollution and the representativeness of urban monitoring stations have been investigated on distance scales ranging from street segments to citywide. Review of the research suggests that spatial variability may be scale-invariant and may sometimes modulate characteristic temporal patterns.

Governmental agencies in many countries have created networks of fixed point monitoring stations to record airborne concentrations of major environmental air contaminants, such as carbon monoxide (CO), sulfur dioxide (SO_2), ozone (O_3), nitric oxide (NO), nitrogen dioxide (NO_2), mixed oxides of nitrogen (NO_x), lead (Pb), and particulate matter. The pollutant concentrations measured at monitoring stations located in urban areas have been used as surrogate measures of human exposure. For example, monitoring sites which are set up primarily for the purpose of determining compliance with air quality standards are often selected to be "representative" of population exposure on an urban or local scale. Another important use of these surrogate measures has been in epidemiological studies in which conclusions about health effects of air pollutants have been based on apparent relationships between concentration time series and health effect incidence time series (*1*). Because health effects such as death or disease episodes are manifested on the level of the individual member of the population, and personal exposure is usually dominated by indoor sources (*2*), the relationship between exposure and effect should ideally be modeled using measurements of personal exposure to airborne contaminants. However, measurement of personal exposure is a practical possibility only in research studies involving limited numbers of individuals for limited durations. Fixed-point monitoring data are also used in the development and validation of mathematical models for predicting air pollution concentrations.

Air pollutant concentrations vary in time and space under the influence of many factors, which are summarized in Table I. Discussion of these factors may be found in basic texts on air pollution (*3*). Potential determinants of ambient exposure (such as meteorological factors, traffic volume, and industrial emissions) and suspected health effects also show spatial and temporal variation.

Table I. Factors influencing air pollutant levels

Generation of primary pollutants	
Process	Examples of pollutants emitted
combustion	NO, NO_2, CO, soot, SO_2
vaporization	hydrocarbons
mechanical comminution	coarse particles, e.g. dusts, sea spray
Generation of secondary pollutants	
Process	Examples of pollutants generated
tropospheric chemical reactions (mainly photodissociations and oxidations)	O_3, OH radical, NO_2, aldehydes, peroxyacetyl nitrate (PAN), H_2SO_4
condensation and coagulation	ultrafine and fine particles ($PM_{2.5}$)
Transport	
diffusion	
wind flow	
turbulence	
mixing height	
Removal processes	
chemical reactions	
wet deposition	
dry deposition	

The existence of variations makes possible the use of powerful statistical methods for identifying apparent relationships – possibly causal – among air pollution data, determinants of exposure and health effects. On the other hand, variation presents a challenge in the design of monitoring strategies; the number and location of monitoring sites and the frequency and duration of sampling events must be selected carefully to yield data that will be representative of pollutant concentrations on the distance and time scales of interest.

Temporal and Spatial Monitoring Strategies

Air pollution sampling methods may be classified in three categories based on the temporal patterns of the sampling process: automated continuous sampling, cumulative sampling, and grab sampling.

Automated monitoring devices sample the air on a continuous or repetitive short-term batch basis, analyze the pollutant level *in situ*, and record the results sequentially, thus providing detailed information on temporal variation. The U.S. EPA reference methods for determining compliance with National Ambient Air Quality Standards (NAAQS) include automated methods for CO, O_3, and NO_2 (*4*). Automated methods are also widely used for SO_2 and particulate matter (*5*). Typically particulate matter is sampled using size-selective sampling methods to collect the particulate fraction having aerodynamic diameter less than 10 μm (PM_{10}) or less than 2.5 μm ($PM_{2.5}$). Volatile organic compounds (VOCs) and peroxyacetyl nitrate (PAN) have been measured with hourly resolution using automated gas chromatographs, and airborne formaldehyde concentrations have been determined with half-hour resolution by automated derivatization with fluorescence detection (*6*).

For many automated monitoring devices the sensitivity to temporal variation comes at the price of portability: they are used principally in fixed-point monitoring applications because they are expensive to purchase, large, heavy, and require a temperature-controlled housing. However, a number of researchers have overcome these constraints to some extent by using mobile air monitoring laboratories to sample spatial as well as temporal variation (*7, 8, 9*).

Relatively inexpensive portable datalogging devices with electrochemical sensors have been used to measure spatial and temporal variation in CO levels (*10*). Though low-cost datalogging electrochemical monitors are also available commercially for other pollutants, including NO and NO_2, these devices may not be sensitive enough to quantify concentrations below 100-500 parts per billion (ppb), limiting their usefulness for ambient urban air monitoring. Portable fine particulate counters have been applied to personal exposure monitoring studies (*11*), and could be used to study spatial-temporal variation.

Cumulative sampling is the collection of air contaminants over a defined sampling period, which may be on the order of hours or days, for subsequent laboratory analysis. In the sampling process, the contaminants may be separated from the air and captured on a suitable filter or sorbent medium, or an entire volume of air may be collected in an evacuated container. Sampling may be active, air being drawn through the collection medium by a calibrated pump, or passive, relying on diffusion to transport the contaminant into and through the collection medium.

Quantification of the mass of contaminant collected by cumulative sampling yields a measure of the ambient concentration at the sampling point, time-averaged over the sampling period. Therefore, cumulative sampling methods are insensitive to variation on time scales shorter than the sampling period. The minimum sampling period sufficient to allow quantification of ambient contaminant concentrations in all samples collected during a sampling campaign depends upon the anticipated lowest average ambient concentration, the limit of detection of the analytical method, and the rate of transport (active or passive) of the contaminant onto the collection medium. For example, passive sampling tubes for measurement of NO_2 have been used with sampling durations ranging from three days (7) to two weeks (12, 13) to ensure sufficient sensitivity and precision in the measurement of ambient NO_2 levels.

Cumulative samplers can be used to greatest advantage in determination of spatial variations in pollutant concentration; their low cost and small size relative to automated air monitoring devices makes it practical to deploy samplers at multiple locations simultaneously throughout a study area. Cumulative sampling is also the only feasible method currently available to make quantitative measurements of personal exposure to most air contaminants at typical ambient concentrations in the general urban environment.

Grab sampling is the collection of short-term samples (on the order of minutes) to provide a "snapshot" of air contaminant concentrations at a particular time and place. Grab sampling may be used to assess the influence of short-term events, such as a contaminant release episode or an unusual meteorological condition, on contaminant concentrations.

Siting Criteria for Monitoring Stations

Air monitoring data generated by public authorities are a readily accessible and potentially valuable resource for researchers. However, an understanding of the criteria for siting monitoring stations is necessary for appropriate interpretation of these data. Two alternative approaches to siting are (1) to locate monitoring sites in a regular grid pattern or (2) to select monitoring sites that are believed to be representative of important features of the concentration profile of an urban area, e.g. "maximum", "background", "residential", etc. (*14*)

Air quality standards developed by the U.S. EPA under the Clean Air Act of 1970 mandated the creation of air monitoring networks, the State and Local Air Monitoring Stations (SLAMS). The design objectives for SLAMS networks with respect to urban air quality are to determine (1) "highest concentrations expected to occur in the area covered by the network"; (2) "representative concentrations in areas of high population density"; (3) "the impact on ambient pollution levels of significant sources or source categories"; (4) "general background concentration levels." (*15*). The concentration of an air pollutant monitored at a site that was selected to meet one of these four objectives is considered to be "representative" on a spatial scale over which concentrations are reasonably uniform. EPA defines four spatial scales relevant to urban air monitoring: *microscale* refers to areas with dimensions up to about 100 meters, e.g. about the size of a city block; *middle scale* refers to areas with dimensions of 100 to 500 meters, e.g. several city blocks; *neighborhood scale* refers to areas on the scale of 0.5 to 4.0 kilometers with fairly uniform land use; and *urban scale* refers to the city as a whole, with dimensions of about 4 to 50 kilometers. (*15*). The scale represented by a given monitoring site depends upon the site objective and the characteristic spatial variation of the target contaminant. For example, a monitoring site selected to detect the highest concentration of CO might be located in an urban canyon, i.e. a street in which the height of buildings on both sides is similar to the width of the street. Concentrations measured at the midpoint between intersections (*16*) may be considered representative on the microscale, and possibly on the middle scale as well, if traffic intensity and land use is uniform over the length of several blocks.

Assessments of Spatial Variation in Urban Air Quality

Relative importance of spatial and temporal variation

Spicer *et al.* (*6*) collected 16 consecutive 3-hour cumulative samples of VOCs, trace elements, and particle-bound organics on three different occasions over a period of four weeks at each of six sampling locations in different sections of Columbus, Ohio. The sampling sites included a downtown commercial location and several residential areas with varying proximity to major roads, highways, and business areas. Analysis of variance performed on the sampling results indicated that variability between sampling locations was generally small relative to temporal variability for hydrocarbons. Low spatial and temporal variability and large spatial-temporal interaction were found for halogenated hydrocarbons and potassium, reflecting sporadic releases from local sources at some sampling sites.

In the Small Area Variations in Air Quality and Health (SAVIAH) study (*13*), 40 to 80 simultaneous sampling sites, intended to reflect regional background, urban background, and street level concentrations of NO_2, were set up in each of four European cities (Amsterdam, Netherlands; Huddersfield, U.K.; Poznan, Poland; Prague, Czech Republic). Two-week cumulative passive sampling was conducted in each city in four separate seasonal surveys. In each city, both spatial and seasonal variations were statistically significant, but spatial variation accounted for most of the overall variability among samples. The spatial coefficient of variation ranged from 22% in Amsterdam to 42% in Prague. Site concentrations tended to be highly correlated across seasonal surveys. However, the site-survey interaction was also statistically significant, indicating that seasonal variations were more pronounced in some sites. The difference between the Columbus study and the SAVIAH study in the relative importance of spatial variation probably reflects the different time scales used in the two studies and the more complex role of NO_2 in photochemical cycles.

Urban scale variation

Kuttler and Strassburger (*8*) took continuous measurements of NO, NO_2, and O_3 in a mobile laboratory during several 50 to 60 km long round-trip traverses of the city of Essen, Germany, spanning green areas, residential areas, secondary streets, main roads, and highways. Traverses were made between the daily rush hours, when traffic emissions were expected to be relatively stable. NO and NO_2 concentrations varied by as much as 20-fold between land use areas. O_3 concentration and the NO_2/NO ratio showed the opposite trend, reflecting the typical suppression of O_3 by NO in high traffic areas. Studies of NO_2 concentrations in Lancaster, U.K. (*12*) and Seattle-Bellevue, Washington (*9*) found a similar relationship between NO_2 concentration and proximity to major traffic routes. Despite significant differences between sampling sites in the Seattle-Bellevue study, in which four-month average concentrations ranged from 13 ppb to 26 ppb between sites, successive 3-week passive sampling measurements at the highest concentration site were highly correlated over time with measurements at all other sites.

McCurdy *et al.* (*17*) took 24-hour samples of acid aerosols at roughly 2-3 day intervals for three months at four locations in the Pittsburgh, Pennsylvania metropolitan area. Sulfate and ammonium ion concentrations were found to be similar at all sites, and their temporal patterns were also very similar. Strong acid, measured as H^+, was also temporally correlated across sites, but the concentration at an "upwind" suburban site was significantly higher (by a factor of 2) than at the city center, suggesting the importance of regional transport of acid aerosols combined with partial neutralization by local sources of ammonia.

To determine urban and regional scale representativeness of time-series data from urban monitoring stations, the temporal correlation between monitors was evaluated over seven adjacent states in the heavily industrialized north-central region of the United States (18). After statistical removal of seasonal effects and longer-term trends, temporal correlations between monitors within 100 miles of each other were generally higher for PM_{10}, O_3, and NO_2 than for SO_2 and CO. Correlations decreased more markedly with distance for PM_{10} and NO_2 than for the other pollutants, dropping from about 0.7 (zero distance intercept) to about 0.5 at 30 miles separation. Correlation between monitors for some pollutants was also influenced by land use (residential vs. industrial vs. commercial), location (urban vs. suburban), and/or monitoring objective (population exposure vs. maximum concentration).

Prediction of spatial variation in air quality on the urban scale is an important feature of mathematical models that use emissions inventories or estimates from stationary and mobile sources within an urban area in simulations of multiple source dispersion processes. McNair et al. (19) used the Carnegie/California Institute of Technology (CIT) photochemical airshed model to calculate the volume-averaged concentrations of NO_2, CO, O_3, and other pollutants in 5 km by 5 km grid cells covering the Los Angeles basin. The calculated spatial distribution pattern for O_3 was consistent with measured data from 37 monitoring stations, though the model was biased downward or upward depending upon the motor vehicle emissions estimates used as an input to the model. The magnitude of bias in the model was comparable to the magnitude of observed inhomogeneity (about 10-20%) between monitoring sites on a spatial scale similar to the model grid cell spacing, i.e. 5 km. Georgopoulos et al. (20) evaluated the ability of the Urban Airshed Model (UAM-IV) to reproduce spatial and temporal exposure patterns throughout the state of New Jersey and neighboring areas during two ozone episodes. Normalized bias and error was comparable to that found by McNair et al. (19). However, comparison of population-weighted exposure estimates derived from interpolation of monitoring data with population-weighted exposure estimates derived from detailed or interpolated UAM calculations suggested that spatial variation in population-weighted exposure depends more on population distribution than on detailed concentration patterns (20).

Neighborhood and middle scale variation

Research results indicate that neighborhoods or land use areas with high average concentrations of pollutants also tend to have higher spatial variability within the neighborhood (*8, 12*). This scaling of variability with average value is characteristic of lognormal concentration distributions. The coefficient of variation of annual mean NO_2 concentrations within neighborhoods in the Lancaster study (*12*) was about 30-40%; this was comparable to the variation between neighborhoods.

Variation in pollutant concentrations within neighborhoods, like variation between neighborhoods, may be associated with proximity to traffic. As part of the SAVIAH study, sampling for PM_{10}, $PM_{2.5}$, particle-bound polycyclic aromatic hydrocarbons (PAHs), soot, and VOCs was conducted outside homes on high-traffic main streets and low-traffic side streets in central Amsterdam (*21*). At each residence, the cumulative samplers were placed on a balcony at one floor above street level, facing the street. The mean concentrations of PAHs, soot, and VOCs were about twice as high on high-traffic streets compared to low traffic streets, and particulate concentrations were about 15-20% higher. Contaminant concentrations measured simultaneously inside the homes were also significantly higher on high-traffic streets, potentially affecting personal exposure.

Chan and Hwang (*7*) assessed the spatial representativeness of a fixed point hourly monitoring station in Taipei, Taiwan, by comparing the fixed-point station results to NO_2 measurements taken using 3-day passive samplers at 22 sites within the same radius. Representativeness was evaluated in terms of a statistic which quantified the relative absence of bias between fixed-point and satellite site measurements. The fixed-point monitoring station was found overall to be highly representative within a 500 meter radius. Representativeness decreased at larger distances. The monitoring station was also found to be more representative of surrounding high-traffic locations than low-traffic locations. To assess the representativeness of time-series data from the fixed-point monitoring station, a mobile monitoring laboratory was used to measure hourly concentrations of PM10, CO, NO, NO2, NOx, SO2, total hydrocarbons (THC) and non-methane hydrocarbons (NMHC) for three days each at six locations within a radius of 750 meters from the station. Monitoring results from the fixed-point and mobile stations were moderately correlated for all pollutants, except THC and NMHC, which were only weakly correlated, apparently due to the influence of local sources.

Microscale variation

A number of researchers have measured or modeled spatial variation on the scale of a city block. Variability within urban canyons has been of particular interest. Laxen and Noordally (*22*) used 1-week cumulative sampling to investigate variation of NO_2 in three orthogonal directions in urban canyons. NO_2 concentrations were found to be highest at the center line of the road, decreasing to near local background level within 10-15 meters transversely from the center. Concentrations measured longitudinally at curbside tended to increase in the direction of traffic flow upstream of traffic lights, a result that is consistent with higher local emissions due to increased vehicle density, idling, and acceleration after traffic stops. NO_2 concentrations decreased with height to near background level at the top of the canyon.

In a study of horizontal and vertical microscale variations in PM_{10} and total suspended particulates in Taipei (*23*), particulate concentrations measured at open windows of a high-rise building decreased with height between the second and seventh floors but showed no consistent change between the seventh and fourteenth floors. At street level, no consistent patterns in spatial variation were found between the roadside, sidewalk, and covered walkway along a high-traffic main road, nor between main streets, side streets, and alleys. The absence of typical dispersion patterns could be due to emissions from sources such as motorcycles which were not confined to streets.

In symmetrical urban canyons, cross-canyon wind flow at roof level creates a vortex such that the wind direction at street level is opposite that at roof level. Consider, for example, a street canyon running north-south, with roughly equal building heights on both sides. If the wind at roof level is from the west, air from roof level will sweep down the front of buildings on the east side of the street, across the street, and up the front of buildings on the west side of the street. Pollution concentrations would tend to be higher on the west side and lower on the east side of the street. Studies of the influence of roof top wind direction on continuous monitoring data in central London found as much as a threefold difference in simultaneous CO concentrations at mid-block monitors on opposite sides of a street (*24*), and about a two-fold variation in concentration (normalized to remove the effect of wind speed) at a single CO monitor located near an intersection of two canyons (*25*).

Temporal Variations

Seasonal and diurnal patterns in urban air pollution concentrations have been widely reported. Average concentrations of the primary combustion products NO and CO tend to be higher in the winter than in the summer (*26-28*), while average concentrations of the secondary photochemical product O_3 show the opposite seasonal behavior (*26-29*). Diurnal patterns of urban NO, NO_2, and CO concentrations typically show daytime peaks corresponding to rush hours (*8, 25-27, 30*). In contrast, hourly ambient concentrations of benzene and toluene measured in Columbus, Ohio in the summers showed a broad peak during night-time hours, probably the effect of nocturnal inversions (*6*).

To provide a more detailed overview of temporal variation patterns in fixed-point monitoring, we will present analyses of several time series of urban air pollution concentration measurements. Hourly concentration data on NO, NO_2, and O_3, recorded at a SLAMS in central Oklahoma City between October 1998 and June 1999, were obtained from the Oklahoma Department of Environmental Quality. The station, which was designed to be representative of population exposure on the neighborhood scale, was located on a sprawling campus not immediately adjacent to high-traffic streets. The probe was at the unusually high elevation of 15.5 meters above street level.

Daily average concentrations of NO and O_3 were calculated from the hourly data. Daily average NO concentrations were lognormally distributed with a geometric mean of 5 ppb and a geometric standard deviation of 2.75. The distribution of daily average O_3 concentrations was less skewed, with a geometric mean of 24 ppb and a geometric standard deviation of 1.65. Daily average NO_2 concentrations had a geometric mean of 11 ppb and a geometric standard deviation of 1.65. As shown in Figure 1, the winter season was marked by frequent high daily concentrations of NO, with large day-to-day variation above a very low baseline. O_3 daily concentrations showed a complementary seasonal dependence and a higher baseline compared to NO. This result is not surprising, because NO concentrations tend to decrease with horizontal or vertical distance from roadways due to dispersion and oxidation, whereas O_3 tends to build up due to photochemical conversion.

The distributions of hourly concentrations were more strongly right-skewed than the daily averages, with geometric standard deviations of 2.88 for NO, 2.43 for O_3, and 2.18 for NO_2. To investigate diurnal and weekly periodicity, the autocovariance functions of the hourly concentration time series were calculated as:

$$V(\tau) = \sum_i c_i c_{i-\tau} - \bar{c}^2$$

where the c_i are the sequential hourly concentrations, $\bar{c}$ is the concentration averaged over the entire time series, and τ is the lag between data points. The autocovariance functions, plotted in Figure 2, showed a strong 24-hour

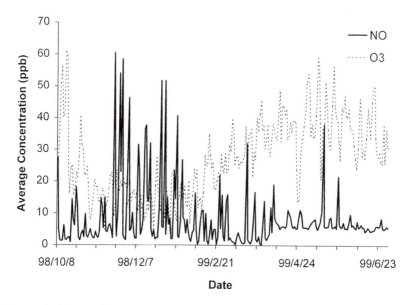

Figure 1. Seasonal variation of 24-hour average NO and O_3 concentrations at a monitoring station in Oklahoma City.

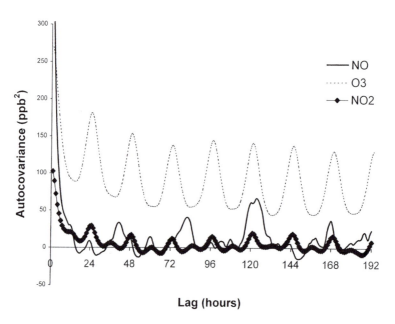

Figure 2. Autocovariance functions of NO, NO_2, and O_3 time series.

periodicity in the O_3 time series and 12- and 24-hour periodicity in the NO_2 time series. On the other hand, the periodic component in the NO autocovariance function was obscured by random associations on time scales longer than 24 hours. Similarly, spectral analysis of an autoregressive model of the time series, performed using S-Plus 2000 statistical software, showed peaks for NO_2 and O_3 corresponding to 6, 8, 12, and 24 hour components, but yielded no discernible periodic components for NO.

The diurnal relationships between NO, NO_2, and O_3 are illustrated in Figures 3 and 4 for weekdays and weekends, respectively. To obtain these curves, the hourly concentrations were first normalized by dividing by the maximum hourly concentration for each day. Normalization removed the effect of day-to-day fluctuations in the peak concentration. The normalized hourly concentrations for each hour of the day were then averaged over the entire monitoring period, consisting of 176 weekdays and 72 Saturdays and Sundays. The average normalized NO_2 diurnal curve showed strong morning and evening rush hour peaks on weekdays, while on weekends the morning peak was attenuated. Similar weekday and weekend diurnal patterns were seen in CO and $PM_{2.5}$ data from other SLAMS in Oklahoma City and Tulsa. The average normalized NO diurnal curves showed only a morning peak. Late afternoon or evening rush hour NO peaks, which occurred at inconsistent times and in only 30% of the days monitored, were almost completely smoothed away in the averaging process. The afternoon peaks occurred mostly in the late autumn and winter when O_3 formation was lower. A similar seasonal effect has been reported in urban green areas (8). This suggests that NO emissions from afternoon rush hour traffic were extinguished by the high O_3 concentrations around the monitoring station.

Implications

Monitoring studies indicate that time-averaged concentrations of some air pollutants often vary by factors of two or more between monitoring points on the same street segment, the same neighborhood, or the same city. This apparent scale-invariance may suggest that methods for analysis of fractals could be fruitfully applied to the problem of spatial variability, as they have already been applied to temporal variability of pollution (28).

The existence in some cases (6, 7, 13) of significant interaction between spatial and temporal variability renders the use of fixed-point data as surrogates for human exposure in time series analysis highly problematic. The uncertainties inherent in using an Eulerian average (i.e. from a stationary sampler) to estimate a Lagrangian average (i.e. for a mobile person) are

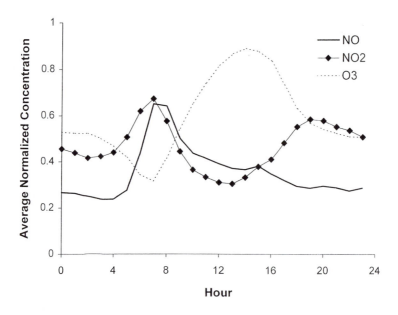

Figure 3. Weekday diurnal variation of NO, NO₂, and O₃ in Oklahoma City.

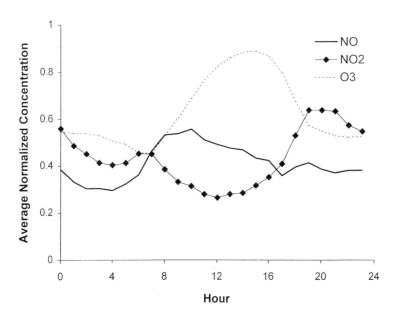

Figure 4. Weekend diurnal variation of NO, NO₂, and O₃ in Oklahoma City.

compounded under conditions of temporal variability at the monitoring locale, because the time average over the period spent by the person in the concentration milieu spatially "represented" by the sampler is not necessarily representative of the overall time average recorded by the sampler (*31*). Empirical studies using simultaneous automated personal and fixed-point monitoring can help elucidate the relationship between Eulerian and Lagrangian exposure patterns.

Acknowledgements

We gratefully acknowledge the assistance of Leon Ashford and co-workers at the Air Quality Division, Oklahoma Department of Environmental Quality, in providing air monitoring data and supporting information. This research was funded by the U.S. EPA National Center for Environmental Research and Quality Assurance under Agreement No. R82-6786-010.

References

1. Dockery, D. W.; Pope, C. A. *Ann. Rev. Public Health.* **1994**, *15*, 107-132.
2. Wallace, L.; Nelson, W.; Ziegenfus, R.; Pellizzari, E.; Michael, L.; Whitmore, R.; Zelon, H.; Hartwell, T.; Perritt, R.; Westerdahl, D. *J. Expos. Anal. Environ. Epidemiol.* **1991**, *1*, 157-192.
3. Seinfeld, J. H.; Pandis, S. N. *Atmospheric Chemistry and Physics*; John T. Wiley & Sons: New York, NY, 1998.
4. Code of Federal Regulations; July 1, 1999; Title 40, Part 50.
5. Allegrini, I. In *Urban Air Pollution – European Aspects*; Fenger, J.; Hertel, O.; Palmgren, F., Eds.; Environmental Pollution 1; Kluwer Academic Publishers: Dordrecht, Netherlands, 1998; pp. 279-296.
6. Spicer, C. W.; Buxton, B. E.; Holdren, M. W.; Smith, D. L.; Kelly, T. J.; Rust, S. W.; Pate, A. D.; Sverdup, G. M.; Chuang, J. C. *Atmos. Environ.* **1996**, *30*, 3443-3456.
7. Chan, C.-C.; Hwang, J.-S. *J. Air & Waste Manage. Assoc.* **1996**, *46*, 755-760.
8. Kuttler, W.; Strassburger, A. *Atmos. Environ.* **1999**, *33*, 4101-4108.
9. Norris, G.; Larson, T. *J. Expos. Anal. Environ. Epidemiol.* **1999**, *9*, 586-593.
10. Croxford, B.; Penn, A.; Hillier, B. *Sci. Total Environ.* **1996**, *189/190*, 3-9.
11. Brauer, M.; Hirtle, R. D.; Hall, A. C.; Yip, T. R. *J. Expos. Anal. Environ. Epidemiol.* **1999**, *9*, 228-236.

12. Hewitt, C. N. *Atmos. Environ.* **1991**, *25B*, 429-434.

13. Lebret, E.; Briggs, D.; van Reeuwijk, H.; Fischer, P.; Smallbone, K.; Harssema, H.; Kriz, B.; Gorynski, P.; Elliott, P. *Atmos. Environ.* **2000**, *34*, 177-185.

14. Larssen, S. In *Urban Air Pollution – European Aspects*; Fenger, J.; Hertel, O.; Palmgren, F., Eds.; Environmental Pollution 1; Kluwer Academic Publishers: Dordrecht, Netherlands, 1998; pp. 297-316.

15. Code of Federal Regulations; July 1, 1999; Title 40, Part 58, Appendix D.

16. Code of Federal Regulations; July 1, 1999; Title 40, Part 58, Appendix E.

17. McCurdy, T.; Zelenka, M. P.; Lawrence, P. M.; Houston, R. M.; Burton, R. *Atmos. Environ.* **1999**, *33*, 5133-5145.

18. Ito, K.; Thurston, G. D.; Nadas, A.; Lippmann, M. *J. Expos. Anal. Environ. Epidemiol.* **2001**, *11*, 21-32.

19. McNair, L. A.; Harley, R. A.; Russell, A. G. *Atmos. Environ.* **1996**, *24*, 4291-4301.

20. Georgopoulos, P. G.; Purushothaman, V.; Chiou, R. *J. Expos. Anal. Environ. Epidemiol.* **1997**, *7*, 191-215.

21. Fischer, P. H.; Hoek, G.; van Reeuwijk, H.; Briggs, D. J.; Lebret, E.; van Wijnen, J. H.; Kingham, S.; Elliott, P. E. *Atmos. Environ.* **2000**, *34*, 3713-3722.

22. Laxen, D. P. H.; Noordally, E. *Atmos. Environ.* **1987**, *21*, 1899-1903.

23. Chen, M.-L.; Mao, I.-F. *Sci. Total Environ.* **1998**, *209*, 225-231.

24. Croxford, B.; Penn, A. *Atmos. Environ.* **1998**, *32*, 1049-1057.

25. Scaperdas, A.; Colvile, R. N. *Atmos. Environ.* **1999**, *33*, 661-674.

26. Mayer, H. *Atmos. Environ.* **1999**, *33*, 4029-4037.

27. Bogo, H.; Negri, R. M.; San Román, E. *Atmos. Environ.* **1999**, *33*, 2587-2598.

28. Lanfredi, M; Macchiato, M.; Ragosta, M.; Serio, C. *Fractals*, **1998**, *6*, 151-158.

29. Varshney, C. K.; Aggarwal, M. *Atmos. Environ.* **1992**, *26B*, 291-294.

30. Hao, J.; He, D.; Wu, Y.; Fu, L.; He, K. *Atmos. Environ.* **2000**, *34*, 453-465.

31. Esmen, N. A.; Hall, T. A.; *Appl. Occup. Environ. Hyg.* **2000**, *15*, 114-119.

Chapter 18

Trophic Transport of Metals in Birds: Birds as Indicators of Exposure and Effect

B. M. Adair, T. J. McBride, M. J. Hooper, and G. P. Cobb

The Institute of Environmental and Human Health, Texas Tech University, Lubbock, TX 79409–1163

The ubiquitous nature of metals from both natural and anthropogenic sources combined with their necessity in biological processes produce a complex system for study. Metal distributions in abiotic and biotic systems must be examined to accurately assess impact on ecosystems. Wildlife studies of exposure and effect can be challenging, but the results are more complete than evaluation of only metal concentrations. Birds are good sentinel species because they visible, sensitive to toxins, and occupy different trophic positions. Therefore, studies to assess avian population status, reproductive success, and toxicological significance of metal exposures can be extrapolated to other wildlife and possibly humans.

In order to understand metals in the environment, we must first examine their sources, fate, and availability. Once the major characteristics of metals are understood, we can assess exposure scenarios of birds to metals. Life history of birds is important, because it influences metal exposure via ingestion and the consequent effects from that exposure. The specific metals and bird species being examined dictate the endpoints and design of exposure and effect studies.

Metals in the Environment

Metal Sources

Metals are released into the environment by numerous anthropogenic and natural sources. Primary anthropogenic sources of metals are fossil fuel combustion, reservoir flooding, mining and smelting, plating and agriculture. Marshes, volcanoes and erosion are representative natural sources that mobilize metals into the environment. It is often difficult to determine if a metal is mobilized from a natural or anthropogenic source, because both sources generally deposit metals into the air and water. Metals tend to be distributed on a global scale via atmospheric transport (*1*), while metals are transported on a regional or continental scale in rivers (*2*). As the geographic scale of an assessment is reduced, metal concentration gradients can be refined, allowing better differentiation between anthropogenic and natural sources. On or near a single site, refinement of concentration gradients is sufficient to determine likely source(s) of metal release(s) into the environment (*3*).

Processes that release metals dictate the species entering the environment. Chemical speciation is often a critical factor in risk assessments at metal contaminated sites. For example, leather tanning facilities release hexavalent chromium (Cr(VI)), which moves through the environment far differently than does the more common trivalent chromium. Further, Cr(VI) is a carcinogen while Cr(III) is essential for good health. Arsenical herbicides are alkylated arsenates, which behave somewhat differently from inorganic arsenite/arsenate resulting in more significant risks than do inorganic arsenic species (*4*). Study designs to evaluate metal exposure and effects must consider the quantity, form, transport and possible transformation of metals in the environment.

Metal Fate

Ultimately soil and sediment serve as sinks for toxic metals (*5*), while water and air are primarily transport media. Metal contaminants in air are largely found in the particle fraction. These particles eventually settle into water bodies or deposit onto soil or biotic surfaces. Metal transport in water is largely dependent on solubility. Parameters such as pH, alkalinity, salinity, and dissolved organic matter greatly influence metal solubility in natural waters (*6, 7*). Metals in the water column may also sorb to suspended particles thereby influencing the transport and transformation (*8*). Particle flocculation in slow flowing water bodies such as lakes or estuaries cause increased metal deposition into sediments. Increases in ionic strength also accentuate metal deposition in estuarine sediments (6, 9).

Soils from different areas often contain dissimilar metal content (10). Deposition of metals into soils of a given area may have occurred from geologic activity, aerial deposition, flooding, or anthropogenic sources. Microbes, plants, and macro invertebrates are the primary receptors of metals from soil and other abiotic compartments. Ingestion of these types of food transfers metals into higher-level organisms.

Metal Availability

The key to understanding metal toxicity to wildlife is understanding the availability of metals from various exposure media. Metal availability to organisms is largely determined by the metal compound or species present, the environmental compartment in which it exists, and the other constituents present in that compartment largely determine metal availability to organisms. Changes in natural conditions by human activities often alter metal availability. For instance, acid rain mobilizes aluminum from soils, which may in turn cause chlorosis in plants (9). The relationship between acidified soil and metal uptake has also been demonstrated in the laboratory (11). Topography alterations due to mining can bring metals into closer contact with surface water and air; therefore, acid mine drainage can more readily transport metals through soil and into water systems (12, 13).

Metal speciation and availability in sediment and soil are dependent on parameters such as pH, alkalinity, salinity, and organic matter. In soil and sediment systems, clays and organic matter form a cation exchange capacity (CEC) that allows the soil/sediment to bind cations. CEC and oxidation state are critical for accurate evaluation of metal availability from a given soil (14). While CEC is important in the fate of cationic metals, retention of anionic species, such as arsenate, arsenite, selenate, selenite, and chromate, in soil is not influenced by CEC. Acid volatile sulfides represent another metal binding ligand, acting to sequester metals primarily in sediments (15). The binding (formation) constants of metals or metal containing species control their solubility and stability in water, which ultimately controls the thermodynamics of their uptake by organisms (8).

Inorganic forms of metals do not biomagnify through food chains/webs because of their reactivity and availability. Organisms in most intimate contact with the contaminated food source or abiotic compartment generally contain the highest concentration of metal (16, 17). However, biota may transform metals into more toxic species. For instance mercury alkylation, most often methylation, produces the biomagnifying cation, methyl mercury ($MeHg^+$). Due to MeHg$^+$ accumulation in fish, advisories against fish consumption exist in most coastal watersheds from Texas to Massachusetts. (*18*). In cases where stable organometallic species are formed, metal speciation is essential for evaluation of metal availability to fish and wildlife of interest.

Avian Characteristics

Since many birds migrate, and numerous species reside in the same geographic location, diverse bird populations may be at risk of exposure to environmental contaminants. Metal contamination in one area can have a detrimental affect on individuals throughout a region, particularly in active feeding or breeding areas, i.e. migratory flyways, staging, wintering and land fall areas (*19*).

Most avian studies emphasize ingestion as the primary route of exposure to toxic metals, because other routes of exposure are extremely uncommon. Because risk of exposure is greatly influenced by bird feeding and nesting habits, the characteristics of birds need to be taken into account for any study and its subsequent results. Age and sex differences in uptake and distribution of metals also influence metal exposure. Ultimately, even the same species of bird may behave differently in different environments, creating variations in risk of exposure (*20*).

Feeding Guilds

Feeding guilds differentiate bird species by what they eat, which also influences their risk of exposure to contaminants. The feeding guilds discussed will be defined as oligivores, omnivores, raptors, and piscivores. Oligivores include insectivores and herbivores that feed on specialized food sources. Omnivores are usually distinguished by including both plant and animal items in their diet. Raptors and piscivores tend to choose prey items higher in trophic webs. Many omnivorous birds subsist on distinct, oligivorous diets for certain parts of the year based on food availability and nutrient requirements. For example, high protein diets of invertebrates are important during breeding and rearing of young, due to increased protein requirements for adults in production of the egg and in nestling rearing. Similarly, intensive feeding on high calorie food items occurs before migration to increase lipid stores needed for sustained migratory flight. Each food source accumulates different metals at varying rates. Feeding guilds are extremely important when discussing exposure and possible effects to birds, as the food sources contain different metals.

Oligovores

Uptake, compartmentalization, and toxicity of metals in plants are important for exposure assessment of plant eating birds (*21-25*). Structures of plants, such as fruits, seeds, leaves and shoots, will accumulate metals at differing concentrations based on soil characteristics and air quality (*12, 24, 25*), thus controlling accessibility to herbivores.

Some bird species restrict their diet to grains, nuts, fruits and seeds for food. This is important since the reproductive portions of plants (seeds, fruits) generally accumulate lower concentrations of metals than the roots, stems, and leaves (*21, 25*). Alternately, browsers utilize such structural components of plants, as grasses, leaves, and shoots. Metal exposure to browsers can occur through ingestion of the contaminant incorporated by the plant and via accidental ingestion of soil. Willow leaves and buds (Salix spp.) biomagnify Cd up to two orders of magnitude above background concentrations, exposing species that utilize the plants as a primary food source to dangerous levels of Cd, most notably the Willow Ptarmigan (*26*). Aluminum can also be a concern for herbivores, because it replaces calcium in plants causing reduced nutrient uptake with contaminant exposure (*27*). Browsing species tend to have a greater likelihood of exposure to metals than selectors as soil-incorporated metals are more easily accumulated into the shoots, leaves and grasses, than in seeds and fruits. Filter feeders, such as geese, cranes and some species of duck utilize aquatic plants as a considerable food source, subsequently eating large amounts of associated sediment (*28*).

Several studies have compared avian accumulation of metals from different food sources. Scheuhammer (*12*) demonstrates in his review that cadmium accumulation in target organs of herbivores was greater than in insectivores of the same or similar species, because cadmium concentration in plants was higher than in prey items. Other studies have revealed that game birds accumulate variable amounts of lead following exposure to lead shot, while uptake by raptors is more consistent (*29*). The variability of uptake between species has been explained by different exposure and depuration mechanisms (*19, 29*). Comparisons between terrestrial and aquatic herbivores reveal that birds feeding on terrestrial plants accumulate lower concentrations of cadmium into kidney and liver than do aquatic herbivores due to less metal accumulation in terrestrial plants than in aquatic plants (*30*). Seeds and plants from a selenium-contaminated reservoir contained lower concentrations of selenium than did invertebrates from the same system causing insectivores to be exposed to higher concentrations of selenium than were herbivores or omnivores (*31, 32*).

Terrestrial and aquatic invertebrates exhibit similar propensity for metal accumulation. Larvae, beetles, and centipedes accumulate metals above background levels but tend to depurate them quickly (*33-35*). Metal concentrations in invertebrates inhabiting contaminated sites can fluctuate seasonally (*21*). Therefore, invertebrate food samples collected for exposure assessment must be collected at times relevant to bird exposure. Soil properties are important in uptake and accumulation of heavy metals in oligochaete invertebrates (*36*). Uptake of metals can vary between similar insect species at the same site, making accurate identification of insects as food items for birds extremely important (*34, 35, 37*).

Omnivores

Many species are seasonal omnivores to compensate for changes in food availability. Nevertheless, several families including gallinaceous birds (Quail, Pheasant, Turkeys), corvids (Jays, Ravens, Crows), and some waterfowl are generally regarded as omnivores. Omnivorous birds may be at increased risk of exposure due to a greater variety of food items. Numerous studies have examined the exposure and effects of metals to omnivorous waterfowl. Accumulation, dietary effects, and reproductive effects after metal exposure have been studied in mallards (*38-40*) that utilize aquatic plants, insects and mollusks as food. Lead exposure from lead shot and food is a concern for many birds, particularly waterfowl (*19, 29, 41, 42*). Studies have shown that birds inhabiting the same sites do accumulate different metal concentrations because of omnivorous versus insectivorous eating habits (*26, 43*).

Raptors

Raptors, or birds of prey, include falcons, hawks, owls, and vultures, among others. They consume a wide variety of invertebrates, fish, mammals, reptiles, and other birds exposing them to a diversity of potential contaminants (*44-47*). In light of their position at the top of trophic food webs, raptors have proven useful in investigating the bioaccumulation of environmental contaminants in toxicological studies (*48*).

Raptors may remove indigestible fur and feathers before consumption or regurgitate pellets of undigested material, which means they may not be exposed to total body burden concentrations (*48, 49*). Studies have shown little association between metal accumulation in Tengmalm's owls and body burden of prey items (*49*) compared to more significant association between lead in kestrels and prey (*50*). Therefore, species-specific differences of digestion and prey item consumption should be considered (*51*) in contaminant studies. Considerable research has examined health effects on raptors due to elevated concentrations of metals in the environment (*44, 46, 52-55*), further illustrating the importance of accumulation of these toxicants from contaminated prey species.

Piscivores

Piscivores include a large array of species: raptors (bald eagle, osprey), shore birds and waterfowl (heron, egret, merganser, loon, crane, cormorant, pelican), seabirds (penguin, eider, auklet), penguins, and kingfishers. Many piscivores are exposed to the entire body burden of their prey items, because they eat their prey whole, without dismemberment or pellet regurgitation. Additionally, piscivores are at greater risk of exposure to metals that biomagnify in aquatic environments. Numerous investigations have demonstrated effects on birds exposed to mercury (*56-62*), selenium (*63, 64*) aluminum (*27, 65*), and lead (*41, 66*) in aquatic systems. Significant differences in metal accumulation in feathers exist between insectivorous and piscivorous egret species, with the fish-eating species showing the highest concentrations of Cd and Hg (*56*) demonstrating the different risk from terrestrial and aquatic systems.

Age, Sex, and Habitat Differences

Food preference is only one reason for differences in exposure and consequent effects in birds. As with many contaminants, metal exposure and effect can be influenced by age and sex of birds (*20*). Studies suggest the sex differences in metal exposure are from different eating habits of males and females (*26, 67*). However, Furness (*30*) emphasizes that sex differences in Cd uptake may not be biologically significant.

Metal accumulation in birds varies with age because of exposure duration and depuration rate differences between different life stages. Some metal concentrations increase with time, because they are slowly accumulated, while other metals decrease between early nestling age and pre-fledge as a result of molting. Studies of sea birds have shown age differences in feather, tissue and blood metal concentrations (*68, 69*). Effects from exposure at different ages also vary. Therefore, determining the most sensitive life stage is critical when evaluating heavy metal exposure and effects in avian populations (*70*). There is evidence that exposure to metals pre-hatching can effect development and reproduction (*71*). Age dependent effects have also been documented from aluminum exposures (*65*).

The same species of birds inhabiting different locations can accumulate and assimilate metals differently for several reasons. Diet, climate, contaminant distribution, and habitat can influence life history of birds that will in turn alter exposure and affects from contaminants (*20, 56*).

Study Endpoints

Residue Endpoints

Analysis of liver, kidney and brain provide the most widely accepted diagnosis of metal concentrations that may produce toxicosis (*26, 30, 72*). Liver and kidney accumulate most inorganic forms of metals, and organometallic forms of compounds such as methyl mercury may concentrate in the brain (*12*). The route of exposure can be evaluated through food item analysis. Food samples can be collected, identified, and separated for residue analysis. The concentration data will provide a distribution of metal(s) in the dietary components (*73*). Regurgitation samples from some species can also provide food samples to quantify metal exposure (*57, 60*).

A number of non-lethal techniques with foundations in human forensics are available for metal determination. Analysis of feathers or blood can be used, much as one would use human hair or blood, to evaluate metal exposure (*74, 75*). Feathers can provide a historic residue profile for some metals that represents metal loads during feather growth. Feathers are primary metal excretion routes for juvenile birds and serve as good biomarkers of metal exposure during this early life stage. Blood residue analyses provide a sensitive residue-based diagnostic for lead exposure (*76*). Metal concentrations in avian excreta have proven to be useful non-lethal endpoints in exposure assessments (*70, 77*). Heavy metal concentrations in eggshells have been indicated as a potential excretion method for females (*78*) and have been used as indicators of metal exposure to adults. While metal concentrations in eggs have been correlated with environmental metal concentrations surrounding nesting sites (*79, 80*).

Frequently, samples collected during environmental studies are small and contain low concentrations of metals. Preparation of such samples for residue analysis should include few sample transfers and reagents and should result in the smallest possible volume in order to minimize metal dilution and contamination. Many methods in the literature do not accommodate both small sample volumes and low metal concentrations. Therefore, it has become increasingly important to develop methods to analyze environmental samples at relevant concentrations (*81*). Graphite furnace atomic absorption (GFAA) can provide the requisite detection limits, but multi-element analyses are quite time consuming using this technique. Inductively coupled plasma atomic emission (ICP-AES) spectrometers alleviate the time constraints of GFAA analyses, but instrument limits of detection are higher for most elements. Current ICP-Mass Spectrometers provide limits of detection that are equivalent to or better than AA and provide rapid sample analyses (*82*).

Biomarkers

Physiological biomarkers are useful indicators for assessing exposure or detrimental effect from a toxicant. Biomarkers can be specific for a particular contaminant, or as a general indicator of physiological condition. Alterations in cellular structure can be linked to contaminant exposure. Other biological markers measure alterations in chemical reactivity of biological processes. Numerous biomarkers can be measured in easily collected sources such as blood and excreta. Bird studies using biological endpoints in combination with residue endpoints produce data that can be used to assess not only metal exposure but also effects of such exposures to the birds and overall population health.

Cell and Tissue Level Effects

Histopathology provides a detailed assessment of detrimental affect by allowing for examination of specific morphological disturbances of target organs in response to metals. The technique does require extensive training and knowledge of normal cell structure for each species examined. The structural and functional nature of the kidney and liver tend to put them at high risk for contaminant accumulation, and subsequently structural and cellular disturbance. Histopathological examination revealed that 57% of Willow Ptarmigans living in high cadmium areas had renal damage (*26*). In addition, selenium toxicosis in mallards gave rise to hepatic histological lesions (*63*). The central and peripheral nervous systems can also show cellular damage from metal exposure. Scheuhammer (*83*) and Burger (*19*) both review negative effects on central and peripheral nervous systems of birds exposed to metals, particularly methyl mercury and lead.

Heme-Related Markers

Effects of metals on the heme biosynthesis pathway have been examined using porphyrin profile changes and ALAD inhibition as endpoints. Porphyrins are the molecular precursors of heme, the prosthetic group in several oxygenating mechanisms of most eukaryotic cells. Eight distinct enzymes catalyze the formation of porphyrin intermediates and heme, and different metals react with specific enzymes producing characteristic alterations of the porphyrin profile (*84, 85*). Porphyrin concentrations have historically been analyzed in liver and kidney tissues; however, urine and fecal matter may offer a reliable non-lethal substrate (*86*) that can be easily collected. Porphyrin increases were significantly correlated with hepatic arsenic concentrations of spectacled eiders (*87*) in western Alaska. In American kestrels, considerable decreases in hemoglobin concentration were shown, indicating a significant alteration in porphyrin biosynthesis (*53, 88*) due to lead intoxication. Starlings inhabiting a metal contaminated site in Montana show porphyrin increases in kidneys of starlings that have accumulated lead. High concentrations of zinc in kidneys, however, appear to modulate the porphyrin increases (*89*). These results emphasize the necessity for more dosing studies of metal mixtures to determine the resulting effects on birds.

Delta aminolevulinic-acid dehydratase (ALAD) is perhaps the most widely studied heme-related enzyme that is altered by metal contamination. ALAD conjugates two aminolevulinic acid molecules critical to heme biosynthesis. The zinc-dependant enzyme is easily inhibited by lead substitution and has been extensively characterized as a sensitive indicator of lead exposure. Blood is the ideal matrix for assessing ALAD activity in wildlife, and sampling is easy and non-lethal. Decreased ALAD activity has been correlated to lead accumulation in a wide variety of birds such as waterfowl, game birds, raptors, and passerines (*28, 42, 53, 54, 90-92*).

Other Markers

Metallothionein is utilized for natural cellular metal ion maintenance and sequestration. This protein is highly inducible as a defense mechanism by cellular presence of some non-essential metals (*93*). Metallothionein induction has been seen in several avian species after cadmium exposure (*30, 72, 94*). No significant metallothionein induction occurred in spectacled eiders, despite mean hepatic copper concentrations of 546ug/g (*87*). Glutathione (GSH) is extremely important in cellular disposition of metals (*95*). The effects of mercury and selenium on GSH function are most commonly discussed. Several studies have examined the effects of mercury and selenium on glutathione metabolism in waterfowl (*96, 97*). Hepatic (GSH) levels have also been examined in starlings and were found to increase 52% in starlings fed high levels of cadmium chloride (*98*).

Cholinesterase and ATPase are enzymes also effected by metal exposure, though few avian studies have investigated the mechanisms of response. Dieter indicated an increase (22 to 28%) in butyl-cholinesterase activity in mallards dosed with a commercial lead shot (*90*). Hoffman found lead intoxication (*88*) did not exhibit an effect on ATPase in American kestrels.

Ecological Endpoints

Residue and biomarker endpoints can be measured in field and lab studies. Ecological endpoints, however, can only truly be examined in the field. Survival, reproduction and population endpoints are the most studied ecological endpoints (*70, 80*). Careful monitoring of nestling health and parental breeding behavior may reveal responses to a contaminant that are not identified by overt physiological changes.

Avian nesting behavior permits investigation of contaminant effects on reproduction. Egg production, hatching success, nestling development, and survival are valuable endpoints to assess metal contamination effects on populations. High concentrations of cadmium may suppress egg production, and Cd and Pb have both been studied for risks of eggshell thinning (*30, 99*).

Metal uptake has been linked to developmental effects in several bird studies (*71, 97, 100*). Brain lesions have been documented in lead exposed chick embryos (*101*). One study revealed that nestling deformities in pied flycatchers were linked to metal exposure while great tits inhabiting the same metal contaminated site did not show nestling deformaties (*70*). Slow growth rates from heavy metal exposure have also been demonstrated in natural populations (*70, 80*). Nestling survival rates of several bird species exposed to metals have been examined in the lab and field (*66, 70, 102*) yielding different results for different metals.

Interpreting the significance of reproductive endpoints to a population can be difficult and requires collaboration between toxicologists and ornithologists to better interpret population changes (*13*). The health, population demographics, and cadmium concentrations observed in white-tailed ptarmigan from contaminated and non-contaminated sites revealed a possible population shift as a result of increased cadmium exposure to females (*26*).

Study Designs

The 12 toxic metals and metalloids on the Resource Conservation and Recovery Act (RCRA) list are investigated most often. Of these, arsenic, lead, chromium (VI), and cadmium top the list of toxicants most frequently found above toxicity thresholds at Superfund sites (*103*). Due to the prevalence of metals in the environment and the fact that many metals are essential nutrients

(*104*), exposure to a single metal on site is unlikely. Even so, most studies of toxic metal effects evaluate the uptake of one or, at most, a few RCRA metals.

Laboratory studies have evaluated the effects of metal mixtures in some vertebrates, but laboratory studies of avian responses to metal mixtures are essentially absent from the literature. Interestingly, more field studies have evaluated interactive effects of metals in birds, but there are too few studies to adequately describe interactive effects of metal mixtures. (*19, 83*). The following sections will discuss general study designs used for both laboratory and field studies.

Laboratory Studies

Dosing studies provide the foundation for determining toxic effects following metal exposure. Metal exposure becomes a concern when essential metal concentrations in tissues exceed levels required for normal homeostasis or when non-essential metal concentrations exceed normal background levels (*70, 74, 76*). Therefore, interactions between toxic metals and nutrients have been evaluated in several controlled dosing studies (*85, 105-107*). Numerous studies demonstrate combinations of metals yielding synergistic and antagonistic effects in mammals (*108*). The studies only examined two or three metals in combination. Also, physiological differences between birds and mammals may yield different metal interactions.

Several lab studies have examined effects from single metal exposure to birds. Eggs from adults dosed with arsenic have been examined for accumulation and effects (*109*). Eggs have been dosed with lead followed by embryo examination for effects (*101*). Although Beyer et al. (*42*) did not examine multiple metals, they did examine the uptake and biological effects of six avian species dosed with lead. The dose dependent distributions and effects of lead, selenium and cadmium have been examined in several species (*72, 98, 110*). Scheuhammer (*83*) reviews numerous studies examining chronic toxicity of several metals.

Field

Well-designed field studies can define exposure routes and adverse effects in study populations. Problems associated with these studies are the areas required to support sufficiently large avian populations and the manpower required to monitor the study of populations (*111*). Many birds migrate, which leads to multiple exposure scenarios for the same bird. Birds can be exposed pre-migration, during stopovers, or at study sites. Prey item availability may shift within a season, forcing adults to feed at opportune times and places during the breeding season. Therefore they may be exposed at sites other than the site of interest.

An ideal study design allows scientists to examine long-term exposure and effects from short term monitoring. Feeding ranges for birds vary from less than one to hundreds of square miles (*45*). Spatial scales and natural population fluctuations must be considered for designs of field studies (*70*). Burger discusses scale, feeding, and biomarker issues for risk assessment of lead in birds (*19*), which can be related to other metals as well.

Single time point studies are useful when assessing current conditions of birds in a small area, and results can serve as preliminary data for long-term studies. Some birds are collected by shooting (*87*), thus it is important to make sure that ammunition used does not contaminate the bird. Opportunistic collection of dead birds is also used (*112*). When birds are shot or collected dead to examine one time point, life history information is limited.

Mist nets are often incorporated into studies to catch birds for multiple time point and life history assessment. Population statistics, such as species, age, sex, count, size, can be monitored from birds that can be tagged and released for further study. Non-lethal samples used for contaminant quantification or biomarker analysis can also be collected from mist netted birds. Larger birds such as waterfowl and sea birds are difficult to trap in nets, however, they can be herded into corrals (*75*) where samples can be collected.

Visual observations and radio telemetry are used to examine foraging patterns and success, parental care, and reproductive behavior of adult birds. When monitored adults are associated with nests, further data on egg and nestling survival, food item identification, and development can also be assessed. Radio telemetry studies of adults breading in nest boxes yield valuable information of nestlings and adults (*113*). Radio telemetry allows adult birds to be monitored beyond visual inspection. However, the position must be triangulated by multiple, simultaneous readings, necessitating substantial investigator efforts. If the readings are taken mechanically, then expensive monitoring equipment is needed.

Incorporation of artificial nest boxes in avian studies attracts cavity-nesting birds to the site and provides an easily monitored environment. Some species of birds are more easily examined from boxes than others. Blue birds and tree swallows tend to nest sporadically throughout a site, while starlings nest in groups and breed synchronously. Arrays of boxes can be set up on different sites in an area, thereby allowing birds to be monitored at the same time, which yields consistent changes of prey item differences within a season. Birds that do not breed synchronously can be useful in identifying exposure differences over time in the same area.

Larger birds, such as kestrels and owls, have much larger home and foraging ranges, making nest box studies challenging (*49*). Individual boxes have to be separated by a greater distance than they do for passerines. However, the raptor species are larger and eat food samples higher in the food chain making residue quantification easier.

Conclusions

Toxicological studies of metals are difficult, because metals are ubiquitous and essential to biological processes. Exposure to a single metal is highly unlikely, and accumulation of metal mixtures can produce either antagonistic or synergistic effects that are dictated by the metals present. Avian field studies have examined metal exposure, accumulation, and depuration using lethal and non-lethal techniques for residue analysis. Biological markers have been combined with residue endpoints to assess reproductive and population effects from metal exposure to wild populations of birds. Few dosing studies of metal mixtures have been performed with little, if any, focus on birds. For these reasons effects of metal mixture accumulation in birds are poorly defined in controlled lab studies.

Avian field studies have provided exposure and effect data at metal contaminated sites. Trophic transport through different exposure scenarios can be examined simultaneously, because multiple avian species at different trophic positions can occupy the same site. Reproductive effects can be effectively monitored using nest box studies. Temporal and spatial scale are important for a well designed avian field study. Birds are often more mobile than other indicator species; therefore, spatial scale for the study species must represent the home and foraging ranges of that species. If results of reproductive and developmental effects from metal are to be extrapolated to population effects, the study must be long enough to assess some long-term consequences.

It is crucial that more bird dosing studies with environmentally relevant mixtures of metals be performed to aid in interpretation of field study results. Results from such dosing studies combined with well designed avian field studies can produce reliable toxicological data on metal exposure and effects.

References

1. Hanisch, C. *Environ. Sci. Technol.* **1998**, 176a-179a.
2. Ebinghaus, R.; Hintelmann, H.; Wilken, R. D. *Fresenius J. Anal. Chem.* **1994**, *350*, 21-29.

3. Icme *Environmental risk assessment methodologies for metals and inorganic metal compounds*; International Council on Metals and the Environment, Montreal, Canada: Monpellier, FR, 2000.
4. Aposhian, H. V. *Ann. Rev. Pharmacol. Toxicol.* **1997**, *37*, 397-419.
5. Goyer, R. A. In *Casarett and Doull's Toxicology*; M.O. Amdur, J. D., and C.D. Klassen, Ed.; Mc Graw-Hill: New York, NY, 1995.
6. Stumm, W.; Morgan, J. J. *Aquatic Chemistry*; Wiley Interscience: New York, NY., 1992.
7. Ownby, D.; Robillard, K.; Lapoint, T. W.; Cobb., G. P. *Toxicology and Environmental Chemistry* **2001**, *in press*.
8. Janes, N.; Playle, R. C. *Environ. Toxicol. Chem.* **1995**, *14*, 1847-1858.
9. Schlesinger, W. H. *Biogeochemistry: An Analysis of Global Change.*; Academic Press, 1991.
10. Friske, P.; Coker, W. B. *Water, Air, and Soil Pollution* **1995**, *80*, 1047-1051.
11. Waters, M. MS, Clemson University, Clemson, SC, 1996.
12. Scheuhammer, A. M. *Environ. Pollut.* **1991**, *71*, 329-375.
13. Gard, N. W., M.J. Hooper In *Ecology and Management of Neotropical Migratory Birds: A Synthesis and Review of Critical Issues*; Martin, T. E., D.M. Finch, Ed.; Oxford University Press: New York, 1995, pp 294-310.
14. Salomons, W.; Branch, H.; Haren, R. A. *Dredged material and mine tailings: Similarities in behavior and impact.*; International Conference on Heavy Metals in the Environment: Athens, Greece., 1985.
15. Ditoro, D. M.; Mahoney, J. D.; Hansen, D. J. *Environ. Toxicol. Chem.* **1990**, *9*, 1487-1502.
16. Mccoy, J. MS, Clemson University, Clemson, SC, 1994.
17. Beyer, W. N., et al. *Environ. Pollut.* **1985**, *38A*, 63-86.
18. Usepa *Listing of Fish Consumption Advisories*; United States Environmental Protection Agency, 1999.
19. Burger, J. *J. Toxicol. Environ. Health* **1995**, *45*, 369-396.
20. Burger, J.; Gochfeld, M. *Environ. Res.* **2000**, *82A*, 207-221.
21. Hunter, B. A.; Johnson, M. S.; Thompson, D. J. *J. Appl. Ecol.* **1987**, *24*, 587-599.
22. Zenk, M. H. *Gene* **1996**, *179*, 21-30.
23. Vazquez, M. D.; Lopex, J.; Carballeira, A. *Ecotoxicol. Environ. Saf.* **1999**, *44*, 12-24.
24. Kozlov, M. V., et al. *Environ. Pollut.* **2000**, *107*, 413-420.
25. Cobb, G. P., et al. *Environ. Toxicol. Chem.* **2000**, *19*, 600-607.
26. Larison, J. R.; Likens, G. E.; Fitzpatrick, J. W.; Crock, J. G. *Nature* **2000**, *406*, 181-3.
27. Hui, C. A.; Ellers, O. *Environ. Toxicol. Chem.* **1999**, *18*, 970-975.

28. Beyer, W. N.; Day, D. L.; Melancon, M. J.; Sileo, L. *Environ. Toxicol. Chem.* **2000**, *19*, 731-735.
29. Kendall, R. J., et al. *Environ. Toxicol. Chem.* **1996**, *15*, 4-20.
30. Furness, R. W. In *Environmental Contaminants in Wildlife*; CRC Press, 1996, pp 389-404.
31. Hothem, R., L. ; Ohlendorf, H. M. *Arch. Environ. Contam. Toxicol.* **1989**, *18*, 773-786.
32. Alaimo, R. S. O.; Knight, A. W. *Arch. Environ. Contam. Toxicol.* **1994**, *27*, 441-448.
33. Kozlov, M. V.; Haukioja, E.; Kovnatsky, E. F. *Environ. Pollut.* **2000**, *108*, 303-310.
34. Kramaz, P. *Bull. Environ. Contam. Toxicol.* **1999**, *63*, 531-537.
35. Kramaz, P. *Bull. Environ. Contam. Toxicol.* **1999**, *63*, 538-545.
36. Peijnenburg, W. J. G. M., et al. *Ecotoxicol. Environ. Saf.* **1999**, *43*, 170-186.
37. Rabitsch, W. B. *Arch. Environ. Contam. Toxicol.* **1997**, *32*, 172-177.
38. Heinz, G. H.; Pendleton, G. W.; Krynitsky, A. J.; Gold, L. G. *Arch. Environ. Contam. Toxicol.* **1990**, *19*, 374-379.
39. Hoffman, D. J., et al. *Arch. Environ. Contam. Toxicol.* **1992**, *23*, 163-171.
40. Stanley, T. R.; Spann, J. W.; Smith, G. J.; Rosscoe, R. *Arch. Environ. Contam. Toxicol.* **1994**, *26*, 441-451.
41. Pain, D. J. In *Environmental Contaminants in Wildlife*; CRC Press, 1996, pp 251-264.
42. Beyer, W. N.; Spann, J. W.; Sileo, L.; Franson, J. C. *Arch. Environ. Contam. Toxicol.* **1988**, *17*, 121-130.
43. Eens, M., et al. *Ecotoxicol. Environ. Saf.* **1999**, *44*, 81-85.
44. Lincer, J. L.; Mcduffie, B. *Bull. Environ. Contam. Toxicol.* **1974**, *12*, 227-231.
45. Craig, E. H.; Craig, T. H. *Lead and Mercury Levels in Golden and Bald Eagles and Annual Movements of Golden Eagles Wintering in East Central Idaho, 1990-1997*; Idaho Bureau of Land Management: Tendo, ID, 1998.
46. Erry, B. V., et al. *Environ. Pollut.* **1999**, *106*, 91-95.
47. Dement, S. H.; Jr, J. J. C.; Barber, J. C.; Strandberg, J. D. *J. Wildl. Dis.* **1986**, *22*, 238-244.
48. Stendell, R. C.; Beyer, W. N.; Stehn, R. A. *J. Wildl. Dis.* **1989**, *25*, 388-391.
49. Hornfeldt, B.; Nyholm, N. E. I. *J. Appl. Ecol.* **1996**, *33*, 377-386.
50. Custer, T. W.; Franson, J. C.; Pattee, O. H. *J. Wildl. Dis.* **1984**, *20*, 39-43.
51. Kim, E. Y., et al. *Environ. Toxicol. Chem.* **1999**, *18*, 448-451.
52. Garcia-Fernandez, A. J., et al. *Arch. Environ. Contam. Toxicol.* **1997**, *33*, 76-82.
53. Henny, C. J.; Blus, L. J.; Hoffman, D. J.; Grove, R. A. *Environ. Monit. Assess.* **1994**, *29*, 267-288.

54. Henny, C. J., et al. *Arch. Environ. Contam. Toxicol.* **1991,** *21*, 415-424.

55. Heath, J. A.; Dufty Jr, A. M. *Physiol. Zool.* **1998,** *71*, 67-73.

56. Burger, J.; Gochfeld, M. *Arch. Environ. Contam. Toxicol.* **1997,** *32*, 217-221.

57. Frederick, P. C., et al. *Environ. Toxicol. Chem.* **1999,** *18*, 1940-47.

58. Bouton, S. N.; Frederick, P. C.; Spalding, M. G.; Mcgill, H. *Environ. Toxicol. Chem.* **1999,** *18*, 1934-39.

59. Sepulveda, M. S.; Frederick, P. C.; Spalding, M. G.; Williams Jr, G. E. *Environ. Toxicol. Chem.* **1999,** *18*, 985-992.

60. Gariboldi, J. C.; Jagoe, C. H.; Bryan Jr, A. L. *Arch. Environ. Contam. Toxicol.* **1998,** *34*, 398-405.

61. Beyer, W. N.; Spalding, M.; Morrison, D. *Ambio* **1997,** *26*, 97-100.

62. Van Der Molen, A. J.; Blok, A. A.; Graaf, G. J. D. *Ardea* **1981,** *70*, 173-184.

63. Green, D. E.; Albers, P. H. *J. Wildl. Dis.* **1997,** *33*, 385-404.

64. Paveglio, F. L.; Kilbride, K. M.; Bunck, C. M. *J. Wildl. Manage.* **1997,** *61*, 832-839.

65. Sparling, D. W.; Lowe, T. P.; Campbell, P. G. C. In *Research Issues in Aluminum Toxicity*; Yokel, R. A., Golub, M. S., Eds.; Taylor and Francis, 1997.

66. Grand, J. B.; Flint, P. L.; Petersen, M. R.; Moran, C. L. *J. Wildl. Manage.* **1998,** *62*, 1103-1109.

67. Burger, J.; Rodgers Jr, J. A.; Gochfeld, M. *Arch. Environ. Contam. Toxicol.* **1993,** *24*, 417-420.

68. Burger, J.; Gochfeld, M. *Arch. Environ. Contam. Toxicol.* **1997,** *33*, 436-440.

69. Honda, K.; Min, B. Y.; Tatsukawa, R. *Arch. Environ. Contam. Toxicol.* **1986,** *15*, 185-197.

70. Eeva, T.; Lehikoinen, E. *Oecologia* **1996,** *108*, 631-639.

71. Hoffman, D. J.; Ohlendorf, H. M.; Aldrich, T. W. *Arch. Environ. Contam. Toxicol.* **1988,** *17*, 519-525.

72. Scheuhammer, A. M. *Toxicol. Appl. Pharm.* **1988,** *95*, 153-161.

73. Mellott, R. S.; Woods., P. E. *J. Field Ornithol.* **1993,** *64*, 205-210.

74. Burger, J. In *Reviews in Environmental Toxicology*; Hodgson, E., Ed.; Toxicology Communications Inc.: Raleigh, NC, 1993; Vol. 5, pp 203-311.

75. Franson, J. C.; Schmutz, J. A.; Creekmore, L. H.; Fowler, A. C. *Environ. Toxicol. Chem.* **1999,** *18*, 965-969.

76. Blus, L. J.; Henny, C. J.; Hoffman, D. J.; Goyer, R. A. *Environ. Pollut.* **1995,** *89*, 311-318.

77. Dauwe, T., et al. *Arch. Environ. Contam. Toxicol.* **2000,** *39*, 541-546.

78. Burger, J. *J. Toxicol. Environ. Health* **1994,** *41*, 207-220.

79. Gochfeld, M. *Arch. Environ. Contam. Toxicol.* **1997,** *33*, 63-70.

338

80. Nyholm, N. E. I. *Arch. Environ. Contam. Toxicol.* **1998**, *35*, 632-637.

81. Adair, B. M.; Cobb, G. P. *Chemosphere* **1999**, *38*, 2951-2958.

82. Lobinski, R.; Szpunar, J. *Anal. Chim. Acta* **1999**, *400*, 321-332.

83. Scheuhammer, A. M. *Environ. Pollut.* **1987**, *46*, 263-295.

84. Woods, J. S. In *Toxicology of Metals*; Chang, L. W., Ed.; CRC Press, Inc.: Boca Raton, New York, London, Tokyo, 1996.

85. Fowler, B. A.; Mahaffey, K. R. *Environ. Health Perspect.* **1978**, *25*, 87-90.

86. Martinez, G.; Cebrian, M.; Chamorro, G.; Jauge, P. *Proc. West. Pharmacol. Soc.* **1983**, *26*, 171-174.

87. Trust, K. A., et al. *Arch. Environ. Contam. Toxicol.* **2000**, *38*, 107-13.

88. Hoffman, D. J., et al. *Comp. Biochem. Physiol.* **1985**, *80C*, 431-439.

89. Adair, B. M. *T.J. McBride, M.J. Hooper, G.P. Cobb-Texas Tech University: D. Hoff-USEPA, W. Olsen- USFWS, unpublished.*

90. Dieter, M. P.; Finley, M. T. *Environ. Res.* **1979**, *19*, 127-135.

91. Johnson, G. D., et al. *Environ. Toxicol. Chem.* **1999**, *18*, 1190-1194.

92. Hoffman, D. J.; Pattee, O. H.; Wiemeyer, S. N.; Mulhern, B. *J. Wildl. Dis.* **1981**, *17*, 423-431.

93. Sanders, B. M.; Goering, P. L.; Jenkins, K. In *Toxicology of Metals*; Chang, L. W., Ed.; CRC Press: Boca Raton, New York, London, Tokyo, 1996.

94. Elliott, J. E.; Scheuhammer, A. M. *Marine Pollut. Bull.* **1997**, *34*, 794-801.

95. Zalups, R. K.; Lash, L. H. In *Toxicology of Metals*; Chang, L. W., Ed.; Lewis Publishers: Boca Raton, New York, London, Tokyo, 1996.

96. Hoffman, D. J.; Ohlendorf, H. M.; Marn, C. M.; Pendleton, G. W. *Environ. Toxicol. Chem.* **1998**, *17*, 167-172.

97. Hoffman, D. J., et al. *Arch Environ Contam Toxicol* **2000**, *39*, 221-32.

98. Congiu, L., et al. *Arch. Environ. Contam. Toxicol.* **2000**, *38*, 357-361.

99. Grandjean, P. *Bull. Environ. Contam. Toxicol.* **1976**, *16*, 101-106.

100. Kuiken, T.; Fox, G. A.; Danesik, K. L. *Environ. Toxicol. Chem.* **1999**, *18*, 2908-2913.

101. Narbaitz, R.; Marino, I.; Sarkar, K. *Teratology* **1985**, *32*, 389-396.

102. Hoffman, D. J., et al. *Arch. Environ. Contam. Toxicol.* **1985**, *14*, 89-94.

103. Atsdr *CERCLA List of Priority Hazardous Substances*; Agency for Toxic Substance and Disease Control: Atlanta, GA, 1999.

104. Eaton, D. L.; Klaussen, C. D. In *Casarett and Doull's Toxicology.*; M.O. Amdur, J. D., and C.D. Klassen, eds., Ed.; Mc Graw-Hill.: New York, NY., 1995.

105. Weltje, L. *Chemosphere* **1998**, *36*, 2643-60.

106. Yanez, L., et al. *Toxicol.* **1991**, *67*, 227-34.

107. Kostial, K., et al. *Toxicol. Ind. Health* **1989**, *5*, 685-98.

108. Krishnan, K.; Brodeur, J. *Arch. Complex Environ. Studies* **1991**, *3*, 1-106.

109. Holcman, A.; Stibilj, V. *Arch. Environ. Contam. Toxicol.* **1997**, *32*, 407-410.

110. Yamamoto, J. T.; Santolo, G. M.; Wilson, B. W. *Environ. Toxicol. Chem.* **1998,** *17,* 2494-2497.

111. Dickerson, R. L., et al. *Environmental Health Perspectives* **1994,** *102,* 65-69.

112. Debacker, V.; Jauniaux, T.; Coignoul, F.; Bouquegneau, J.-M. *Environ. Res.* **2000,** *84A,* 310-317.

113. Reynolds, K. D., et al. *Environ. Toxicol. Chem.* **in press, 2001.**

Chapter 19

Genetic Diversity Provides a Useful Measure of Environmental Impacts

Dan E. Krane

Department of Biological Sciences, Wright State University,
Dayton, OH 45435–0001

Environmental insults diminish an ecosystem's ability to maintain productive and adaptable populations of organisms. We have analyzed the DNA profiles of naturally occurring populations of organisms within freshwater and terrestrial sites with varying degrees of exposure to stressors and find that changes in the underlying genetic diversity of these populations are significantly correlated with the extent to which they have been exposed to anthropogenic stressors. Since it is a population's genetic diversity that is largely responsible for its vigor and ability to adapt to subsequent stressors, these results suggest a generally applicable and sensitive means of directly assessing the impact of stressors upon individual species within an ecosystem.

Long-standing concern over the deteriorating quality of water resources within the United States has resulted in intervention though several far-reaching pieces of federal legislation (*1*). Emphasis on the maintenance of the biological integrity of the nation's waters has resulted in significant improvements in their vigor and economic productivity (*2*) though serious questions have been raised regarding the appropriate means of assessing progress toward that goal (*3*). Toxicity and biologically based indices have proven to be among the most sensitive and useful measures of environmental impacts through their direct assessment of the extent to which water resource systems can harbor adaptable biological communities (*4*). These indices on the biochemical, individual, population and community levels have demonstrated that both acute and chronic exposures to stressors can exert selective pressures upon organisms and that the bioavailability of pollutants is strongly correlated with decreased survival and reproductive success within populations (*5*).

While species diversity and population densities often return to normal levels shortly after remediation in both terrestrial and acquatic habitats (*6*), the population bottlenecks imposed by anthropogenic stressors can result in significant reductions in genetic diversity that are only slowly restored in nature through the processes of migration and mutation. The consequences of long-term diminished genetic diversity can be profound. Free-living populations with high levels of genetic diversity utilize resources more broadly and efficiently (*7*), and selection by either natural or anthropogenic stressors is less likely to result in local extinction of genetically diverse populations (*8*).

As a direct result, short-term acquatic field studies using allozyme-based measures of genetic diversity appear to be more sensitive and reliable means of measuring environmental impacts than assays of more transiently variable features of populations such as their densities or variances in their morphologies (*9*). However, many currently used allozyme markers are from loci that are under direct selection by anthropogenic stressors (*10*) and may not reflect overall levels of genetic variability within a population (*11*). This, in conjunction with practical limitations in their sensitivity (*12*), has caused allozyme assays of genetic diversity to be poorly correlated with generally accepted biotic indices of water quality (*13*).

An approach that more generally and sensitively assesses the extent that anthropogenic stressors have altered levels of overall genetic diversity in populations should be of great utility. RAPD-PCR (random amplified polymorphic DNA polymerase chain reaction) (*14*) is a good candidate method for the assessment of population health in that it has been found to generate very sensitive measures of genetic relatedness within populations of organisms (*15*). RAPD-PCR differs from conventional PCR in that it utilizes short (typically 10 nucleotide long) primers to amplify anonymous and often polymorphic loci (*16*) from genomic DNA isolates (*17*). The resulting markers are generally free of selective constraint (*18*) and can be: 1) readily resolved on agarose gels due to differences in their sizes; 2) reproducibly amplified; and 3) used to distinguish between even closely related individuals of the same species (*19*).

Preliminary results

Site selection

To test the usefulness of RAPD-PCR-based measures of genetic diversity as an augmentation of existing measurements of the effect of environmental impacts, we have used relatively small numbers of primers to generate sets of DNA profiles such as those shown in Figure 1. The RAPD-PCR profiles shown in Figure 1 have been made using the genomic DNA of representative populations (N=18) of native crayfish (*Orcorectes rustics*) collected from three impacted stream systems with closely associated unaffected reference sites within Ohio. Each of the three streams investigated have well documented histories of exposure to anthropogenic stressors and depressed aquatic communities. Specifically, the Little Scioto River at Marion, Ohio is subject to severe sediment contamination from polycyclic aromatic hydrocarbons and receives metal and organic inputs from combined sewer overflows, municipal and industrial effluents, and agricultural runoff. The reference site upstream of the affected region of the river however is free of contamination. Similarly, Dick's Creek at Middletown, Ohio receives several discharges from a large steel plant and non-point source inputs from neighboring hazardous waste landfills. A nearby reference stream (Elk Creek) of similar order provided a suitable reference comparison. Lastly, the impacted sites of the Ottawa River at Lima, Ohio receives refinery and municipal waste water effluents, combined sewer overflows, and agricultural runoff unlike the immediately upstream reference site that has a good fish and benthic community.

Methodology

All genomic DNA isolations used in the generation of RAPD-PCR profiles by our research group are made with QIAquick PCR purfication kits (Qiagen). The resulting pellets are washed with 70% ethanol, dried and resuspended in 50 µL of TE [10mM TRIS (*p*H 8.3), 1 mM EDTA]. The quantities of DNA isolated for each sample are estimated by electrophoresis on a 1% agarose yield gel. Upon dilution to make DNA concentrations consistent between samples, all isolates are either immediately utilized or frozen at -20° for later use.

RAPD-PCR profiles are routinely generated from total genomic DNA using the following conditions: final reaction volumes were 10 µL and contained 2 µL of diluted genomic DNA, 1.5 units of KT1 KlenTaq (Wayne Barnes, Washington University, St. Louis, MO), 20 mM TRIS, *p*H 8.0, 2.5 mM MgCl$_2$, 16 mM (NH$_4$)$_2$SO$_4$, 150 µg/mL bovine serum albumin, 0.2 µM of a single primer (i.e. A-01: 5'-CAGGCCCTTC-3' or A-02: 5'-TGATCCCTGG-3') and

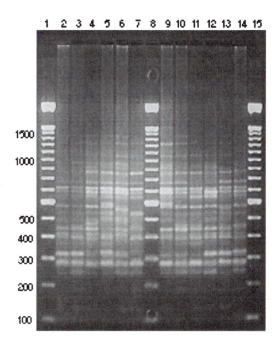

Figure 1. RAPD-PCR profiles of crayfish collected from an impacted site and one of its associated reference sites. PCR products generated using the A-01 primer (Operon Technology) on genomic DNA isolated from six crayfish collected at a reference site (IBI value: 36) on the Ottawa River (lanes 2-7), and from six crayfish collected at a downstream impacted site (IBI value: 21) on the same river (lanes 9-15). Molecular size markers are included in lanes 1, 8 and 16 and are indicated at left.

60 μM dNTP. MJ Research thermocyclers (PTC-100 and Mini-cycler models) are used for amplifications for 45 cycles consisting of the following steps: 92° for 1 minute, 36° for 1 minute, 68° for 2 minutes. An additional extension at 68° for 5 minutes follows the last round of amplification. All samples are held at 4° until RAPD-PCR products can be resolved by gel electrophoresis. RAPD products are then electrophoresed in 2% agarose gels in TBE buffer (10 mM TRIS, pH 8.3, 10 mM boric acid, 1 mM EDTA) at 4°. Gels are stained in ethidium bromide for one half hour and destained in water for one hour. Bands can then be visualized with a UV lamp and documented using a Gel Print 1000i imaging system (BioPhotonics Corp.).

Scoring and data analysis

Each RAPD-DNA profile is routinely scored twice, independently, and the RAPD-PCR amplification of a sample is repeated in cases where scorings were not in complete agreement. A measure of the genetic similarity of individual organisms within a population to others collected at the same site was obtained by determining the fraction (f) of markers it shared with other individuals from the same site using the following equation:

$$f = 2\left\{ \frac{mxy}{mx + my} \right\}$$

where mxy is the number of bands any two samples share and mx and my are the number of bands amplified from each organism (20). For example, pairwise similarities generated in this way for crayfish populations ranged from 0.62 to 0.86 (x=0.739; s=0.056) for reference sites and were significantly less than those observed at impacted sites (p<0.01) where pairwise similarities ranged from 0.72 to 0.92 (x=0.848; s=0.044) (21). Average pairwise genetic similarities between crayfish at reference sites were consistently lower than those observed among individuals collected at otherwise similar impacted sites (Table 1) (21).

Biologically based indices such as the Index of Biotic Integrity (IBI) (22) and the Invertebrate Community Index (ICI) (23) are presently the method of choice for assessing the impact of anthropogenic stressors at aquatic sites. While both of these indices reduce a large number of metrics (ranging from measures of population density to species diversity) to a final summary value, neither utilize direct measures of the genetic diversity of organisms resident to a site. In addition to utilizing smaller quantities of sample material and being less expensive to perform, the average pair-wise genetic similarities generated with RAPD-PCR profiling generate a measure of genetic diversity that is significantly correlated (for crayfish, N=144; r=-0.770; p<<0.001 and N=144;

r=-0.707; p<<0.001 for IBI and ICI, respectively) with these other accepted indicators of biotic integrity (Figure 2) unlike allozyme-based approaches (*13*).

A nonparametric measure, the Jonckheere test for ordered alternatives (*24*), also suggested that a monotonic trend (not necessarily linear) existed between diminished genetic diversity in crayfish and reported IBI values for all the sites included in this study (p=0.0002). The probability value was determined based on an analysis of 5,000 random permutations of the data set with the null hypothesis stating that there is no difference among the mean pairwise genetic similarities from the eight sites used in this study (H_O: $\mu_1 = \mu_2 = \mu_3 = \ldots = \mu_n$) and the alternative hypothesis that there is a monotonic trend based on a priori information (H_A: $\mu_1 \geq \mu_2 \geq \mu_3 \geq \ldots \geq \mu_n$, where at least one of the inequalities is a strict inequality).

On-going work with snails (*Physella gyrina*) from the same sites used in the crayfish study indicate that a similar inverse relationship between genetic diversity and environmental quality exists for populations of these non-migratory organisms.

Prospects

Environmental stress is just one of many potential causes for reduction of standing variation within a population and genetic diversity measures based on RAPD-PCR are unlikely to be able to distinguish between naturally occuring and anthropogenic stressors. In addition, a reduction of genetic diversity is not necessarily the only selective response to anthropogenic stressors.

Trends contrary to those seen for crayfish and snails provide practical examples of the consequences of these other effects. For instance, preliminary studies of adult damsel fly populations actually show an increase in genetic diversity at polluted sites. Damsel flies are popular ecoindicator species because of the sensitivity of their aquatic larvae to sediment contamination. Migrant adults from surrounding unaffected sights appear to dominate sampling at polluted sites and result in genetically diverse populations whose progeny are likely to not survive development. Similarly, other on-going studies within the laboratory on pill bug populations also suggest an increase in genetic diversity that is correlated with anthropogenic stressor levels. Pill bugs are known to be unusually resistant to a wide range of pollutants, and decreased predation and increased refuge availability at polluted sites allows them to maintain larger, more diverse populations than at unaffected sites.

Nonetheless, RAPD-PCR appears to be a sufficiently sensitive measure of population health in at least four different species to detect significant differences in genetic diversity between sites affected by common anthropogenic stressors and very similar but unaffected reference sites. As such, RAPD-PCR based measures of a population's genetic diversity have the potential to be the basis of a valuable alternative or augmentation to conventional assessments of environmental insults.

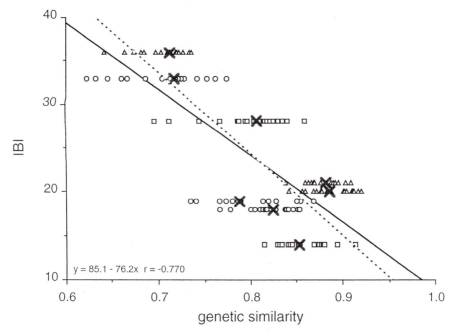

*Figure 2. Index of Biotic Integrity (IBI) values are inversely correlated with RAPD-PCR based measures of genetic similarity. Average pairwise similarity of each organism relative to all others collected at its site are plotted against the independently obtained IBI value (21) for that site (N=144, r=-0.770, p<<0.001) and a solid line shows the best fit linear regression. Crayfish collected from the impacted and reference Ottawa River sites are displayed as triangles (Δ) those collected from sites along the Little Scioto River are displayed as circles (O) and those collected from Elk Creek and its reference stream, Dick's Creek, are shown as squares (□). Large **X**'s correspond to the mean pairwise genetic similarity of crayfish at each site (N=8, r=-0.804, p<0.01) and a dashed line corresponds to the best fit linear regression for those points.*

Unlike the metrics that are considered in assigning IBI and ICI values, genetic diversity measures should be relatively insensitive to naturally occurring seasonal and habitat variability as well as other complicating factors. Instead, genetic diversity measures directly assess the biological basis of a population's ability to adapt and respond to both natural and anthropogenic stressors and, as a result, to its vigor and economic potential. Since RAPD-PCR profiles simultaneously survey a relatively large set of random genetic markers that are very likely to be free from selection by stressors, they are also more likely to provide a truer estimate of a population's genetic diversity than other molecular-based systems. As a direct result, they should allow detection of both chronic and acute exposures to stressors within populations of organisms with low migration rates and provide earlier warnings of impending local and even global extinctions. Choice of organisms with long life spans (i.e. trees) or at high trophic levels or with large geographic ranges (i.e. eagles) might also allow characterizations of habitat quality over long time spans and/or large geographic ranges. And, unlike other biologically based indicators, genetic diversity approaches also have the additional advantage of being potentially useful in a wide variety of systems ranging from freshwater, marine and terrestrial.

Just as measures such as IBI and ICI rely upon information derived from a variety of species, measures of genetic diversity should also be broadly based since the genetic diversity of any single species can also be influenced by a variety of natural factors such as competition and niche size. However, our analyses using crayfish and on-going work with snails (*Physella gyrina*), pill bugs and damsel flies from the same sites used in this study indicate that measures of the genetic diversity of even a single well-chosen sentinel species are highly indicative of the extent to which a site has been impacted by anthropogenic stressors.

Acknowledgements

David Sternburg was instrumental in the collection and generation of the crayfish and snail RAPD-PCR data while Scott Rousch and Billy Grunwald, Jr. played major roles in the collection and generation of profiles for damsel flies and pill bugs. Benjamin Collins kindly provided a custom-written FORTRAN program for Jonckheere testing. Ze' Ayala and Keith Grasman provided helpful suggestions about the manuscript. This work is support by an EPA STAR grant to DEK.

References

1. The Water Pollution Control Act of 1966. The Federal Water Pollutions Control Act Amendments of 1972. The Clean Water Act of 1977.

348

2. E. A. Starbird, *Natl. Geogr.* **141,** 816 (1972); National Research Council, in *Restoration of Aquatic Ecosystems,* (National Academy Press, Washington, DC, 1992).
3. National Research Council, in *Testing of the Effects of Chemicals on Ecosystems* (National Academy Press, Washington, DC, 1981).
4. J. F. McCarthy and L. R. Shugart, in *Biomarkers of Environmental Contamination,* J. F. McCarthy and L. R. Shugart, Eds.(Lewis Publishers, Boca Raton, FL, 1990), pp.3-14.
5. G. A. Burton, *Eniron. Toxicol. Chem.* 10, 1585 (1991).
6. J. E. Thorpe and J. F. Koonce, *Can. J. Fish. Aquat. Sci.* 38, 1899 (1981); F. W. Allendorf and R. F. Leary, in *Conservation Biology: The Science of Scarcity and Diversity,* M. E. Soule, Ed. (Sinauer Associates, Inc. Sunderland. MA, 1986), pp. 57-76.
7. J. A. Beardmore, T. H. Dobzhansky, D. Paulovsky, *Heredity* **14,** 49 (1960); J. A. Beardmore and L. Levine *Evolution* **17,** 121 (1963); T. Long, *Genetics* **66,** 401 (1970); J. F. McDonald and F. J. Ayala *Nature* **250,** 572 (1974); A. Minawa and A. J. Birley, *Nature* **255,** 702 (1975); J. R. Powell and H. Wistrand, *Am. Naturalist* **112,** 935 (1978); T. F. C. Mackay, *Genetical Res.* **37,** 79 (1981).
8. F. W. Allendorf and R. F. Leary, in *Conservation Biology: The Science of Scarcity and Diversity,* M. E. Soule, Ed. (Sinauer Associates, Inc. Sunderland. MA, 1986), pp. 57-76.
9. J. B. Mitton and R. K. Koehn, *Genetics* **79,** 97 (1975); E. Nevo, T. Shimony, M. Libni, *Nature* **267,** 699 (1977); E. Nevo, R. Noy, B. Lavie, A. Bieles, S. Muchtar, *Biol. J. Linn. Soc.* 29, 139 (1986); Smith et al., *Copeia* 1983, 183 (1983); B. Lavie and E. Nevo, *Marine Poll. Bul.* 17, 21 (1986); M. J. Benton, S. A. Diamon and S. I. Guttman, *Ecotoxicol. Env. Saf.* 29, 20 (1994).
10. B. Lavie and E. Nevo, *Marine Biol.* **71,** 17 (1982); B. Lavie and E. Nevo, *Marine Poll. Bul.* 17, 21 (1986); B. Lavie, E. Nevo, U. Zoller, *Environ. Res.* 35, 270 (1984); R. B. Gillespie and S. I. Guttman, *Environ. Toxicol. Chem.* **8,** 309 (1989); R. B. Gillespie and S. I. Guttman, in *Environmental Toxicology and Risk Assessment, 2nd Volume,* J. E. Gorusch, F. J. Dwyer, C. G. Ingersoll and T. W. LaPoint Eds.(American Society for Testing and Materials, Philadelphia, 1993) pp.134-145; M. Mulvey et al. *Enivron. Toxicol. Chem.* **14,** 1411 (1995); M. A. Schlueter, S. I. Guttman, J. T. Oris, A. J. Bailer *Environ. Toxicol. Chem.* **14,** 1727 (1995).
11. J. L. Gooch and T. J. M. Schopf, *Biol. Bull.* **138,** 138 (1970).
12. R. B. Gillespie, C. F. Facemire, S. I. Guttman, *Biochem. Syst. Ecol.* **19,** 541 (1991).
13. S. A. Fore, S. I. Guttman, A. J. Bailer, D. J. Altfater, B. V. Counts, *Ecotoxicol. Env. Saf.* **30,** 24 (1995); _____ *ibid.,* 30, 36 (1995).

14. J. G. K. Williams, A. R. Kubelik, K. J. Livak, J. A. Rafalski, S. V. Tingey *Nucl. Acids Res.* **18**, 6531 (1990).

15. J. G. K. Williams, A. R. Kubelik, K. J. Livak, J. A. Rafalski, S. V. Tingey *Nucl. Acids Res.* **18**, 6531 (1990); J. G. K. Williams, A. R. Kubelik, K. J. Livak, S. V. Tingey, J. A. Rafalski, in *More Gene Manipulations in Fungi*, J. W. Bennett and L. L. Lasure, Eds. (Academic Press, San Diego, CA, 1991), pp. 433-439.

16. N. A. Tinker, M. G. Fortin, D. E. Mather, *Theor Appl Genet* 85, 976 (1993).

17. D. R. Levitan and R. K. Grosberg *Mol. Ecol.* **2**, 315 (1993).

18. R. K. Grosberg, D. R. Levitan and B. B. Cameron, in *Molecular Zoology: Advances, Strategies and Protocols*, J. D. Ferraris and S. R. Palumbi, Eds. (John Wiley & Sons, Inc., New York, 1995), pp.67-100.

19. M. Lynch, *Mol. Biol. Evol.* **7**, 478 (1990).

20. A. G. Clark and C. M. S. Lanigan, *Mol. Biol. Evol.* 10, 1096 (1993).

21. D. E. Krane, D. C. Sternberg, G. A. Burton, Envir. Toxic. Chem., **18**, 504 (1999).

22. J. R. Karr, K. D. Fausch, P. L. Angermier, P. R. Yant, I. J. Schlosser, *Ill. Nat. Hist. Surv.Spec. Publ.* **5**, 28 (1986); J. R. Karr *Environ. Toxicol. Chem.* **12**, 1521 (1993).

23. Ohio EPA, *Biological Criteria for the Protection of Aquatic Life, Volume I, The Role of Biological Data in Water Quality Assessment* (Columbus, OH, 1987); Ohio EPA, *Biological Criteria for the Protection of Aquatic Life: Volume II: Users Manual for Biological Field Assessment of Ohio Surface Waters* (Columbus, OH, 1987); Ohio EPA, *Biological Criteria for the Protection of Aquatic Life: Volume III: Standardized Biological Field Sampling and Laboratory Methods for Assessing Fish and Macroinvertebrate Communities* (Columbus, OH, 1987); Ohio EPA, *Addendum to: Biological Criteria for the Protection of Aquatic Life: Volume II: Users Manual for Biological Field Assessment of Ohio Surface Waters* (Columbus, OH, 1989).

24. M. Hollander and D. A. Wolfe, *Nonparametric Statistical Methods* (John Wiley & Sons, New York, 1973), pp. 120-3.

Chapter 20

Plant Biomonitors: Pollution, Dandelions, and Mutation Rates

Steven H. Rogstad, Brian Keane, Matthew Collier, and Jodi Shann

Department of Biological Sciences ML6, University of Cincinnati, Cincinnati, OH 45221–0006

We are exploring the use of the common dandelion (*Taraxacum officinale* Weber; Asteraceae) as a plant biomonitor that may reflect the degree to which communities have been impacted by anthropogenic effects, especially pollutants. Specifically, we are investigating whether dandelion populations that have experienced increased pollution exposure exhibit higher mutation rates or changes in genetic diversity at variable-number-tandem-repeat (VNTR) loci (also called microsatellite or minisatellite loci). Here we review the characteristics of dandelions that suggest why they might be of use in such biomonitoring, outline the types of information we are extracting from dandelion populations to that end, and present some of our initial findings.

Plant biomonitors ideally would provide us with information on environmental anthropogenic influences that are otherwise difficult to detect. While monitoring for numerous individual factors of concern (e.g., surveys for the presence of metals or organic contaminants, Uvb or other radiation, etc.) is possible, it is not often practical, necessarily meaningful, or efficient. For example, assessing air or soil lead concentrations at a given point in time may not reflect fluctuations over time or changing bioavailability or bioimpact. Further, the suite of anthropogenic influences at a site may involve numerous, independently fluctuating factors (e.g., several different organic compounds), and measuring each alone may not reflect how different members of the suite may be interacting through time.

An ideal plant biomonitor, on the other hand, would provide us with information that "integrates the signal" presented by all of the anthropogenic factors affecting a site such that a determination could be made as to whether those factors, over the growth period of the plant, are in some way affecting the biological functioning or integrity of the population or, better yet, the community. Although model animal biomonitors for such purposes are being investigated, plants offer several advantages, including: 1) plant growing tissues are constantly exposed to the environment at the site of interest in contrast to animals that may move offsite, or feed on materials from offsite; 2) many plants have short reproductive cycles compared to many animals; and 3) plants can produce large offspring arrays compared to many animals. The importance of the latter two factors will be emphasized when the method we are using for detecting mutation rates is explained below.

Dandelions (*Taraxacum officinale*) have several characteristics that make them of particular interest as plant biomonitors. *T. officinale* has a very broad ecological amplitude and thus is widely distributed across the United States, and indeed, throughout the world. This species commonly grows at a variety of mesic to drier sites, and ranges from tropical sea level to alpine habitats. Dandelions can be found in both synanthropic and exanthropic habitats. If dandelions prove to be good environmental biomonitors, they can be employed across a very broad range of community types and at several geographical scales of organization. In addition, these plants are opportunistic, and have a very fast rate of establishment and growth under a broad range of ecological conditions. Therefore, dandelions can be introduced into target sites for rapid monitoring of possible pollution stressor presence, or rapidly grown under conditions where the presence of potential stressors could be experimentally manipulated.

Another characteristic of dandelions making them a logical choice for investigation as plant biomonitors is prior research demonstrating that a variety of pollutants can be taken up and sequestered in dandelion tissues (*1,2*; our results - see below). It has been demonstrated that dandelions can grow across a range of polluted conditions, and that these plants may be better environmental biomonitors for pollutants (in the sense that tissue samples can be analyzed for the bioaccumulation of pollutants) than several other plant species already in use as tissue-concentration biomonitors (*1*). Despite the well-known tolerance of dandelions to anthropogenic contaminants, we have been unable to find any information on the genetic effects of the accumulation of pollutants in dandelion tissues, or in tissues of other plants, for that matter. Thus, our research approach described below will provide new information on genetic effects in plants in relation to the presence of environmental pollutants.

Yet another reason for examining dandelions as a possible plant biomonitor is that we have determined that this species can readily be surveyed for a large number of genetic loci of a type previously shown to exhibit relatively high rates of mutation. These loci are DNA sequences including tandem repeats of a short "core" sequence. Many different core sequences have been identified, ranging in size from 2-4 base pairs (so-called simple sequence repeats, e.g., *3*, or microsatellites; *4,5*), to core sequences of up to 30-60 base pairs (minisatellites; *6*), or even larger (macrosatellites; *7*). Loci at which such tandem repeats occur often exhibit relatively high genetic variation detectable as numerous different alleles existing in populations. These alleles are thought to differ primarily in the number of tandem repeats of the core sequence, and thus have been termed variable-number-tandem-repeat (VNTR) sequences. Several different VNTR loci have been shown to be hypervariable with heterozygosities in excess of 90% in several different organisms, and in general, VNTR sequences provide genetic markers that are among the most variable known for eukaryotes (e.g., *8,9*). VNTR loci may be surveyed at the individual locus level (*4,5*), or probes may be used to survey several loci simultaneously (multilocus probes; e.g., *6,10*). VNTR sequences are now known to exist in a wide range of eukaryotes, and in some cases, the same core sequence repeat probe provides variable genetic markers for humans, other vertebrates, and plants (*11,12*).

The high variability of VNTR markers suggests that they should have relatively high, detectable mutation rates. One problem in surveying mutation rates using morphological features or allozymes is that both typically have low levels of variation, and thus have such low mutation rates that very large samples sizes must be surveyed to characterize them. The notion that mutation rates are probably higher at VNTR loci comes from comparison of their genetic diversity with allozymes. For example, a survey (*13*) of allozyme analyses of 468 plant taxa produced a mean H (heterozygosity) of 0.113, with one of the highest heterozygosities being 0.481 for a cactus species. The average H for six plant species surveyed with VNTR markers (*9*) was 0.557, the highest being quaking aspen (H=0.760). The higher level of VNTR variation, and the implication that it is generated through higher rates of mutation, make VNTR sequences leading candidates for objects of study as monitors for the presence of mutagens. Indeed, that VNTR sequences do have readily detectable rates of mutation in dandelions is shown in the Results of Preliminary Research section below.

Another reason for hypothesizing that VNTR sequences may be useful in detecting increased mutation rates is that a few intriguing studies have now demonstrated that increased VNTR mutation is correlated with increased exposure to different types of mutagens in a handful of organisms. For example, two tumor lines exposed in vitro to the mutagen PhIP exhibited increased mutation rates at VNTR loci relative to control cultures (*14*). Exposure to radiation increased VNTR mutation rates in medaka fish (*15*) and in mice (*15,16,17*). Comparing mutation rates at VNTR loci in children born to parents living in areas heavily polluted by the Russian Chernobyl nuclear power station accident of April, 1986, with mutation rates in children from unpolluted locations, Dubrova et al. (*16*) found mutation rates were approximately twice as high in the polluted areas. Although the increase in mutation rates was positively correlated with Caesium-137 surface contamination, it was noted that a wide array of environmental contaminants, radioactive as well as non-radioactive (including heavy metals) were also released in the Chernobyl fallout, so the exact causative agents remain unclear. We are aware of only one study where VNTR mutation rates have been examined in relation to exposure to known non-radioactive mutagens in a natural population, in this case animals. Yauk and Quinn (*18*) found elevated VNTR mutation rates in a gull population inhabiting an industrial site contaminated with a variety of anthropogenic pollutants relative to populations at more "pristine" sites.

A final reason to examine the mutation of VNTR markers in response to mutagens is their involvement in human genetic diseases and cancer (e.g., *19*). Several human diseases, including fragile X syndrome, spinal and bulbar muscular atrophy, Huntington's disease, myotonic dystrophy, and several hereditary ataxias have been found to be caused, at least in part, by expansions of trinucleotide core repeats (*20*). Reports of VNTR sequence or microsatellite mutation involvement in various cancers number in the hundreds (e.g., *21*, *22*). The demonstration that increases in VNTR mutation rates are detectable in pollution-exposed plants would thus not only provide a monitoring system for the presence and activity of environmental mutagen stressors, but would also provide an alarm system reflecting the presence of stressors that might increase the occurrence of VNTR related genetic diseases in humans. Further, since the mechanisms involved in the development of VNTR related diseases are largely unknown, the demonstration that VNTR mutation rates can be stimulated in plants would provide non-human, non-animal models that are easy to manipulate and rapidly generate high sample sizes for future research into the causative mechanisms of such VNTR mutation and associated diseases.

A limited number of studies have demonstrated that various types of pollutants can affect aspects of the genetic diversity of populations (e.g., *23,24,25*). Genetic diversity often decreases in such populations, which is generally viewed as a negative outcome since the maintenance of genetic diversity is thought to be beneficial for the long-term survival and adaptability of most species (*26,27*). What is needed is a collection of biomonitor species that can be utilized to determine if and how genetic diversity changes in response to anthropogenic influences. Again, VNTR markers are the markers of choice for such studies since they provide much higher genetic marker variation, and thus greater sensitivity for detecting changes in genetic diversity. In previous studies, we have demonstrated how VNTR marker variation can be used to investigate whether measures of genetic diversity of plant populations growing at pollution- impacted versus non-impacted sites differ (e.g., *28,29*). We demonstrate below that dandelions have extensive VNTR genetic variation facilitating such comparative analyses of genetic diversity within and between populations, and further, that the mode of seed production of dandelions is especially well suited for the analysis of VNTR mutation rates relative to mutagen exposure.

Results of Preliminary Research

Site Selection and Analysis

With regard to dandelion populations, the anthropogenic influence we are investigating is exposure to metal pollutants, some of which have been shown to be mutagens with other organisms. To identify sites that reflect a broad range of contaminant exposure, we have used two strategies. First, we used USEPA air quality monitoring information available at http://www.epa.gov/airsweb/monvals.htm where data collected at hundreds of stations across the United States are listed. The data on PM_{10} (particulate matter ≤ 10 micrometers, measured as micrograms per m^3 recovered on a daily basis) was used since metal contamination may be positively correlated with PM_{10} (*30*). Based on PM_{10} data averaged over 4-6 years, 21 stations were chosen that span a range from the most "pristine" (PM_{10} <20) to the most "contaminated" (PM_{10} >40). Second, we have identified five sites where an industrial point source resulted in metal contamination of the soil. On these 26 sites, located in the central United States, that vary in contamination composition and degree, we have sampled soils, dandelion population tissues, and seeds.

We extracted and analyzed samples of soil (two samples per site) and dandelion leaf tissue (leaf tissue from at least 25 plants was pooled) from each site for eight metals using inductively coupled plasma mass spectrometry for Cd, Cr, Cu, Fe, Mn, Ni, and Zn, and atomic absorption spectrometry for Pb. Two replicates of each sample were analyzed. Metal concentration values for both samples at a site and replicates were highly correlated (*31*). Figure 1 is an example of mean values from one such analysis (for Zn). In Figure 1, it can be seen that the sites surveyed span a range of Zn exposure. Our results suggest that soil metal concentrations are positively and significantly (P<0.05 criterion) correlated with

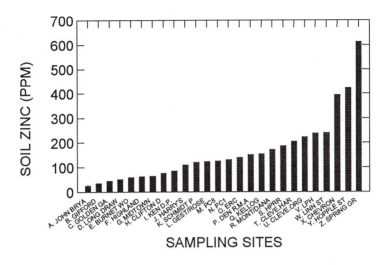

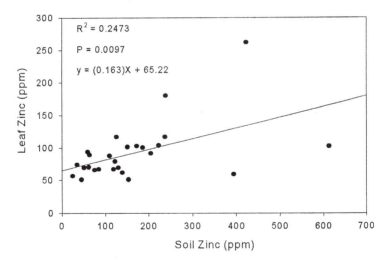

Figure 1. Above. Zn detected in soils sampled at 26 sites spanning a range of pollution exposures. Below. Concentrations of Zn in soils versus in leaf tissues from > 20 dandelion plants collected at each of these 26 sites. More details on these and other analyses may be found in 31.

PM_{10} values (31). These findings may have relevance for studies showing that human death rates increase with increasing $PM_{2.5}$.

However, our results also suggest that metal presence in tissues may be only weakly positively, or not at all, correlated with soil metal concentrations (note positive, significant correlation with relatively low R^2 value in Figure 1). In other words, the bioavailability of a metal may not be strongly correlated with its presence at a site. Metal uptake by dandelions may be influenced by a host of factors, including seasonality (31), moisture fluctuation, soil pH, presence of other elements or chemicals including natural or anthropogenic organics, intraspecific differences in plant genotypes, varying routes of exposure and uptake (from soil versus from atmosphere, transport mobility, etc.).

Development of Widely Applicable VNTR Probes

A method has been developed by which the polymerase chain reaction (PCR) is used to generate long, synthetic tandem repeat (STR) probes (10). To summarize the method for generating a particular PCR-STR probe (see 10, 11 for more details), a short oligomer (the 20-30 bp template) of tandem repeats of a selected core sequence is created with a DNA synthesizer. A second sequence (the 20-30 bp complement) composed of repeats of the complementary core sequence is also synthesized. Equimolar mixes of the template and complement sequences are then subjected to several, more or less standard, PCR cycles. During the annealing step of each cycle, some of the template and complement strands will anneal out of register in a manner (with DNA double strand 5' overhangs) permitting extension at internal 3' strand ends (the notations 5' and 3' refer respectively to the phosphate end and the ribose free hydroxyl end of the DNA strand). This extension produces new longer strands composed of more core repeats. Over several PCR cycles, longer and longer strands are created. For each core sequence (template-complement combination), a series of reactions using different cycling conditions (Mg concentrations; annealing temperatures) are explored to determine the most stringent conditions under which extension will take place. These optimal conditions for a particular template-complement combination need only be determined once, after which they can be used in all subsequent PCR reactions.

To generate a PCR-STR probe, the optimized reaction and cycling conditions are used to evenly produce strands over a broad size range, and strands of 300-700 bp are then excised and recovered from an agarose gel. A fraction of the recovered DNA is then used in a one-sided PCR reaction to produce ^{32}P-labeled probes. These labeled PCR-STR probes are used individually in Southern blot hybridization protocols. Using longer repetitive DNA probes may yield more stable, clearer results than obtained under similar conditions with end-labeled, shorter oligomer repeats (e.g., 32; personal observations). Several studies have shown that VNTR genetic markers are generally nuclear (as opposed to mitochondrial or chloroplast DNA), and that they obey Mendelian inheritance (e.g., *11,33,34,35,36*).

All of the 20 core repeat sequences we have thus far used as PCR-STR probes (e.g., *11*) were chosen based on previously reported minisatellite sequences shown to reveal genetic variation in some organism, except for three that are randomly chosen sequences. We have shown that all of these probes reveal genetic variation in more than one plant species, and several have detected extensive variation in several different species (*9,11*). While for any given species it is rather unpredictable exactly which probes will be most effective at revealing genetic variation, in our experience, variation in all species examined has been detected with at least some probes (*9*). The wide taxonomic applicability of PCR-STR probes in revealing VNTR genetic markers indicates that the methods employed here for investigating the presence of anthropogenic stressors that may affect mutation rates or the genetic diversity of dandelion populations can readily be transferred to other species.

VNTR Genetic Markers in Dandelions

Yet another reason that dandelions are of interest as possible biomonitors for mutation stressors is their mode of reproduction via seeds. Most angiosperms produce seeds via sexual means in which a zygote is produced through the fusion of two different gametes, the zygote then developing to become the embryo and the new seedling. However, in dandelions, each seed bears an embryo produced asexually through a process known as agamospermy in which a diploid cell of the "maternal" plant fails to undergo meiosis and becomes the first cell of what develops into the embryo (*37,38*). In other words, all seeds produced by one dandelion plant should be genetically identical at all loci to the maternal plant (barring mutation).

We have shown that most dandelion offspring are genetically identical to their maternal parental plant when surveyed across more than 70 VNTR loci using PCR-STR probes (*9*). However, we have also determined that novel mutation of new markers in offspring also occurs at a detectable rate. Discerning such mutations is easy since they appear as new bands not present in the maternal plant, or the lack of a maternal band. The fidelity of transmission of most VNTR markers to dandelion offspring, and the detection of a new mutant marker are demonstrated in Figure 2.

To date, surveying eight dandelion populations for 76 adults and 540 of their offspring we have detected 191 mutant VNTR bands among 24,845 transmitted, yielding an average mutation rate of 0.0077. This is equivalent to 1 mutant band for every 130 bands transmitted, being one of the highest mutation rates known for plants. These initial results thus confirm that VNTR markers should be among the most informative in biomonitoring for mutation stressors.

Figure 2 also demonstrates another important feature of dandelion VNTR genetic markers that makes dandelions useful as ecological indicators. Note that for each of the two maternal parent-offspring sets of individuals, while in each case the offspring are identical to the maternal parent (except for the indicated mutation), the two parent-offspring sets differ from each other at numerous VNTR markers. These two parents were collected approximately 1 m apart at the same natural field site. Other initial results (not shown) indicate that when 22 dandelion individuals are sampled from 100 m² plots in natural populations, most of the individuals sampled have different VNTR fragment profiles, with individuals differing from one another similarly to what is depicted for the two maternal parents in Figure 2. A computer program, GELSTATS, has been designed to statistically analyze data sets composed of such VNTR fragment profile information for various population genetic diversity parameters (*39*). The VNTR genetic marker diversity we have detected in dandelions, combined with these analyses, will permit us to examine whether dandelion populations at pollution-impacted sites differ from non-impacted sites in several population genetic parameters (e.g., average band number per individual, levels of similarity within and between populations, whether genetic diversity differs among sites, etc.). For an example of the application of these same methods demonstrating that VNTR diversity can be used to discern genetic differences in polluted versus non-polluted, adjacent conspecific plant populations of *Rubus idaeus* and *Typha latifolia*, see *28,29*.

Prospects

There are different types of biomonitors. Our initial results, and the results of others (*40,41*), suggest that dandelions and other plants may not always be good biomonitors in the sense that pollutant amounts in tissues should be highly correlated with the degree of site contamination. This implies that contaminant uptake by plants is a complex process, some of the variously interacting factors involved being noted above. Again, environmental presence is not necessarily equivalent to bioavailabililty. Instead, however, dandelions may be good biomonitors in the sense that they do reflect contaminant bioavailability, and thus, exposure risk. Further, since it is clear that dandelions do sequester metals (e.g., Figure 1), some of these metals may contribute to increased mutation rates (*42,43*). Assessing VNTR mutation rates in dandelion populations resident across a range of pollution exposures will provide the first information as to whether this species can be used as a biomonitor for mutation stressors. As noted earlier, the suite of stressors and their fluctuation through time can be very complex, and involve

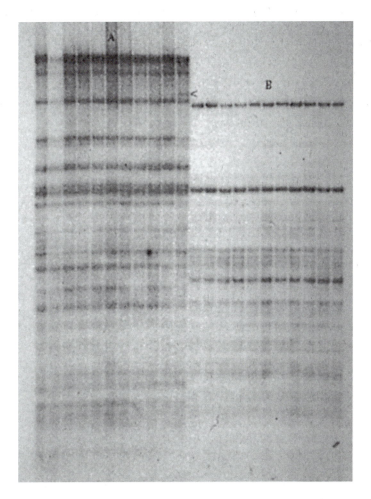

Figure 2. Maternal plant to offspring transmission of VNTR genetic markers in dandelion detected with the TTCCA (= core sequence) PCR-STR probe (prepared according to 10,11). Markers for two different maternal parents are shown (lanes A and B), with 10 offspring of each maternal parent also shown (5 offspring on each side of each parent). More than 30 VNTR markers are shown across both parents. One new mutant marker is detectable in the eleventh lane from the left (marked with a <). If an autoradiograph is interpreted to have a transmission of approximately 30 markers to 20 offspring (= 600 markers transmitted), with one novel marker, the detectable mutation rate is 1/601 = 0.00166 for that particular gel-probe autoradiograph. 3 ug genomic DNA per lane was digested with a five-fold excess of TaqI restriction endonuclease.

factors other than chemical pollution (e.g., UVb or other radiation). *In situ* stressor bioindicators would act to integrate the combined signal coming from this complex and varying array of anthropogenic influences, providing a measure of whether this signal is having an impact on the genetics of populations. Such ecological indicators may even provide us with a new type of the proverbial "miner's canary," warning us of previously undetected, synergistic effects of combinations of anthropogenic factors that, when measured alone, seem below levels known to induce detrimental consequences.

Another area for future research concerns the use of plants as tools for the phytoremediation and reclamation of contaminated environments. Although numerous plants are being explored for their capability to remove or stabilize contaminants at polluted sites, to our knowledge, no studies have been conducted on the genetic consequences. If such exposed plants are experiencing increased rates of mutation, then increased rates of novel mutations may be spreading into adjacent conspecific populations, altering their genetic characteristics. The demonstration of increased VNTR mutation rates in response to increased contamination would thus have implications for the planning of phytoremediation projects.

As previously noted, VNTR loci have been involved in certain types of genetic diseases, primarily due to expansion of triplet cores, and carcinogenesis in humans. It would seem likely that deleterious triplet-like expansions at VNTR loci in plants may also occur, contributing to increased genetic loads. Knowledge that mutation rates increase in natural populations with increasing pollution may thus have implications for the maintenance of the biological integrity of populations. Further, establishing a link between VNTR mutation in plants and in mammalian or human cell cultures relative to mutagen exposure becomes of interest: if a relationship exists, plants would become powerful surrogate monitors for environmental mutagens (humans obviously cannot be directly used as *in situ* monitors) that could be used to survey for mutagenic risks to humans across a wide array of habitats.

If successful, the approach being used with dandelions should be applicable to numerous other plant and animal species that reproduce in a similar asexual manner. Agamospermy, the production of the new seed embryo from a maternal cell, occurs widely in the dandelion family, and is also taxonomically widespread among the angiosperms (*37*). Thus, VNTR mutation rates can be explored in a broad range of plants in a fashion similar to that suggested here for dandelions. The ultimate goal is to determine a series of plant species for which changes in genetic diversity inform us about changing anthropogenic influences that are otherwise difficult to detect.

Acknowledgements

We thank J. Abel, C. Daley, K. Gomez, C. Hennessy, M. Krichko, M. Miller, N. Osswald, K. Ridel, E., N., and L. Rogstad, and J. Smith. We also thank the J. Caruso lab for use of the ICP-AES. Although we are grateful for research funding provided by the U.S. Environmental Protection Agency STAR program (grant R826602-01-0 to SHR and BK), it does not necessarily reflect the views of the Agency, and no official endorsement should be inferred. We are also grateful to partial funding from the Ohio Plant Biotechnology Consortium to SHR.

References

1. Djingova, R.; Kuleff, I. In *Plants as Biomonitors*; Markert, B., Ed.; VCH Publishers: New York, NY, 1993; pp 435-460.
2. Mitchell, L. K.; Karathanasis, A. D. *Environ. Geochem. Health.* **1995,** *17*, 119-126.
3. Ali, S.; Muller, C. R.; Epplen, J. T. *Hum. Genet.* **1986,** *74*, 239-243.
4. Litt, M.; Luty, J. A. *Am. J. Hum. Genet.* **1989,** *44*, 397-401.
5. Weber, J. L.; May, P. E. *Am. J. Hum. Genet.* **1989,** *44*, 388-396.
6. Jeffreys, A. J.; Wilson, V.; Thein, S. L. *Nature* **1985,** *314*, 67-73.
7. Jabs, E. W.; Goble, C. A.; Cutting, G. R. *Proc. Natl. Acad. Sci. USA* **1989,** *86*, 202-206.
8. Scribner, K. T.; Arntzen, J. W.; Burke, T. *Mol. Biol. Evol.* **1995,** *11*, 737-748.
9. Rogstad, S. H. In *Genomes (Proceedings of the twenty-second Stadler Genetics Symposium)*; Gustafson, J. P.; Flavell, R. B., Eds.; Plenum Publishing: New York, NY, 1996; pp 1-14.
10. Rogstad, S. H. *Meth. Enzymology* **1993,** *224*, 279-294.
11. Rogstad, S. H. *Theoret. Applied Genet.* **1994,** *89*, 824-830.
12. Weising, K.; Nybom, H.; Wolff, K.; Meyer, W. *DNA Fingerprinting in Plants and Fungi*; CRC Press: Boca Raton, LA, 1995.
13. Hamrick, J. H.; Godt, M. J. W. In *Plant population genetics, breeding and genetic resources*; Brown, A. H. D.; Clegg, M. T.; Kahler, A. L.; Weir, B. S., Eds.; Sinauer: Sunderland, MA, 1990; pp 43-63.
14. Kitazawa, T.; Kominami, R.; Tanaka, R.; Wakabayashi, K.; Nagao, M. *Molec. Carcinogen.* **1994,** *9*, 67-70.
15. Sadamoto, S.; Suzuki, S.; Kamiya, K.; Kominami, R.; Dohi, K.; Niwa, O. *Internat. J. Radiation Bio.* **1994,** *65*, 549-557.
16. Dubrova, Y. E.; Jeffreys, A. J.; Malashenko, A. M. *Nature Genet.* **1993,** *5*, 92-94.

362

17. Fan, Y. J.; Wang, Z.; Sadamoto, S.; Ninomiya, Y.; Kotomura, N.; Kamiya, K.; Dohi, K.; Kominami, R.; Niwa, O. *Internat. J. Radiat. Bio.* **1995,** *68*, 177-183.
18. Yauk, C. L.; Quinn, J. S. *Proc. Natl. Acad. Sci. USA* **1996,** *93*, 12137-12141.
19. Krontiris, T. G. *Science* **1995,** *269*, 1682-1683.
20. Warren, S. T. *Science* **1996,** *271*, 1374-1375.
21. Fujiwara, Y.; Chi, D. D.; Wang, H.; Kelemen, P.R.; Morton, M. L.; Turner, R.; Hoon, D. S. B. *Cancer Res.* **1999,** *59*, 1567-1571.
22. *Microsatellites: Evolution and Applications*; Goldstein, D. B.; Schlotterer, C., Eds.; Oxford University Press, Oxford, Great Britain, 1999; 368 pp.
23. Gillespie, R.B.; Guttman, S. I. *Environ. Toxicol. Chem.* **1989,** *8*, 309-317.
24. Nevo E.; Shimony, T.; Libni, M. *Nature* **1977,** *267*, 699-701.
25. Krane, D. E.; Sternberg, D. C.; Burton, G. A. *Environ. Toxicol. Chem.* **1999,** *18*, 504-508.
26. Guttman, S.I. *Environ. Health Perspect.* **1994,** *102*, 97-100.
27. Lynch, M. In *Conservation genetics: case histories from nature;* Avise, J. C.; Hamrick, J. L., Eds.; Chapman & Hill: New York, NY, 1996; pp 471-501.
28. Keane, B.; Smith, M. K.; Rogstad, S. H. *Environ. Toxicol. Chem.* **1998,** *17*, 2027-2034.
29. Keane, B.; Pelikan, S.; Smith, M. K.; Toth, G.; Rogstad, S. H. *Amer. J. Bot.* **1999,** *86*, 1226-1238.
30. *Analysis of Industrial Air Pollutants*; MSS Information Corporation: New York, NY, 1974; pp 36-52.
31. Keane, B., Collier, M. H., Shann, J. R., Rogstad, S. H. *Sci. Total Environ.*, *in press*, **2001**.
32. Arens, P.; Odinot, P.; van Heusden, A. W.; Lindhout, P.; Vosman, B. *Genome* **1995,** *38*, 84-90.
33. Hanotte, O.; Bruford, M. W.; Burke, T. *Heredity* **1992,** *68*, 481.
34. Menotti-Raymond, M.; O'Brien, S. J. *Proc. Natl. Acad. Sci. USA* **1993,** *90*, 3172-3176.
35. Hogan, N. C.; Slot, F.; Traverse, K. L.; Garbe, J. C.; Bendena, G.C.; Pardue, M. L. *Genetics* **1995,** *139*, 1611-1621.
36. Taylor, E. B. *J. Heredity* **1995,** *86*, 354-363.
37. Richards, A. J. *Plant Breeding Systems*; George Allen & Unwin: London, UK, 1986.
38. King, L. M.; Schaal, B. A. *Proc. Natl. Acad. Sci. USA* **1990,** *87*, 998-1002.
39. Rogstad, S. H.; Pelikan, S. *Biotechniques* **1996,** *21*, 1128-1131.
40. Marr, K.; Fyles, H.; Hendershot, W. *Can. J. Soil Sci.* **1999,** *79*, 385-387.
41. Normandin, L.; Kennedy, G.; Zayed, J. *Sci. Total Environ.* **1999,** *239*, 165-171.
42. Hansen, K.; Stern, R. M. *Toxicol. Environ. Chem.* **1984,** *9*, 87-91.
43. Gebhart, E. *Toxicol. Environ. Chem.* **1984,** *8*, 253-265.

Chapter 21

Molecular Identification of Chironomid Species

E. Newburn and D. Krane

**Department of Biological Sciences, Wright State University,
Dayton, OH 45435–0001**

Despite their utility as indicators of aquatic ecosystem health, chironomid species are generally difficult to identify using the morphological features of their larvae. Molecular identification should significantly improve the reliability of species identification of known chironomid larvae that have important indicator roles. It may also aid in the discovery and phylogenetic placement of new chironomid species with different indicator potentials. Preliminary analyses of the internal transcribed spacer regions (ITS) of chironomid rDNA provide sequence data that suggests high amounts of variation between species of this group and at the same time the intraspecific variation is low. The polymerase chain reaction (PCR) amplification and restriction enzyme digests of these ITS regions produce characteristic fragments for each species tested to date. Nucleotide sequence analysis of these regions confirmed the interspecific diversity found in both the ITS-1 and ITS-2 region.

Introduction

Many species of chironomid larvae (Diptera: Chironomidae) are morphologically indistinguishable from one another and careful microscopic scrutiny is needed to accurately identify to the species level for most members of this group (*1*). Anatomic features such as labial plates, mandibles, and antennal structure are of particular utility in such classification efforts. Drastic changes in many morphological characters during the transition from immature to mature stages in some species further compound the difficulty associated with proper identification of these organisms. As a result, existing larval keys for chironomids only allow characterization to the level of generic groups or subgroups in many cases. Constantly changing nomenclature and unstandardized references cause further difficulty.

Despite these challenges, these macroinvertebrates are known to be very valuable in biomonitoring efforts (*2*). Midge (chironomid) larva are particularly useful as freshwater ecoindicators because of their extensive distribution and wide range of known tolerances to environmental disturbances (*3*). In addition, the immobility of midge larvae and their relatively stable population dynamic also make them well suited for assessing conditions at single sites. Taken as a whole, these characteristics make chironomid larvae exceptionally valuable indicators of freshwater ecosystem integrity (*4, 5*).

In practice, the extent to which a freshwater system has been stressed by pollutants can be determined through an examination of the presence or absence of specific chironomid species with known pollution tolerances. Finer scale analyses of relative abundances can also provide clues to the presence of specific stressors. For instance, high abundance of a chironomid species such as *Dicrotendipes nervosus* frequently indicates the presence of abundant decomposable organic matter, whereas species such as *Cricotopus bicinctus* are found more commonly in systems with high levels of inorganic contaminants (*1*). Since even closely related species can have distinct pollution tolerances, the need for proper identifications is particularly important and useful (*6*).

Macroinvertebrates such as chironomid larvae have been used as integral parts of freshwater monitoring systems by the Ohio EPA since 1973 (*7*). The principal chironomid-based measurement of macroinvertebrate integrity is the Invertebrate Community Index (ICI) (*8*). The ICI considers the status of a total of 150 different chironomid species commonly found in North American streams and was developed in a fashion similar to that used for the more broadly based Index of Biotic Integrity (IBI) (*9*). These standardized systems have avoided many of the complications encountered with more subjective approaches (*3*).

Improvements in the ability to resolve and distinguish chironomid species has the potential to clarify the taxonomic status of many specimens of this diverse group that are currently unknown. With more reliable and complete identifications, a detailed survey of species composition and range could be made with these significant ecoindicator species. Phylogenetic characterization of this group should also yield important insights into their evolution, range and sensitivities.

Furthermore, molecular identification markers may provide a more reliable and efficient means of midge larvae identification because of a wide range of advantages. In fact, a growing number of population ecologists are investigating the use of DNA-based techniques in their field studies due to their ease of use and cost effectiveness (10). Small quantities of material (even from incomplete organisms) can be used since DNA is found in all cells and is typically needed in only nanogram quantities. Also, usable DNA samples can be easily extracted from both living, dead and even preserved tissue (10).

The selective constraint upon different regions of an organism's genome varies considerably and the best molecular markers for species identification correspond to those unconstrained sequences that accumulate numerous substitutions after species divergence. The ITS (intergenic transcribed spacer) regions between rRNA encoding regions within eukaryotic genomes corresponds to just such a locus (11). In most eukaryotic organisms, rRNA genes are found in tandem arrays hundreds of repeat units in length. While the rRNA coding sequences themselves are typically under strong selection and therefore well-conserved evolutionarily, some spacer regions within them and all the intergenic sequences between them appear to be free to accumulate mutations at a rapid rate (11). Further, the many gene conversion events that appear to occur in these tandem arrays causes them to be exceptionally homogenous within species since mutations are either quickly corrected or sweep through each organism's tandem arrays (12). The unusual combination of :1) rapid divergence between species in these regions, 2) homogeneity within species, and 3) availability of highly conserved sequences flanking the variable regions makes the ITS sequences extremely well-suited for use as a phylogenetic marker in closely related species (13).

These regions rapidly diverge in reproductively isolated populations and yet are exceptionally homogenous within species (14). In previous work, the ITS-1 region of ribosomal DNA has been used to effectively distinguish between morphologically similar species such as many species of parasites (15), yeast (16), and algae (17). The labor intensive and slow microscope procedures for differentiating the eggs of *Ostertagia ostertagi* from other nematode genera were defined through amplification of ITS-1 sequences (15). The need for species differentiation was indispensable since specific GI infections cause major loss to the cattle industry. Also, specific diagnosis of hookworm species is necessary to effectively treat infection in humans. *Necator americanus* and *Ancylostoma duodenale* were distinguished through PCR amplifications of the ITS-1 (18). Again, small size and morphological limitations of filth fly larva, *Muscidifurax* species, caused identification difficulty. PCR - RFLP analysis of the ITS regions allowed easy differentiation with specific enzymes (14). A rapid and reliable PCR method was also used to amplify the ITS region of strains of *Saccharomyces*. It was effective and beneficial to industries such as winemaking and brewing (*16*). The green algae *Chlamydomonas reinhardtii* has been analyzed through rDNA ITS-1 and ITS-2 to determine genetic relatedness (*17*).

The conserved flanking sequences of the ITS regions, 18S and 28S subunits, greatly facilitate amplification of this region through the polymerase chain reaction (PCR) (*19*, 20). Specific oligonucleotide primers have been designed to facilitate the amplification of the chironomid ITS-1 and ITS-2 region by PCR. Upon confirmation that a single amplification product is obtained, restriction fragment length polymorphism (RFLP) analyses, similar to those used to characterize bacterial communities (*21*), have been used to distinguish between species. Due to the high variability of the ITS-1 and ITS-2 region between chironomid species, treatments with restriction enzymes generate DNA fragments of distinctive lengths that can be used to determine the presence or absence of any known chironomid in a sample.

Methods

Grab sampling of chironomid species was conducted at four sites in southwestern Ohio. (Caesar Creek; Dick's Creek in Middletown; Little Miami River in Sugarcreek Township; and Twin Creek in Germantown) Upon recovery, midge larva were sorted and stored in 95% ethanol until DNA extraction. A permanent slide of each specimen was made to confirm identification. Species were cleared by using a 9% solution of KOH. Larvae were kept in the solution for two days unheated and then put through a series of baths: glacial acetic acid for a minimum of 5 minutes, 70% ETOH for 15 minutes and 100% ETOH for 15 minutes. Specimens were then mounted on slides with Euparal mounting medium (BioQuip Products, Inc). Permanent slides of specimen have been deposited with the Ohio Biological Survey (Columbus, OH).

DNA was extracted from individual organisms through use of QIAamp DNA Mini Kit (Qiagen). Instructions of the manufacturer were followed with one exception : an extended lysis of 48 hours was used to increase the DNA yield due to small quantity of tissue from immature larvae. DNA concentrations were determined by spectroscopy and diluted to between 10-50 ng/uL with total yields typically being a total of 750 ng of high quality genomic DNA per specimen.

PCR was performed using specific primers designed from the conserved 18S and 28S subunits of rDNA of *Chironomus tentans* from Genbank. 18S primer sequence, 5' - GAT GTT CTG GGC GGC ACG CG 3', and 28S primer sequence, 5'- TTG GTT TCT TTT CCT CCC CT 3', were used. Both primers were used in 20 pmol concentrations. PCR was carried out in 20 uL volumes using 50 ng of template, 20 pmol primer, 2mM dNTP, 25 mM MgCl$_2$, and 1 U Taq DNA polymerase (Promega). Reactions were carried out on a MJ Research Thermocycler Model PTC-150 under the following conditions 95°C, 35 seconds (denaturing); 63.5°C, 35 seconds (annealing); 72°C, 1 min 20 seconds (extension) for 30 cycles.

Gel electrophoresis was performed with 1.8% agarose gels (Agarose DNA grade (high melting), Fisher Scientific) prestained with 0.5 uL of Ethidium bromide (10 mg/mL). Gels (60mm X 55 mm) were ran at 100V using TE buffer at 22° C. After a single amplification product was confirmed, restriction digests using *Hin*fI and *Rsa*I restriction enzymes were carried out using buffers provided by the supplier (Gibco). A water bath of 37 °C was used to restrict PCR products to completion during eight hour incubations. (A protocol of 3.5 uL ddH20, 5 uL PCR product, 1 uL buffer, and 0.5 uL enzyme (10 U/uL) was used for restriction).

All cloning procedures were carried out as described in TOPO TA Cloning Kit for Sequencing (Invitrogen). Colonies were grown overnight in LB culures of 50 ug/ml ampicillin. Plasmid DNA was isolated using QIAprep Miniprep (Qiagen) with no modifications to manufacturer's protocol. Plasmids were analyzed for inserts by restriction analysis using *Eco*RI (Gibco). Glycerol stocks of plasmid cultures were kept at 70 °C in 2 parts culture: 1 part glycerol.

Sequencing reactions were carried out using an ABI Prism 310 Genetic Analyzer. (Dye Terminator Cycle Sequencing Ready Reaction Kit, Perkin Elmer). 0.2 ug/uL of template was used and a primer concentration of 3.2 pmol. Products were purified using spin column purification (Centri-Sep).

Multiple sequence alignments with hierarchical clustering, were generated with the help of the computer program Multalin version 5.4.1 (*22*).

Results

Morphological features of all collected chironomids were microscopically scrutinized for species identification. Labial and paralabial plates (Figure 1) were some of the key structures that were among the most helpful determinants for the species collected in this study. A total of six species (*Glyptotendipes lobiferous*, *Cardiocladius obscurus*, *Chironomus riparius*, *Dicrotendipes fumidus, Cricotopus bicinctus,* and *Polypedilum convictum*) were identified in this way and two independent isolates of each species were subjected to molecular characterization.

PCR amplifications of the rDNA of the 18S and 28S subunits of representatives of these six species each generated a single amplification product (Figure 2). Size of the amplification product for each species was itself distinctive (*Dicrotendipes fumidus*: 1210 bp; *Dicrotendipes lobiferous*: 1250 bp;

A

B

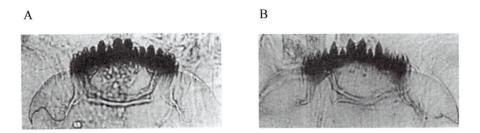

Figure 1. Morphological structures are slightly varied between chironomid species (Simpson and Bode 1980). 1(A) The labial plate of Polypedilum illinoense 1(B) The labial plate of Polypedilum convictum.

A   B   C   D   E   SM

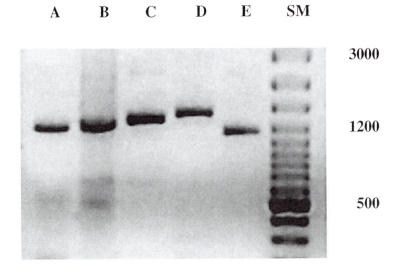

3000

1200

500

Figure 2. PCR amplification of the internal transcribed spacer regions of five Chironomid species. Using primers specific to the 18S and 28S regions of rDNA, five different Chironomid species yield amplification products of characteristic lengths. Lane A - Glyptotendipes lobiferous, Lane B - Cardiocladius obscurus, Lane C - Chironomus riparius, Lane D – Dicrotendipes fumidus, Lane E - Cricotopus bicinctus, and Lane SM - 100 bp size marker.

Chironomus riparius: 1350 bp; *Cardiocladius obscurus*: 1440 bp; *Cricotopus bicinctus*: 1180 bp).

Nucleotide sequence variation between all six collected species in the amplified regions was confirmed through the detection of restriction fragment length polymorphisms (RFLPs). Restriction enzymes were selected by using GenBank sequences of *C. tentans* (accession # x99212) the software program Webcutter (Webcutter 2.0, copyright 1997, Max Heiman). Restriction patterns with *Hin*f I generated distinctive banding patterns for all six chironomid species (Figure 3). Similarly distinctive banding patterns were also seen after digestion with *Rsa*I (Figure 4B). While interspecific PCR-RFLP patterns were distinctive, PCR-RFLP patterns were invariant between individuals of the same species (Figure 4). RFLP analyses were replicated twice for each species.

The complete nucleotide sequence of the amplification products from *Dicrotendipes fumidus* and *Cricotopus bicinctus,* respectively, were also determined (Figure 5). Comparison to homologous sequences in the ITS-1 and ITS-2 regions of all other chironomid species currently available in Genbank confirm the distinctive nature of these regions in these chironomids as well as others (Figure 5B). In contrast, functionally constrained 18S and 28S rDNA sequences (including those used for amplification primers) were almost invariant (Figure 5).

Discussion

Exposure to anthropogenic stressors in freshwater ecosystems is commonly assessed by determining the presence or absence of particular chironomid species. Distinguishing between the large number of recognized, closely related chironomid species on the basis of their morphological differences requires detailed analysis by relatively uncommon experts. Low cost, high throughput and objectivity make molecular typing an appealing alternative or supplemental approach.

We have found that PCR amplification of the rDNA ITS-1 and ITS-2 followed by RFLP analysis of those amplicons yields DNA fragments whose lengths alone can serve as species-specific markers. In fact, the size of the amplification products themselves were distinctive for all six chironomid species collected in this preliminary survey. While it is unlikely that product length alone would allow a unique determination of all 150 taxa of midge larvae known, the PCR-RFLP approach seems to represent a promising alternative or supplement to current morphologically based assignments. Genomic DNA can be easily isolated from individual specimens in quantities that allow many amplifications to be performed.

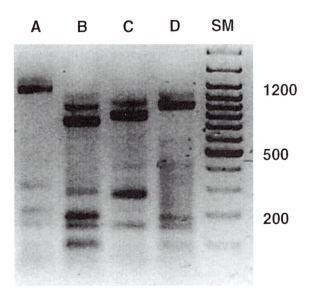

Figure 3. Hinfl RFLPs of PCR products from internal transcribed spacer regions. After amplification, the PCR product of four Chironomid species is restricted using Hinfl to generate distinctive restriction fragment length polymorphisms (RFLPs). Sequence variability results in different, characteristic, banding patterns for each species. Lane A - Polypedilum convictum, Lane B - Cricotopus bicinctus, Lane C - Glyptotendipes lobiferous, Lane D -Chironomus riparius, and Lane SM - 100 bp size marker.

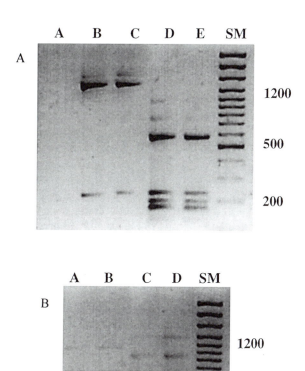

Figure 4. Individuals of the same species yield indistinguishable PCR-RFLP banding patterns. Two different PCR amplifications of the rDNA of the same species give the similar banding patterns. Intraspecific variation is apparently low in the amplified region. (4A) A - Negative control in which amplification was performed with no genomic DNA template, HinfI RFLPs of PCR products: B,C – two different Cardiocladius obscurus individuals, D,E – two different Dicrotendipes fumidus individuals, and Lane SM - 100 bp size marker. (4B) RsaI RFLPs of PCR products: A,B – two different Cardiocladius obscurus individuals, C,D – two different Dicrotendipes fumidus individuals, and Lane SM - 100 bp size marker.

372

```
             1                                                         50
C.tentans    GATGTTCTGG GCGGCACGCG AGTTACAATG AAGCTGACAA CGTGTTACCT
D.fumidus    .......... .......... .......... .......... ..........
C.bicinctus  .......... .......... .......... ...CATA.G. ..........

                                                              100
             TATCCGAGAG GATTGGGAAA TCACTTAGCC AGCTTCCTAG TGGGGATTGT
             .......... .......... ..A...T... .......... ..........
             .G.T...A.. A.......A. ..G....... .......... ........A.

             101                                                       150
C.tentans    GGACTGAAAA AGTTCACATG AACCA--GGA ACTCCTAGTA AGTGTGAGTC
D.fumidus    .......... .......... ......--.. ....-----. ..........
C.bicinctus  .......... ...C..T... TTTTTTTT.. ATAT.T.... .CA.......

                                                              200
             ACTAGCTTGC ATTGATTACG ACCCTGATCT TTGTACACAC CGCCCGTCGC
             .......T.G ..A.....T. .....TGAG. .......... ..........
             .........T .G..A.... T.....TGAG .......... ..........

             201                                       250
C.tentans    TATTACCGAC GAATTATTTA GTGAGATCTC TGGAGGTAAA CATTGCGGTG
D.fumidus    .......... .......... ........T. .......... ....A.A
C.bicinctus  ..G.......G .......... .......... .G.G .TC.A....G
```

A

Figure 5. Alignment of sequence data of rDNA from chironomid species. Sequence data is aligned showing similarities and differences. (5A) Alignment of C. tentans (Chironomus tentans), D. fumidus (Dicrotendipes fumidus,), and C. bicinctus (Cricotopus bicinctus) from the 18S subunit of rDNA. (5B) Alignment of partial /8S, ITS-1, 5.8S, ITS-2, and 28S subunits of rDNA from 14 chironomid species: C.tentans (Chironomus tentans); Genbank accession #X99212, D.fumidus (Dicrotendipes fumidus,); Genbank accession ####, G.salinus (Glyptotendipes salinus,); Genbank accession # AJ296804, G. barbipes (Glyptotendipes barbipes); Genbank accession #AJ296797, G. pallens (Glyptotendipes pallens,); Genbank accession number #AJ296801, C.pallidivittatus (Chironomus pallidivittatus,) Genbank accession #AJ296805, C.halophilus (Chironomus halophilus); Genbank accession #AJ279432, C.luridus (Chironoimus luridus); Genbank accession #AJ296779, C. cingulatus (Chironomus cingulatus,); Genbank accession #AJ296774, C.melanotus (Chironomus melanotus,); Genbank accession #AJ296781, C.plumosus (Chironomus plumosus); Genbank accession #AJ296822, C.nuditarsis (Chironomus nuditarsis); Genbank accession #AJ296783, C.duplex (Chironomus duplex); Genbank accession #AJ296776, and C.bicinctus (Chironomus bicinctus); Genbank accession ####. Periods (.) indicate matches with the nucleotide found within C. tentans at that position, while nucleotides are shown for positions that differ relative to those found in C. tentans. Dashes (-) indicate gaps inserted to improve the alignment.

B

```
                      251                                                                              300                                                      350
                      CCTCGGT-AT CGCGATTGCT TTTGCCAAAG TTGATCAAAC TTGATGATTT GGAGGAAATA AAAGTCGTAA CAAGGTTTCC GTAGTGTGAAC CTGCGGAAGG
C.tentans             .......... .......... .......... .......... .......... .......... .......... .......... .......... ..........
D.fumidus             -.....T..- TT........ .......... .......... .......... .......... .......... .......... .......... ..........
G.salinus             -T........ .......... .......... .......... .......... .......... .......... .......... .......... ..........
G.barbipes            .......... T......... .......... .......... .......... .......... .......... .......... .......... ..........
G.pallens             .......... T..C.T.... .......... .......... .......... .......... .......... .......... '......... ..........
C.pallidivittatus     .......... .A........ .......... .......... .......... .......... .......... .......... .......... ..........
C.halophilus          .......... .A........ .......... .......... .......... .......... .......... .......... .......... ..........
C.luridus             .......... T......... .......... .......... .......... .......... .......... .......... .......... ..........
C.cingulatus          .......... T......... .......... .......... .......... .......... .......... .......... .......... ..........
C.melanotus           .......... TA........ .......... .......... .......... .......... .......... .......... .......... ..........
C.plumosus            .......... TA........ .......... .......... .......... .......... .......... .......... .......... ..........
C.nuditarsis          .......... TA........ .......... .......... .......... .......... .......... .......... .......... ..........
C.duplex              .......... .......... .......... C......... .......... .......... C......... .......... .......... ..........
C.bicinctus           A..TT.C... T.T.G...T. .......... .......... .......... .......... C......... .......... .......... ..........
```

```
                      351                                                                              400                                                      450
                      ATCATTAATG TATG-----T TTTGCACACG CATTTATGC- -TCTTTCATC TTGTTTTTTT ATGGGG---- ---------- ---------- -TGAGAATTA TTAATTA----
C.tentans             .......... .......... .......... .......... .......... .......... .......... ---------- ---------- .......... ..........
D.fumidus             ........C. ..A--TT..- AA.-TT..-- .-T-A----- -........- -......... ----------  ---------- ---------- ----G.CC .CTCC.-
G.salinus             .......... ..A--T--   CA-TT---TT ATA.G..ATG GG.....T..A CAT.C.A.A. G..T.TATAA AGTTTGTGT G..GTTTGA. A...A..AAT
G.barbipes            .......... ..A--T-- CA-TT--GTT ATA.G..ATG GG.....T..A CATA..C.A. G..T--ATAA AAGCTTGTGT G..GTTTGA. A..AC.AAT
G.pallens             .......... ..A--TT-- CA-AT--TTT ATA.G..AA- -....T..G C.T.AAC.AG T.----TAA AAGTTGGTG- --GTTT.A. AGTG..TTTT
C.pallidivittatus     .......... ...--TT. -GCA---G-C AT......- C.T-....- .CA...CT.G. ---------- ---------- ---------- .-..CT.G.- .........
C.halophilus          .......... .A.---TTA CA-A--T--C AT.C---...T C......CA C.T.G....A. T.T.T.GAGT GTAGTAGAAG T..T.TA. ..T.A.TCA
C.luridus             .......... .T---AA C--AA---C AT.-...-..T C......CA C.T.G....A. TG..T.A--- --.T.T.. ..T.T... ---GGAC-
C.cingulatus          .......... .-......-AC AT........- TCT.---CA TCT.---CA .CT.G.---A. G..T.A--- GATGT----- --.G.G.TAG. ----G.AC-
C.melanotus           .......... .......--TT. CG-----AC AT........- TCT.---CA .CT.G.---- .CT.G.--- GATGT---- ----G.G.TAG. ----G.AC-
C.plumosus            .......... .......--TC. CG-A---AC AT........-G TC.---CA .CT.G.---   -A.ATGTT GGTGTTTTTG GG.G...GG AT..A.TCA
C.nuditarsis          ...GTGTT. CAAA---AC AT........-G TC.---CG .CT.G....A. A.A.ATGTT GGTGTTTTTT TG.G.G--- --A.CC-
C.duplex              .G----- ------T-AC AT.A....-  -.....CA .CT.G.---A. T.CAC.A--- ---------- G....A.T A.TTA..TA-
C.bicinctus           .A.A---.T. CA-TT-TG-C TT..AC..A- .....T.CA C.T.G.GAAA .A.------- ---------- ---------- G...A.GAG. AAT.CA.--
```

Figure 5. Continued. *Continued on next page.*

374

```
                     451                                                              500                                                    550
C.tentans          ------AAAT CCTAGGTACT AGAATTGCGA -TATGTGTGC GAT-TA--AT GTCGTAC-AC ATGTTGTTGG TTTTATAAAG GGCTTCGCCT AGGTA-----
D.fumidus          -------... .......... .......... ...TCAC. .AC.C.T--T. .T.C.T-G. .......... .......... .......... ----------
G.salinus          TTG--T.... .......... .......... ..CA.CA TG.G.----- ------I-G. .......... ACT....... .......... ..ATCAT
G.barbipes         TTG--T.... .......... .......... ..CA.CT TG.A.----- ------GT.. .......... ACT....... .......... ..ATCAT
G.pallens          ATG--T.... .......... .......... ...C-CA T.AA.----- ------GT.. .......C.. ACT....... .......... ..ATCAT
C.pallidivittatus  .......... .......... .......... ....TA--- .......... .......... .......... .......... .......... ..T-ATT
C.halophilus       TACACT.... .......... .......... ...C.. .CGCA--TA TGT.C..G. .G........ .......... .......... ..--ATC
C.luridus          -------... .......... .......... .GC... A.GCA----- GC........ .......... .......... .......... ..--ATC
C.cingulatus       -------... .......... .......... ...-TT .TGT.---CA CA..C.-- .......... .......... .......... ..--AAC
C.melanotus        -------... .......... .......... ...-TT .TGT.---CA CA..C.-- .......... .......... .......... ..TAAAC
C.plumosus         TATA-T.... .......... .......... ...CTT .TGTGTCA.A CG..C.-- .......... .......... ..A....... ..---AC
C.nuditarsis       -----T.... .......... .......... ..-C.T T.TA----- ----C.-- .......... .......... .......... ..---AC
C.duplex           -----T.... .......... .......... ...ACAT--- .......... ----C..- .A........ .......... .......... ..A-ACT
C.bicinctus        -------... ...T...... .C.C.--... C.CAA.CAA TCGC.T---- ------GT.. .......C.. .......T.A ..A .G.--AC

                     551                                                              600                                                    650
C.tentans          -TA--TTTTA CTTTTTATGC CAAAAAACAT AAAAAAAAAT TTGTGATTA- --TAATAAAC AGTTTTTTCG ATAAGAAAAA ATGAATAAAC
D.fumidus          -..CT...... ...C...... ------...T. .TTG...C- .G..A----- ...A....... TA.T.T.T. .ACT.GT..T
G.salinus          TACT-...... TGC..ATGAA AT.T.T..C. TTT.T.G.- GTAA.A.TA GA.T..TGTG TA.G-CAATA .TTAG..G. .AA--A....A
G.barbipes         TACT-...... TGC..ATAAA AT.T.---.C. TTT.T.G.- GTCA..AT. AA.T..TGCG TA.G-CGATA .TTAG..G. .AA--A..C-
G.pallens          T.CTG...... T.C.ATAAA TTG-----CA TTT.T..GT- GTTG..ATA .G--.GGTA TA.T.TGTG T..G.C.AGC ..T.CC..T ATTGAG..A
C.pallidivittatus  T..CT...... TGCCAA.AAA A..C.T.A.A ..C..... ...... ...C...... .......... .......... AAT.A....-
C.halophilus       T..CT.C..T T--------- ...C... TTT.T... ...A... ...T----- ...T.AG..A .T-------- AA..A....-
C.luridus          T..CT.----- T--------- ...C... T------ ...A... ...T----- --C..AT..A .T-------- AA..A....-
C.cingulatus       T..CTC...C T--------- T...C... TT----- ...A... -T.GA.A... ...T----- -G..TGGTTT .T.A....TC T......-
C.melanotus        T..CTC...C T--------- T...C... -T----- ...A... -T.GA.A... ...T----- -G..TGGTTT .T.A....TC T.-.T....
C.plumosus         T..CTC...C T--------- T...C... -T----- ...A... -T.AA.A... ...TA... ...T----- T--.CAGTTT .T.A....TC T.T.T.....
C.nuditarsis       T..CTC...C T--------- T...C... ------ ...A... -T.AA.A... .TG..... ...T----- TG.CGGTTT .T.A....TC T......T.A
C.duplex           T..CT.....  TGC...CATA A..C...A.A ..........  ------- ...AA.A... ...TCC.TA AGAA....TC GA......G .AA....CT
C.bicinctus        TC.CTA.ACC T-----.... ATT...G... C--------- ------- .AA..C..T- --A..AC.TA .CGGA.ATC G.T--TC... .A.G.AT.GA
```

```
              651                                                                                          750
C.tentans           AAAAA-CTTA ACCCTAGACA GGGGATCACT TGGCTCATGG GTCGATGAAG ACCGCAGCAA ACTGCGCGTC GCCATGTGAA CTGCAGGACA CATGATCATT
D.fumidus           .T.-T..... .......... .......... .......... .......... .......... .......... .......... .......... ..........
G.salinus           .T..TT.... .......... .A........ .......... .......... .......... .......... .G........ ..-....... ..........
G.barbipes          -T..TT.... .......... .......... .......... .......... .......... .......... .G........ ..-....... ..........
G.pallens           TTG.TT.... ....T..... .T........ .......... .......... .......... .......... .G........ ..-....... ..........
C.pallidivittatus   ....-..... .......... .......... .......... .......... .......... .......... .......... .......... ..........
C.halophilus        T...-..... .......... .......... .......... .......... .......... .......... .......... .......... ..........
C.luridus           T...-..... .......... .......... .......... .......... .......... .......... .......... .......... ..........
C.cingulatus        ..G.T..... .......... .......... .......... .......... .......... .......... .......... .......... ..........
C.melanotus         ---------- .......... .......... .......... .......... .......... .......... .......... .......... ..........
C.plumosus          ...----... .......... .......... .......... .......... .......... .......... .......... .......... ..........
C.nuditarsis        ...----... .......... .......... .......... .......... .......... .......... .......... .......... ..........
C.duplex            ...TT..... .......... .......... .......... .......... .......... .......... .......... .......... ..........
C.bicinctus         G...TT.... .......... .......... .......... .......... .......... .......... .......... .......... ..........

              751                                                                                          850
C.tentans           GACATGTTGA ACGCATATTG CGCCTTATAC ATTTGGTTCT CTTTATAATA TACACAAAAT ----TTATA ATGTGGAACT GTATAAGG-T ACATATGGTT
D.fumidus           .......... .......... .......... .......... ..CGT..--- .......-.. ----..... ..GAA..... .A........ ..........
G.salinus           .......... .......... .......... .......... ......--.. .......-.. ----..... A.GAA..... .......-.. ..........
G.barbipes          .......... .......... .......... .......... ......--.. .......-.. ----..... A.GAA..... .......-.. ..........
G.pallens           .......... .......... .......... .......... ......--.. .......-.. ----..... ..GAA..... ......G.-. ..........
C.pallidivittatus   .......... .......... .......... .......... ......G... AATAT..... .AATAT... .......... .......... ..........
C.halophilus        .......... .......... .......... .......... ......G... ..C....... A---T.... ..G....... .......... ..........
C.luridus           .......... .......... .......... .......... ......G... .TT....... A---T.... ..G....... .......... ..........
C.cingulatus        .......... .......... .......... .......... ......G... ......-... ----..... .......... .......... ..........
C.melanotus         .......... .......... .......... .......... ......G... .-T....... ----..... .......... .......... ..........
C.plumosus          .......... .......... .......... .......... ......G... ..-T...... ----..... ..G....... .......... ..........
C.nuditarsis        .......... .......... .......... .......... ......G... ..CT--.... ----..... ..G....... .......... ..........
C.duplex            .......... .......... .......... .......... T.A....... .TT....... ----..... .......... .......... ..........
C.bicinctus         .......... .......... .......... .......... AAC..A.GCC .CA....... ----..CT. ........T. .......... ..........
```

Figure 5B. *Continued*   *Continued on next page.*

```
      851                                                              900                                                  950
C.tentans         GAGTGTCGT- AATTTCATAT GATTACAACT ATAAGT-ATC TATCGCACAC ATAGTGTTGT TAT---AGTA CATAATAGAG TGTCATCAAA GCCGTCTCAC
D.fumidus         .........- .......CA. A....T.... .C.....-.. CGA..TT.CA G.CA----.. -----C.CG .A.....T.. ......T... .AT.....--
G.salinus         .........- .......TA. A....T.... .C........ ------CA.. T.T.ATA.-A .----GAT. .C.....A.. ......T... ..TA....--
G.barbipes        .........- .......TA. A....T.... .C........ ------CA.. T.T.ATA.-A .----GAT. .C.....A.. ......T... ..TA....--
G.pallens         .........- .......TA. A....T.... .A.....-.T G...TA..TCA T.GTGTG.-G .T......-G. ......A... ..TA....--
C.pallidivittatus ..........                          ...-CACA.. CATAGTG...                    .....A...                   .........G.
C.halophilus      .........G ......G.. .C........ ...C...... .G.GC--ACA CACAGT---- .....C.... ......T...          ......A..
C.luridus         .......... ......G.. .C.....-.. ...C...... .G.GT-TACA CACAGT---- .....C....                              ...GTCT
C.cingulatus      .......... ......A.. .C.C.----.. .G.GTTGTTA .ACACACACA C.C----.CG                              .......T...   ...--G.
C.melanotus       .......... ......A.. .C-G.---.. G.T.TTTGT. TACACACACA C.C----.CG                              .......T...   ...--G.
C.plumosus        .........C ......TA. ..........  ..C-G.G... .G.ACACTCA C.TT...GTG         ......CG                            ...AT.
C.nuditarsis      .......... ......A.. .........  ..C-G.T... .T.ATGTGTA CACAC.----- ......CG                              ...T.T... ..A...GT
C.duplex          .......TA. T....TG..T. .........  ...AG.G.. AGATCTCTCT C.CTGTAGTA .CCC.T....  .C........            ...C.G-T.
C.bicinctus       .......... .........  .........  ...AC----- ---ATCACA. C.CCG..GTG .-----TG.. T.C....TC.T .A.---.CTT
```

```
      951                                                              1000                                                 1050
C.tentans         .TCAAAGATT GATTTCTGCG CG--GTGTGA CGATTTATGA CTAAAATTCT AATCTAA-TG TCAGTT--TA CGCCTATTTT T--AAATAAA T---GGGGGG
D.fumidus         .TG..---- -------.G .A........ ..........  .......... G.GT.T.---  --------    .AC.......  --.T.ATG.  A------.A.
G.salinus         ...TC.T.A TG.A.A.AT -----A... TA.A...... TA........  .......... G..TA.----  --------    ...AC.CTC  CTTGC.CTC CTTAACT.A
G.barbipes        ...TC.T.A T.CA-A.AT -----A... TA........ TA........  .......... G..TA.----  --------    ..AC..A    CTTGC.CTC CTTAACC.A
G.pallens         A.------- ----G..T. -----A... TA........ ......T...  .......... G..TAG----  --------    ...AC....  --C.CCC   -TTCA.T.T
C.pallidivittatus .AA------- .......GT. .........  ........  .....C....  ..........                                    ...ATC.--
C.halophilus      .A.------- -------.C. ---.C....  ........  .....G....  .......... .TA.                                  ........
C.luridus         .A.------ ------- --A-C.... ..........  .....G....  .......... .A.        -AC.                        ........
C.cingulatus      TG.TGCT.CC T--AGTA.T. -----.... ..........  .....G....  .........A ..........              ...G
C.melanotus       TG.T.CT--- T-AGTA.T. -----.... ..........  ..T.......  .....G....            ..........              ...G
C.plumosus        .A.TGCT.C. TG.AG-A.T. -TG-T.... ..........  ..T.......  .....G....            ..........              ...G
C.nuditarsis      TG.TGCT.C. TG.AGTG.T. GTG-..... ..........  ..T.......  .....G....  ..........            .A...        ...G
C.duplex          .G.GT.---- ----.ATG.G ..........  ..........  ...GG.G...  .........A            ..........              ..-GG....A
C.bicinctus       .AGTG.---- -------G A........  ..........  ..........  ..........  ...A.----  ..........  .AAT.A.-   -------
```

```
                    1051                                                                              1150
C.tentans           AAGAGTGAAA AAT-TCAAAA TTCG----CA CATATATGTG ATGAATCTTG TGAGTCTA-- ---------- ---------- ---TTCTCTC TGGCGCTAAC TTTACA----
D.fumidus           .TCTGA.... GG.....TT  CATT------ ..C.------ ......A.-- CATAAT.CGT TTTCAATAGA AA-------- .......T.. .....C.... ..........
G.salinus           TT......G. T.GGAGGG.. GA-------- T..GA.---- --A.TGAG.T CATAAT.CGT TTTCAATAGA AA-------- .......T.. .......... ..........
G.barbipes          TT......G. ..GGAGGG.. GA-------- T..GA.---- --A.TGAG.T CATAAT.CGT TTTCAATAGA AA-------- .......T.. .......... ..........
G.pallens           GG...A..G. TAGAGGG... GA-------- T..GA..A.. AA.TGAG.T  CATAAT.CGT TTTCAATAGA AA-------- TA-------- .......... ..........
C.pallidivittatus   .......... .C.--.TT.. ...CA----- ---------- ......C.-- ........-- ---------- TA-------- .......... .......... ..........
C.halophilus        .......... .G........ ...CGTG... .C.C.CTGC. C.G..C.... .....AT... TTTCATTGAA AA-------- .......... .......... ..........
C.luridus           G......... .G........ ...GG----- .C.C.CTGC. C.G....... ....-T.... TTTCATTGAA AA-------- .......... .......... ..........
C.cingulatus        .......... ..AA...... ---------- .TC.C..... A......... A...-T.TCT TTTCATTGAA AAG.C..... .......... .......... ......GACG
C.melanotus         .......... .A-.A..... ---------- ...C.C.... A......... A.T-.TCT.. TTTCATTGAA AAG.C..... .......... .......... ......GTCA
C.plumosus          .......... ..AA...... --.TA.----  ...C...... A..AT..... ..T.TCT... TTTCATTGAA AAG.C..... .......... .......... ....AAAA
C.nuditarsis        .......... .A-....... ---------- ...C...... ..ACAT.... ..CT.TCT.. TTTCATTGAA AAG.C..... .......... .......... ....AAAA
C.duplex            .......... .C-....... ...CT--.G. .G..CTAT.. .T.---.... .T.ATATGA- ---------- --G....... .......... .......... ..........
C.bicinctus         C.C..AA.T. T..T.T.CT. .A-------- ---------- ...TG..--- ---------- ---------- ---------- .A........ .......... ..........

                    1151                                                                              1250
C.tentans           ---------- ---------- ------TATA TATATAATGT CTCGTTAGTT GCTCCTGATT TATCCGC--- ---------- ---AT----- -----GTG   AATAA-CGAT
D.fumidus           ---------- ---------- -------.G. C..G.----- A.AA..G... .CAAAA.A.. .CGT...TAA ---------- -----A.... ..........  TG.G.GT.TG
G.salinus           ---------- ---------- --ACA.TC.. ...-GTG.A. TAT....... .CGA.A.AA. A..TTCAT-- ---------- -----T..-- ......TGA   T...CG-..
G.barbipes          ---------- ---------- --ACA.T... .TGTG.A... TAT....... .CGA.A.AA. A..TTCAT-- ---------- -----T..-- ......TGA   T...CG-..
G.pallens           ---------- ---------- TGTACC..CT .TATATG.A. ..T....... .CGA.A.AA. A..TTCAT-- ---------- -----T..-- ......TGA   TG.G.CG-..
C.pallidivittatus   ---------- ---------- ---TATA... C..TAT.... ......G... .......... ----TAT--- ---------- --CCGC.... ..........  ..........
C.halophilus        ---------- ---------- --TACA..C. C..TAT.... ......G... .T.C...... CTCATTGT-- ---------- --GCTTG... ......T    GTCC.CG...
C.luridus           ---------- ---------- --TAAT.--- ----TG.... G......... .......... C--GTTGT-- ---------- ---------- ......CT    GTACGCG...
C.cingulatus        CGCGCTTACA CACACTTGTG TGTGTG.... .GC.GG.A.. G.T....... .A.T...... C..ACAATA  --ACTGTT-- ---------- .G....CG-. ..........
C.melanotus         CGC---TTA  CACACTTGTG TGTGTT.GC. .G--GG.A.. G.T....... .A.T...... C..ACAAAA  CTACTGT.-- ---------- .G....CG-. ..........
C.plumosus          -----TATA  -CCTTTGTG  TGTATA.--- -G...TGTC  A.-....... .AG..T.... C.G.ACGAAA -TACTGTG.G TATATGT.-- GG....CG-. ..........
C.nuditarsis        -----TATA  TACCTTCGTG TGTATA.G.T AG..AT..A. A.-....... .AGG.T.... C.G.ACGAAA -AACTGTG.A T--------- GGC...CG-. ..........
C.duplex            ---------- ---------- ------.... -CG.TG.... ......C... .C-------- C--------- ---------- ---------- ......T    T.G.CGTTGAT
C.bicinctus         ---------- ---------- ---------- -G..ATG.AA AG-AA..... .CATAT.A.. .TG.ATAT-- ---------- ---------- ..........  .
```

Figure 5B. Continued   *Continued on next page.*

```
                   1251                                                         1300
C.tentans          TTTG----AG ATAAAATCAT TCTTTCA--A ATGTACTA-- --CTGAAG-- --TAAAA--A
D.fumidus          A.A.-----. GCCG..GT.G AG...TC-. .A...G..-- ...T.CG    AG.T.TG--
G.salinus          ...GACATT  .A....ATG. AT.C.TT-- .....A.T-- ...T.TCA   CA..T.T--
G.barbipes         ...GACATT  GA....-T.  AT.C.TT-- .....AAT-- ...T.TCA   AAC.T.T-T
G.pallens          ...AGCA.A  .A.---TG.  AT..TT--  .....AAT-- ...T..CA   TA.T.GTAT.
C.pallidivittatus  ......A               .......-.  --ACTA     .....TA    AAG..G--T
C.halophilus       ...GAGT.   .A..G.A.   .......T-  .GTACCC    GA.....TG  TAAT.G--
C.luridus          .....---.  .A.C.G.AG. .......C-- GT.CTA     G-......TA AAA.....
C.cingulatus       ...AG--.A  .A.T---.   A.....T-  .GTA.TA     ...-T      GTAT.--
C.melanotus        ...AG--.A  .A.G---.   .....T--  --ACTA      ...G.-T    --AT.--
C.plumosus         ...AG--.A  .A.G.G--.  .....T--  --ACTA      .G.AT      ATAT----
C.nuditarsis       ...AG--.A  .A.G---.   .....T--  --ACTA      ...GT      GTAT.TTAC.
C.duplex           .GAGATAGT  ...G.AG.   .......TCT. ...TTATAA  TA....TA   AT..T--
C.bicinctus        A.T.-----  -A..TGATGA T.AA.CCG.  CA.C.AA.CA ATAC..TCT  CAAT----

                                                          1350
C.tentans          AGTAAAAAA  AAAAAAAGA- ----------  CAATTTCGCG
D.fumidus          TAAT..T.G. .....A.    ----------  T.GAA.....
G.salinus          TAA..T..   CG.G..AG-  --------A   A.T.......
G.barbipes         TAAT.T..   CG.G..AG-  --------A   A.T.......
G.pallens          .A..T.C.G. CG......G- --------A   A.T.......
C.pallidivittatus  .AA......  -------..  --------C   A.T.......
C.halophilus       .A.......  TTT...     --------C   A.T.C...
C.luridus          -A.....    .......    --------C   A.T.C...
C.cingulatus       .T.G..T.T. .T.G.G..   --------C   G.T.......
C.melanotus        T.G.TT.T   .T.G.G..   --------C   G.T.......
C.plumosus         -A.TT.T.   .T.G.G..   --------C   G.T.......
C.nuditarsis       T.G.T.TGT. .T.G.G...  G AGAGAGAGAC G.T.......
C.duplex           TTG.T.T.C. T.TT...    ----------
C.bicinctus        -A.T..TCG  CTTT.TT.-  ----------

                   1351                                           1400
C.tentans          ACCTCAACTC ATGTGAGACT ACCCCCTGAA TTTAAGCATA TTAATTAGGG GAGGAAAAGA
D.fumidus          .......... .......... .......... .......... .......... ..........
G.salinus          .......... .......... .......... .......... .......... ..........
G.barbipes         .......... .......... .......... .......... .......... ..........
G.pallens          .......... .......... .......... .......... .......... ..........
C.pallidivittatus  .......... .......... .......... .......... .......... ..........
C.halophilus       .......... .......... .......... .......... .......... ..........
C.luridus          .......... .......... .......... .......... .......... ..........
C.cingulatus       .......... .......... .......... .......... .......... ..........
C.melanotus        .......... .......... .......... .......... .......... ..........
C.plumosus         .......... .......... .......... .......... .......... ..........
C.nuditarsis       .......... .......... .......... .......... .......... ..........
C.duplex           .......... .......... .......... .......... .......... ..........
C.bicinctus        .......... .......... .......... .......... .......... ..........
```

Figure 5B. *Continued*

While the actual 18S and 28S rDNA sequences themselves are virtually invariant between the chironomid species for which sequence information is available (Figure 5), the ITS-1 and ITS-2 regions associated with them exhibit very high interspecific variation (Figure 3 and 5). Preliminary studies of intraspecific variability using PCR-RFLP (Figure 4) suggests that members of the same species can be readily distinguished from members of different species. Thus far, no detectable variation in PCR fragment length or RFLPs have been shown within chironomid species. Low levels of variation within species might still allow accurate classification, especially if additional restriction enzymes with different recognition sites are used.

On going work will allow the PCR-RFLP identification of an additional ten chironomid species. Sequencing of the amplification products of all these organisms is currently underway as part of a phylogenetic study of important chironomid species. Present taxonomic uncertainties of chironomid species may be diminished as rDNA sequences are analyzed. It is likely that previously unrecorded species will be found, and that previously unrecognized relationships between species in this important group will be elucidated. At present, only chironomids common in southwestern Ohio have been considered as part of this work but future plans include analyses of specimens from a broader geographic range.

Additional work is also underway to determine the presence or absence of individual chironomid species in mixed samples of organisms such as those commonly found as a result of grab sampling in streams with complex chironomid communities. Terminal restriction fragment length polymorphism (TRFLP) in which one of the two amplification primers is radioactively labelled prior to amplification and subsequent restriction digestion appears promising for analysis of such mixed samples. Each species yields a single, characteristic band upon gel electrophoresis and autoradiography. The resulting "bar code" pattern is easily scored by comparison to a set of standards from reference specimens.

Acknowledgements

We thank Scott Roush for assistance in primer design and field collection, Mike Bolton from the Ohio EPA for expertise in chironomid identifications, and Tim Wood and Maria Gonzalez for providing helpful suggestions in collection and slide preparation.

REFERENCES

1. Simpson, K. and Bode, R. 1980. Common Larvae of Chironomidae (Diptera) from New York State Streams and Rivers. New York State Museum, Albany, New York.

2. Cairns, J., Jr. and K.L. Dickson. 1971. A Simple Method for the Biological Assessment of the Effects of Waste Discharges on Aquatic Bottom-dwelling Organisms. J. Water Poll. Contr. Fed. 43(5): 755-772.

3. DeShon, J.E. 1995. Development and Application of the Invertebrate Community Index (ICI), pp.217-243 (Chapter 15). In W.S. Davis and T. Simon (eds.). Biological Assesment and Criteria: Tools for Water Resource Planning and Decision Making. Lewis Publishers, Boca Raton, FL.

4. Gaufin, A.R. 1973. Use of Aquatic Invertebrates in the Assessment of Water Quality. pp. 96-116. In Cairns, J., Jr. and K.L. Dickson, eds. Biological Methods for the Assessment of Water Quality. Amer. Soc. Test. Mat., Spec. Tech. Pub. No. 528.

5. Goodnight, C.J. 1973. The Use of Aquatic Macroinvertebrates as Indicators of Stream Pollution. Trans. Amer. Microsc. Soc. 92(1):1-13.

6. Curry, L. L. 1965. A Survey of Environmental Requirements for the Midge (Diptera:Tendipedidae). In Biological Problems in Water Pollution, 3rd seminar, 1962. U.S. Public Health Service Publication No. 99-WP-25, pp. 127-141.

7. Wilhm, J.L. 1970. Range of Diversity Index in Benthic Macroinvertebrate Populations. Journal of the Water Pollution Control Federation 42: R221-R224.

8. State of Ohio Environmental Protection Agency. 1987. *Biological Criteria for the Protection of Aquatic Life*, Vol 3 - Standardized biological Field Sampling and Laboratory Methods for Assessing Fish and Macroinvertebrate Communities. Columbus, OH, USA.

9. Karr, J.R. 1993. Defining and Assessing Ecological Integrity: Beyond Water Quality. Environ. Toxicol. Chem. 12:1521-1531.

10. Parker, P., Snow,A., Schug, M., Booton, G., and Fuerst, P. 1998. What Molecules Can Tell Us about Populations: Choosing and Using a Molecular Marker. Ecology. 79: 361-382.

11. Marcon, P.C. R.G. Taylor, D.B., Mason, C.E., Hellmich, R.L., and Siegfried, B.D. 1999. Genetic Similarity among Pheromone and Voltinism Races of *Ostrinia nubilalis*(Hubner) (Lepidoptera:Pyralidae). Insect Mol. Biol.,in press.

12. Elder, J.F. and Turner, B. J. 1995. Concerted Evolution of Repetitive DNA Sequences in Eukaryotes. Q. Rev. Biol. 70: 297-320.

13. Hillis,D.M. and Dixon, M.T. 1991. Ribosomal DNA: Molecular Evolution and Phylogenetic Inference. The Quarterly Review of Biology. 66: 411-442.

14. Taylor, D.B. and Szalanski, A.L. 1999. Identification of *Muscidifurax Spp.* by Polymerase Chain Reaction- Restriction Fragment Length Polymorphism. Biological Control. 15: 270-273.

15. Zarlenga, D.S., Gasbarre, L.C., Boyd, P., Leighton, E., Lichtenfels, J.R. 1998. Identification and Semi-quantitation of *Ostertagia* Eggs by Enzymatic Amplification of ITS-1 Sequences. Veterinary Parisitology. 77: 245-257.

16. Josepa, S., Guillamon, J., and Cano, J. 2000. PCR Differentiation of *Saccharomyces cerevisiae* from *Saccharomyces bayanus/Saccharomyces pastorianus* Using Specific Primers. FEMS Microbiology Letters. 193(2000):255-259.

17. Coleman, A.W. and Mai, J.C. 1997. Ribosomal DNA ITS-1 and ITS-2 Sequence Comparisons as a Tool for Predicting Genetic Relatedness. Journal of Molecular Evolution. 45:168-177.

18. Monti, J.R., Chilton, N.B., Bao-Zhen, Q., and Gasser, R.B. 1998. Specific Amplification of *Necator americanus* or *Ancylostoma duodenale* DNA by PCR Using Markers in ITS-1 rDNA, and It's Implications. Mollecular and Cellular Probes. 12: 71-78.

19. Kocher, T.D., W.K. Thomas, A. Meyer, S.V. Edwards, S. Paabo, F.X. Villablanca, and A.C. Wilson. 1989. Dynamics of Mitochondrial DNA Evolution in Animals; Amplification and Sequencing with Conserved Primers. Proc. National Acad. Sci. 86:6196-6200.

20. Simon, C., S. Paabo, T.D. Kocher, and A.C. Wilson. 1990. Evolution of Mitochondrial Ribosomal RNA in Insects as Shown by the Polymerase Chain Reaction. In M. Clegg and S. O'Brien (eds.), Molecular Evolution, UCLA Symposia on Molecular and Cellular Biology, New Series, 1222: 235-244. Wiley-Liss, New York.

21. Clement. B., Kehl, L., DeBord, K., and Kitts, C. 1997. Terminal restriction fragment patterns (TRFPs), a rapid, PCR-based method for the comparison of complex bacterial communities. Journal of Microbiological Methods 31 (1998):135-142.

22. Corpet, F. 1988. Multiple sequence alignments with hierarchical clustering. Nucl. Acids Res., 16 (22):10881-10890.

Remediation

Chapter 22

Integrating Site Characterization with Aquifer and Soil Remediation Design

Jejung Lee, Howard W. Reeves[*], and Charles H. Dowding

Department of Civil Engineering, Northwestern University,
2145 Sheridan Road, Evanston, IL 60208–3109

This review describes various approaches to propagate uncertainty inherent with the hydrogeological, chemical, and biological input parameters for groundwater flow and contaminant transport models employed to design remediation systems. Design reliability is used to guide site exploration, to compare different remedial designs, and to indicate when sufficient data have been collected to evaluate the design. Also described is a computationally efficient framework to integrate site characterization with aquifer and soil remediation design. A simple example is used to illustrate how design reliability is developed and employed to assess both exploration and design.

Site characterization and source identification are crucial to successful design of systems to remediate soil and groundwater contamination. There are, however, several factors that make it difficult to predict contaminant movement, and, therefore, make it difficult to design a remediation system. First, the geologic structure is usually heterogeneous, but only a limited number of data are available to describe the heterogeneity. Second, it is often difficult to completely identify the initial contaminant source. Third, contaminant movement is impacted by chemical and biological reactions that both are chemical and site specific.

While a variety of remediation techniques to mitigate groundwater and soil contamination have been developed, assessment of their effectiveness before construction has lagged. One of the most widely used mitigation techniques is the pump-and-treat method, which has been used at about three-quarters of the Superfund sites where groundwater cleanup was required (1). However, the effectiveness of conventional the pump-and-treat method to reach the restoration goal has been questioned (1). Recently, in-situ treatment techniques such as monitored natural attenuation, passive reactive barriers, and bioremediation have gained popularity for groundwater contaminant remediation. As in-situ treatment techniques are increasingly employed, the recognition of the importance of site characterization is also increasing. However, knowledge of geologic structure and accuracy of model prediction are still uncertain. These uncertainties give rise to model performance uncertainty that affects the remedial design reliability. Therefore the impact of uncertain hydrogeology on the model performance must be assessed during design.

This paper reviews methods to assess design reliability and presents an efficient framework to both evaluate various remedial designs and integrate site characterization into the design process. Within this framework, whether there are sufficient data to base the design, what additional data are required, and where these data should be collected to increase confidence in the design are quantitatively addressed.

Overview

The fundamental assertion of our research is that remediation design must be considered as site characterization activities progress to yield cost-effective design and effective site characterization. This assertion arises from the observation that different designs require different hydrological, chemical, and biological information from the site.

Model predictions, either analytical or numerical, are generally used to design remediation schemes. Due to the lack of information of site, uncertain input parameters are used in the models, therefore, the predicted performance of the design is uncertain. Freeze et al. (2) present a hydrologic decision framework illustrating how various uncertainties can be taken into account by considering the reliability of design and the value of future data. Applicability of this framework was demonstrated in a series of papers (3-5). The approach presented herein builds upon this work.

386

Figure 1 is the basic framework of this methodology. As shown, it is possible to estimate the probability of success for a design based on the current field data and to assess if additional data will improve the reliability of the design. Information obtained through the procedure also indicates which data should be collected and where on the site it should be obtained.

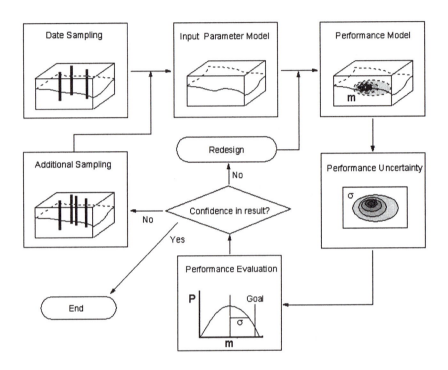

Figure 1. Computational framework to incorporate geologic/chemical input uncertainty with model sensitivity to calculate performance reliability

Review

Many approaches have been developed for each step of the assessment framework shown in Figure 1. Conventional and newly introduced methods based on typical deterministic models are categorized in Figure 2. In this section, each method will be briefly reviewed and discussed.

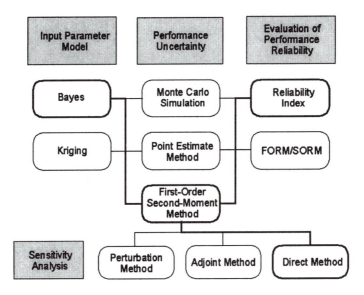

Figure 2. Flowchart of alternative and employed methods for modeling at each step of the framework given in Figure 1. Thick line represents the flow employed in our framework

Input: Parameters and their uncertainties

For the input step, there we consider two basic approaches to interpolate between known geologic data. The most popular method, kriging will be discussed first, followed by the Bayesian approach.

Kriging is the geostatistical method to extrapolate or interpolate point values at unsampled locations within the spatial domain limited data samples. It provides an estimation variance that yields a measure of the uncertainty of its interpolated values. Traditionally, kriging takes the form of linear interpolation based on Gaussian condition. It includes simple kriging, ordinary kriging, universal kriging, and cokriging. Among them, ordinary kriging is the most widely used methods (6).

Ordinary kriging is a "best linear unbiased estimate" of the parameter. It is "linear" because its estimates are weighted linear combinations of the available data, "unbiased" because it tries to have the mean error equal to zero, and "best" because it aims at minimizing the variance of the error (7). The kriging estimator, Z^* is described as,

$$Z^* = \sum_\alpha \lambda_\alpha Z_\alpha \qquad (1)$$

where Z_α is the measured data, λ_α is an estimation weight whose summation should be a unity to satisfy the non-bias condition. Ordinary kriging is applied when the mean is constant and not known. If the mean is not known but is expressed as a polynomial function of spatial coordinate, universal kriging may be used. If the mean of variable is perfectly known, simple kriging is applicable. In most practical situation, however, the mean is not known. It can be possibly known only when the number of data becomes so large as to allow an estimate of the mean.

In many cases, data with more than one variable are obtained, and sometimes they are correlated. Cokriging was developed as a multivariate generalization of kriging (8). It is especially advantageous in cases where the primary measurements are limited and expensive, while auxiliary measurements are available at low cost (9). Hoeksema and Kitanidis (10) compared Gaussian conditional mean and variance with extended cokriging estimations based on point observations of transmissivity and hydraulic head. Harvey and Gorelick (11) applied conditional cokriged estimation to different types of measurement to show that sequential conditioning of data improves linear approximation and brings about a better estimate of hydraulic conductivity.

If the measured data show non-Gaussian distribution or highly skewed histograms, linear kriging is not applicable. For these data, indicator kriging was developed as a non-linear kriging method. In this procedure, the original values are transformed into indicator values, such that they are zero if the datum value is less than the pre-defined cutoff level or unity if greater. Therefore, the estimated value with indicator kriging represents the probability of non-exceedence at a location (9).

The Bayesian approach is beginning to come into favor since the importance of site characterization and sampling strategies has increased. This approach is similar to the geostatistical approach (kriging) in its use to develop statistically realistic descriptions of sites to aid in initial data collection and site exploration when there is a scarcity of hard data (6, 12). The Bayesian approach may provide the general framework to update parameter uncertainty when additional data are available, and to evaluate its effect on estimation or decision (13).

Many researchers have adopted the Bayesian concept for environmental problems. Kitanidis (13) described the Bayesian conditional mean and variance which can be a nonlinear estimator for uncertain covariance parameters. McLaughlin et al. (14) presented a stochastic site characterization procedure that combined field measurements with predictions obtained from mathematical models. In their framework, the model provides prior estimates, and they are updated whenever new measurements of hydraulic conductivity, head, or concentration become available.

James and Gorelick (*15*) suggested a Bayesian data worth framework to improve the cost-effectiveness of data collection in groundwater remediation programs. McLaughlin and Townley (*16*) offered a comprehensive review of the inverse problem and demonstrated how inverse problems may be cast in a Bayesian framework. They described that in most applications of a posteriori estimation, the prior and measurement error probability densities were assumed to be Gaussian. With a multivariate Gaussian assumption, the Gaussian conditional approach can be used to estimate mean and its covariance in Bayesian framework. It is useful even when the Gaussian assumption does not hold, because for known parameters, it gives the minimum variance estimate which is linear function of the observation (*13*). The work discussed here adopts a Gaussian conditional approach to estimate the means and the variance of uncertain input parameters to the design model.

Omre (*17*), Omre and Halvorsen (*18*), and Handcock and Stein (*19*) presented a Bayesian-Kriging approach to use the spatial approximation techniques of kriging while allowing for the uncertainty in specification of the parameters in the geostatistical approach (17). A prior distribution of the expected model is combined with available observations to make a posterior prediction. With an uninformed prior calculation, Bayesian-Kriging is very close to kriging with variogram data. This Bayesian alternative to kriging was compared with cokriging by Le et al. (*20*) and Sun (*21*). They showed that the Bayesian alternative uses all the historical information whereas cokriging only uses the current information for its prediction. They found that in terms of the mean-squared prediction error, prediction by the Bayesian alternative is superior to cokriging although the difference is small. They also found that in estimating the standard errors associated with prediction, the Bayesian alternative is also superior to cokriging.

Design Model : Performance and uncertainty

Next, it is necessary to use the design model to determine an estimate of system performance and the uncertainty in this estimate. As described in this section, this step requires a large amount of computing time by conventionally used Monte Carlo simulation. As an alternative, the first-order second moment (FOSM) method adopted within our framework reduces the computational burden.

Monte Carlo simulation (MC) is the most widely used method for calculation of performance variance. It can be applied when many classes of parameters are simultaneously uncertain, and it is easy to understand (*22*).

This method employs the classical statistics to calculate the variance with a set of model outputs from a set of input parameters that are randomly generated. The number of runs depends on the model and the assumed input parameter distribution. According to Harr (*23*), the required number of MC simulations, N, for, m, independent variables is estimated as, $N=(h^2/4\varepsilon^2)^m$, where h is the standard deviation in a normal distribution corresponding to the confidence interval, and ε is the maximum allowable system error in estimating the confidence interval. For example, if a required confidence interval is 99% with 1% system error, h is 2.58, ε is 0.01, and $(16,641)^m$ is estimated. Therefore computing time is the major disadvantage of this method. Cawlfield and Wu (*24*) required over 400,000 computer runs to achieve a good level of accuracy for a one-dimensional transport code for a reactive contaminant.

The point estimate method (PEM) was developed as an alternative to Monte Carlo simulation and first-order second moment (FOSM) method. In this method, the model is evaluated at a discrete set of points in the uncertain parameter space, with the mean/variance of model predictions computed using a weighted average of these functional evaluations (*25*). Compared with MC, this method has the advantage of low computational requirement, and compared to the more computationally efficient FOSM, this method has the advantage to handle nonlinear models while FOSM requires linear models. Harr (*23*) described applicability of this method for structural problems, and Mishra (*25*) illustrated this method with an analytical model of health risk arising from water-borne radionuclide migration from a repository.

The first-order second moment method (FOSM) is the method adopted within the framework to propagate input parameter uncertainty through numerical models (*26, 27*). FOSM provides two moments, mean and variance of predicted variables. This method is based on Taylor series expansion, of which second-order and higher terms are truncated. The expected value of concentration, $E[u]$ and its covariance, $COV[u]$ are (*23, 27*),

$$E[u(z)] = u(\bar{z}) \tag{2}$$

$$COV[u_i(z), u_j(z)] = \sum_{k=1}^{m} \sum_{l=1}^{n} \frac{\partial u_i}{\partial z_k} \frac{\partial u_j}{\partial z_l} COV(\bar{z}_k, \bar{z}_l) \tag{3}$$

$\frac{\partial u}{\partial z}$ is a Jacobian matrix that relates the change in computed concentration, u to the change in given input parameter, z. Equation (2) states that the best estimate of concentrations at every point in the discritized domain depends on the best estimate of input parameters. The estimate of the covariance of concentration for all the computation points in the domain is given by Equation (3), and it is determined by using the input parameter covariances and the sensitivity.

Wagner and Gorelick (28) used FOSM analysis to propagate parameter uncertainty into contaminant concentration prediction uncertainty for a hypothetical aquifer. James and Oldenburg (29) compared FOSM analysis with MC analysis for three dimensional TCE concentration uncertainty. The comparison indicates that for the large-scale, low concentration problem, the two methods of analysis give similar uncertainty evaluations allowing justification of the use of FOSM analysis for conceptual models. Tiedeman and Gorelick (30) examined the reliability of remediation design and optimal locations for pumping wells used to contain a contaminant plume in an aquifer with uncertain parameters. They used the FOSM to translate input parameter uncertainty into estimated simulation uncertainty and showed that it gave results similar to the MC simulation. McKinney and Loucks (31) applied FOSM to a network design algorithm for improving the reliability of groundwater simulation model predictions. They presented a method to minimize the prediction variance by choice of new aquifer measurement locations with this method.

Sensitivity coefficients from FOSM approaches are essential to predict uncertainty of model performance and design reliability. A perturbation method is generally used method to obtain the sensitivity matrix. However, this method requires a great deal of computing time and a determination of the proper perturbing value. Yeh (32) reviewed three methods: influence coefficient method (33) based on the perturbation concept, sensitivity equation method, and variational method (34, 35). Skaggs and Barry (36) compared the direct method, which involves analytically differentiating the equation, with an adjoint method. In recent years, Graettinger (37), Kunstmann (38), and Graettinger and Dowding (39) introduced the direct computation of sensitivity derivatives within the numerical code rather than numerical approximation of these derivatives. By using an automatic differentiation tool such as ADIFOR (40), this approach can be performed both accurately and efficiently (41, 42). The direct derivative coding method is consistent with the analysis of Yeh (32) and Skaggs and Barry (36) and has the advantage that it requires only a single model run to obtain the whole sensitivity matrix over the perturbation method which requires N+1 model runs where N is the number of input variables.

Output : Performance reliability

Finally the expected value and variation of the performance must be evaluated. Some, as described below have developed the first- and second-order reliability methods for this task. The reliability index approach followed herein provides similar information but appears to be much more efficient.

The first- and second-order reliability methods (FORM and SORM) were initially developed for structural reliability applications to estimate the occurrence of low-probability events. These methods are based on two main concepts: 1) formulation of a performance function describing the behavior of interest in terms of the random variables, and 2) transformation of the problem into standard normal space, where an estimate of probability is obtained (*43*). Through this method, the system reliability index is calculated. Jang et al. (*44*) also showed that sensitivity of the system reliability index identified which parameter has a major influence on the estimate of probability. While FORM and SORM were adopted as more practical alternatives to Monte Carlo techniques. The implementation by Jang et al. (*44*) is quite computationally expensive due to numerical (finite difference) approximation of required derivatives relating changes in the dependent variable to changes in the input parameters within the algorithms. The search for the design point also can be computationally intensive, especially for large numbers of random variables. Details of FORM and SORM are presented in the publications by Cawlfield and Sitar (*43*), Jang et al. (*44*), Der Kiureghian et al. (*45*), and Sitar et al. (*46*).

The reliability index, β, may be calculated as an alternative for obtaining the design point from FORM and SORM. To evaluate the model performance, two measures are considered: the estimated variance in the dependent variable and reliability index β for the remediation design.

β is defined as the difference between the calculated performance, u_i and the allowable performance, $u_{allowed}$ at given locations (x_i, y_i, z_i) at a site divided by the standard deviation of the model behavior, σ_{ui}. It may be written as,

$$\beta = \frac{u_{allowed} - u_i}{\sigma_{u_i}} \qquad (5)$$

By examining the β values for specific target locations at a site, the reliability or probability of success of a modeled remedial design will be evaluated. The probability of success is given by $\Phi(\beta)$ where Φ is the cumulative normal distribution function. If the β values are near zero, the model remedial design implies that the concentrations will be near the allowable maximum concentration. If the β is large and negative, the remedial system does not meet the desired performance, and if β is large and positive, the model predicts that the remedial scheme will be successful. Therefore, the engineer may examine β values and judge if additional sampling may reduce the σ_{ui} and can improve system reliability or if the remedial system must be redesigned to change the expected performance, u_i.

If the examination of the β shows reliability is not sufficient, geological uncertainty and model sensitivity are combined to determine where next data should be collected. The next location will be the data position that creates the greatest variance in the performance model. The estimated covariance matrix has been called the importance matrix by several authors (37,47).

Discussion and Conclusion

Various approaches to propagate uncertainty inherent with the hydrogeological, chemical, and biological input parameters for groundwater flow and contaminant transport models employed to design remediation system were reviewed and discussed. In terms of the framework presented in Figure 1, the results from reliability analysis may give information about (1) which data are required and where these data should be collected to increase confidence in the design, (2) which design alternative is most reliable, and (3) when the acceptable probability of design success is met.

Lee (48) presented the application of the framework to two- and three-dimensional finite element models for various design alternatives to answer the questions (1) through (3). Among methods that are presented in Figure 2, Bayesian conditional calculation for spatial geologic uncertainty, a first-order second moment (FOSM) calculation of model uncertainty, and reliability index calculation are adopted sequentially. For FOSM, the sensitivity matrix is calculated by directly differentiating numerical finite element code. The comparison of the performance uncertainties from multiple uncertain input parameters allowed a determination of the parameter that most influenced the performance model and was the most important for determination of the location for the next sample. Lee (48) demonstrated the use of the reliability index to compare the reliability of different proposed remediation design alternatives and to analyze the impact of additional sampling on the design and performance for the given cleanup goal at the compliance point. If more than one compliance point are examined, a multivariate β should be determined to consider correlation between the compliance points. More research is required in this area.

Fortney (49) applied the framework to a study with field data. He adopted MODFLOW 2000 (50, 51) as the groundwater flow model and incorporated the calculation of performance uncertainty by FOSM. The influence of input parameter uncertainty on the performance uncertainty and the reliability of design alternatives for design goals at different locations were shown. Fortney (49) also presented an alternative way to determine the most influential point in the domain requiring the additional sampling to increase the reliability of the design.

The work by Lee (48), Fortney (49), and Graettinger (37) present the details of integrating site characterization and design. An important and interesting question that arises in the application of this framework is, What probability of success is sufficient ? Baecher (52) discussed socially acceptable probability of success for civil engineering projects, but no groundwater projects. To use our approach in practice, socially acceptable probability of success for groundwater remediation will need to be debated and determined.

Acknowledgement

This work was funded by Grant R-827126-01-0 from the U.S. Environmental Protection Agency. The work has not been subjected to any EPA review, and therefore does not necessarily reflect the views of the Agency.

References

1. *Pump-and-Treat Ground-Water Remediation: A Guide for Decision Makers and Practitioners.* U.S. Environmental Protection Agency (EPA), Office of Research and Development; Washington, DC 1996; EPA/625/R-95/005.
2. Freeze, R.A.; Massmann, J.; Smith, L.; Sterling, T.; and James, B., *Ground Water* **1990**, 28(5), 738-766.
3. Massman, J.; Freeze, R.A.; Smith, L.; Sperling, T.; and James, B., *Ground Water* **1991**, 29(4), 536-548.
4. Sperling, T.; Freeze, R.A.; Massman, J.; Smith, L.; and James, B., *Ground Water* **1992**, 30(3), 376-389.
5. Freeze, R.A.; James, B.; Massmann, J.; Sperling, T.; and Smith, L., *Ground Water* **1992**, 30(4), 574-588.
6. Chiles, J.-P.; Delfiner, P. *Geostatistics: Modeling Spatial Uncertainty*; John Wiley & Sons: New York, 1999.
7. Isaaks, E.H.; Srivastava, R.M. *An Introduction to Applied Geostatistics*; Oxford University Press: New York, 1989, 278-322.

8. Matheron, G. *The Theory of Regionalized Variables and Its Applications*; Ecole des Mines:Fontainebleau, France, 1971.

9. Rouhani, S. Geostatistical Estimation: Kriging, In *Geostatistics for Environmental and Geotechnical Applications*; Srivastava, R.M.;Rouhani, S.; Cromer, M.V.; Johnson, I.; Desbarats, A.J. Eds.; ASTM STP 1283; American Society of Testing and Materials, 1996, pp. 20-31.

10. Hoeksema, R.J.; Kitanidis, P.K., *Water Resour. Res.* **1985**, 21(6), 825-836.

11. Harvey, C.F.; Gorelick, S.M., *Water Resour. Res.* **1995**, 31(7), 1615-1626.

12. Journel, A.G.; Huijbregts, C.J. *Mining Geostatistics*; Academic Press: New York, 1978.

13. Kitanidis, P.K., *Water Resour. Res.* **1986**, 22(4), 499-507.

14. McLaughlin, D.; Reid, L.B.; Li, S.-G.; Hyman, J., *Ground Water* **1993**, 31(2), 237-249.

15. James, B.R.; Gorelick, S.M., *Water Resour. Res.* **1994**, 30(12), 3499-3513.

16. Mc Laughlin, D.; Townley, L.R., *Water Resour. Res.* **1996**, 32(5), 1131-1161.

17. Omre, H., *Mathematical Geology* **1987**, 19(1), 25-39.

18. Omre, H.; Halvorsen, K.B., *Math. Geology* **1989**, 21(7), 767-786.

19. Handcock, M.S.; Stein, M.L., *Technometrics* **1993**, 35(4), 403-410.

20. Le, N.D.; Sun, W.; Zidek, J.V., *J.R.Statist.Soc.B* **1997**, 59(2), 501-510.

21. Sun, W., *Environmetrics* **1998**, 9, 445-457.

22. *Consequences of spatial variability in aquifer properties and data limitations for groundwater modelling practice*; Peck, A.; Gorelick, S.; de Marsily, G.; Foster, S.; Kovalevsky, V., Eds.; IAHS Publication No. 175; IAHS Press: Wallingford, Oxfordshire, UK, 1988.

23. Harr, M.E. *Reliability-Based Design in Civil Engineering*; McGraw-Hill: New York, 1987.

24. Cawlfield, J.D.; Wu, M., *Water Resour. Res.* **1993**, 29(3), 661-671.

25. Mishra, S. In *Calibration and Reliability in Groundwater Modelling: Coping with Uncertainty*; Stauffer, F.; Kinzelbach, W.; Kovar, K.; Hoehn, E. Eds.; IAHS Publication No. 265; IAHS Press: Wallingford, Oxfordshire, UK, 2000.

26. Dettinger, M.D.; Wilson, J.L., *Water Resour. Res.* **1981**, 17(1), 149-161.

27. Townley, L.R.; Wilson, J.L., *Water Resour. Res.* **1985**, 21(12), 1851-1860.

28. Wagner, B.J.; Gorelick, S., *Water Resour. Res.* **1987**, 23(7), 1162-1174.

29. James, A.L.; Oldenburg, C.M., *Water Resour. Res.* **1997**, 2495-2508.

30. Tiedeman, C.; Gorelick, S.M., *Water Resour. Res.* **1993**, 29(7), 2139-2153.

31. McKinney, D.C.; Loucks, D.P., *Water Resour. Res.* **1992**, 28(1), 133-147.

32. Yeh, W.W.-G., *Water Resour. Res.* **1986**, 22(2), 95-108.

33. Becker, L.; Yeh, W.W.-G., *Water Resour. Res.* **1972**, 8(4), 956-965.

34. Carter, R.D.; Kemp, Jr, L.F.; Pierce, A.C.; Williams, D.L., *Soc.Pet.Eng.J.* **1974**, 14(2), 186-196.

35. Sun, N.Z.; Yeh, W.W.-G., *Water Resour. Res.* **1985**, 21(6), 869-883.
36. Skaggs, T.H.; Barry, D.A., *Water Resour. Res.* **1996**, 32(8), 2409-2420.
37. Graettinger, A.J. Ph.D. Dissertation, Northwestern University, Evanston, IL., 1998.
38. Kunstmann, H.G. Ph.D. Dissertation, Swiss Federal Institute of Technology, Zurich, Switzerland, 1998.
39. Graettinger, A.J.; Dowding, C.H., *Journal of Geotechnical and Geoenvironmental Engineering,* ASCE **1999**, 125(11), 959-967.
40. Bischof et al., *IEEE Comp. Sci. & Engng.* Fall, 1996, 18-32.
41. Abate, J.; Wang, P.; Sepehrnoori, K.; Dawson, C., *Commun. Numer. Meth. Engng.* **1999**, 15, 423-434.
42. Park, S.K.; Droegemeier, K.K, *Mon. Wea. Rev.* **1998**, 127, 2180-2196.
43. Cawfield J.D.; Sitar, N. In *Consequences of spatial variability in aquifer properties and data limitations for groundwater modelling practice*; Peck, A.; Gorelick, S.; de Marsily, G.; Foster, S.; Kovalevsky, V. Eds.; IAHS Publication No. 175; IAHS Press: Wallingford, Oxfordshire, UK, 1988; pp. 191-216.
44. Jang, Y.-S.; Sitar, N.; Der Kiureghian, A., *Water Resour. Res.* **1994**, 30(8), 2435-2448.Der Kiureghian, A.; Lin, H.-Z.; Hwang, S.-J., *Journal of Engineering Mechanics,* ASCE **1987**, 113(8), 1208-1225.
45. Der Kiureghian, A.; Liu, P.-L., *Journal of Engineering Mechnics*, **1986**, 112(1), 85-104.
46. Sitar, N.; Cawlfield, J.D.; Der Kiureghian, A.D., *Water Resour. Res.* **1987**, 23(5), 794-804.
47. Tomasko, D.M.; Reeves, M.; Kelley, V.A. In *Proceedings of the DOE/AECL Conference on Geostatistical Sensitivity, and Uncertainty Methods for Ground-Water Flow and Radionuclide Transport Modeling*; Buxton, B., Ed.; Battelle Press: Columbus, Ohio, 1987.
48. Lee, J. Ph.D. Dissertation, Northwestern University, Evanston, IL., 2001.
49. Fortney, M.D. M.S. Dissertation, Northwestern University, Evanston, IL., 2001.
50. Harbaugh, A.W.; Banta, E.R.; Hill, M.C.; McDonald, M.G., *MODFLOW-2000, the U.S. Geological Survey modular ground-water model - User guide to modularization concepts and the Ground-Water Flow Process*: U.S. Geological Survey Open-File Report 00-92, 2000.
51. Hill, M.C.; Banta, E.R.; Harbaugh, A.W.; Anderman, E.R., *MODFLOW-2000, the U.S. Geological Survey modular ground-water model - User guide to the Observation, Sensitivity, and Parameter-Estimation Processes and three post-processing programs*, U.S. Geological Survey Open-File Report 00-184, 2000.
52. Baecher, G. B., *Geotechnical Risk Analysis User Guide*, FHWA/RD-87-011, Fed.Hwy.Admin.,McLean, 1987, p.55.

Chapter 23

Mechanisms Controlling Chlorocarbon Reduction at Iron Surfaces

Tie Li and James Farrell[*]

Department of Chemical and Environmental Engineering,
University of Arizona, P.O. Box 210011, Tucson, AZ 85721

Abstract: This research investigated whether rates of carbon tetrachloride (CT) and trichloroethylene (TCE) dechlorination on iron surfaces are limited by rates of electron transfer. The contributions of direct electron transfer and indirect reduction via atomic hydrogen to the overall dechlorination rates were also investigated. Electron transfer coefficients for CT and TCE were determined from measurements of dechlorination rates over a potential range from –600 to –1200 mV (SHE), and a temperature range of 2 to 42 °C. The transfer coefficient for CT was found to be independent of temperature, and the apparent activation energy was found to decrease with increasingly negative electrode potentials. These observations indicate that the rate of electron transfer controlled the observed rate of CT dechlorination. In contrast, the transfer coefficient for TCE was temperature dependent, and increased with increasingly negative electrode potentials. This indicated that TCE dechlorination was not controlled by an electron transfer step. Comparison of analytically and amperometrically measured reaction rates showed that CT reduction occurred primarily via direct electron transfer, while TCE reduction involved both direct electron transfer, and an indirect mechanism involving atomic hydrogen. Comparison of amperometrically and analytically measured reaction rates for TCE and perchloroethylene (PCE) also supports an indirect mechanism for chloroethene reduction.

Introduction

Chlorinated organic compounds are prevalent groundwater contaminants due to their widespread use as industrial solvents. Because many of these compounds are classified as carcinogens or suspected carcinogens, chlorocarbons must be removed from drinking water supplies prior to potable use. Presently employed remediation technologies include adsorption by activated carbon and air stripping. While these methods are effective for removing chlorocarbons from the aqueous phase, they only transfer the contaminants from water to another medium, which then requires treatment or disposal.

In recent years there has been considerable interest in developing destructive treatment methods for removing chlorinated organic compounds from contaminated waters. The use of zerovalent metals for reductive dechlorination has been a very active research area since Gillham and O'Hannesin (1) proposed that metallic iron filings could be utilized in passive groundwater remediation schemes (2-12). In zerovalent iron remedial systems, the iron serves as an electron donor to reduce chlorinated organic compounds to their nonchlorinated analogs and chloride ions. Redox active metals, water itself, and other groundwater constituents, such as nitrate and carbonate, may also be reduced by reactions with the zerovalent iron (13, 14).

Several possible reaction mechanisms for reductive dechlorination at zerovalent iron surfaces are illustrated in Figure 1. Direct reduction may occur via electron tunneling to a physically adsorbed chlorocarbon, or through a chemical bond to a chemisorbed species (15, 16). Arnold and Roberts postulated that chloroethenes form chemisorption complexes with iron as depicted in Figure 1 (15, 16). Chlorocarbons may also be reduced indirectly via reaction with atomic hydrogen produced from water reduction (17). These reactions may involve formation of hydride-chlorocarbon complexes (15). For reactions involving atomic hydrogen, the iron acts as a catalyst for both the water reduction reaction and the dehalogenation reaction. This electrocatalytic hydrodechlorination mechanism has been found to operate in palladium catalyzed systems used for reductive dechlorination (18-21).

Dechlorination rates and byproducts for most environmentally relevant chlorocarbons have been determined under a wide range of experimental conditions (*1-13, 16*). Chlorinated alkane reduction by zerovalent iron has been found to proceed sequentially, and produces chlorinated products (*3*). For example, dechlorination of carbon tetrachloride (CT) produces primarily chloroform as the first stable product (*3*). In contrast, chlorinated alkene reduction yields primarily completely dechlorinated species as the first stable products. For example, dechlorination of perchloroethylene (PCE) and trichloroethylene (TCE) has been hypothesized to involve a β-elimination mechanism that yields chloroacetylene as the first detectable product (*4, 16*). However, the chloroacetylene is unstable and is rapidly reduced to acetylene and ethene, and has therefore not been reported by most investigators (*4, 5, 16*).

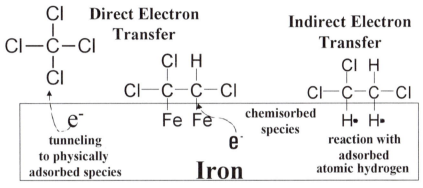

Figure 1. Possible reaction mechanisms for reductive dechlorination at zerovalent iron surfaces.

Although most investigators have reported faster dechlorination rates with increasing degree of halogenation, not all studies are consistent with this trend. For example, Arnold and Roberts (*16*) reported that dechlorination rates for a series of chlorinated alkenes decreased with increasing degree of chlorination, and Farrell et al. (*9*) reported faster rates of TCE dechlorination than for PCE. These observations suggest that there is an indirect mechanism involved in chloroethene reduction.

Rates of multistep electron transfer reactions are dependent on potential according to the Butler-Volmer equation, as given by (*22, 23*):

$$i = i_0 [e^{-\bar{\alpha}F(E-E_{eq})/RT} - e^{\tilde{\alpha}F(E-E_{eq})/RT}]$$ (1)

where i is the net reaction current, i_0 is the exchange current, F is the Faraday constant, R is the gas constant, T is temperature, E is electrode potential, E_{eq} is the equilibrium potential for the overall redox reaction, and $\vec{\alpha}$ and $\overleftarrow{\alpha}$ are the transfer coefficients for the reduction and oxidation reactions, respectively. The first term in brackets represents the rate of the forward reduction reaction, while the second term gives the rate of the reverse oxidation reaction. The transfer coefficients depend on the number of electrons transferred before and after the rate determining step, and whether the rate determining step involves electron transfer (*23*).

One purpose of this investigation was to determine whether the rate determining step for CT and TCE reduction involves electron transfer. A second goal of this research was to determine the contribution of direct and indirect reduction mechanisms to the overall rate of TCE and CT dechlorination.

Materials and Methods

Voltammetric Experiments Voltammetric analyses using chronoamperometry (CA) and Chronopotentiometry (CP) were performed in a custom three-electrode cell using an EG&G (Oak Ridge, TN) model 273A potentiostat and M270 software. All controlled potential or controlled current experiments utilized an EG&G model 616 rotating disk electrode, a Hg/Hg$_2$SO$_4$ reference electrode, and a platinum wire counter electrode encased in a proton permeable membrane (Nafion, Dupont). An iron disk (Metal Samples Company, Mumford, AL) was used as the working electrode. The experiments were conducted in 10 mM CaSO$_4$ background electrolyte solutions that were deoxygenated during and prior to use by purging with argon or nitrogen.

The CA and CP experiments were performed at constant halocarbon concentration by purging the solution with nitrogen gas containing CT or TCE for the duration of each experiment. Prior to each experiment, the iron disk was polished with an EG&G polishing kit, and conditioned at -785 mV with respect to the standard hydrogen electrode (SHE). After a one hour conditioning period, the applied potential was momentarily terminated before applying the fixed potential for each CA experiment. In order to maintain cathodic cell currents, the CA experiments were performed only at potentials below the open circuit potential of the iron disk. The uncompensated resistance between the working and reference electrodes was 20 Ω. Since the cell currents were less than 200 μA in all experiments, the maximum voltage loss due to solution resistance was less than 4 mV. All potentials are reported relative to the standard hydrogen electrode, and cathodic currents are reported as positive.

Experiments with freely corroding iron were performed in a 25 mL, air-tight glass cell. A single iron wire (Aesar, Ward Hill, MA) was used as the reactant. Iron corrosion rates were determined from analysis of Tafel diagrams produced by polarizing the iron wire electrodes ±200 mV with respect to their open circuit potentials. These experiments utilized a platinum wire counter electrode and a Ag/AgCl or Hg/Hg$_2$SO$_4$ reference electrode.

Analyses   TCE, PCE and CT concentrations in all solutions were determined by analysis of 100 μL aqueous samples. The 100 μL samples were injected into 1 g of pentane and analyzed by injection into a Hewlett-Packard 5890 Series II gas chromatograph equipped with an electron capture detector and autosampler. Chloride ion concentrations were determined by analysis of 5 mL samples using a DX 500 ion chromatograph (Dionex, Sunnyvale, CA). An Orion pH probe was used to determine the initial and final pH values of the electrolyte solutions.

Chronoamperometry Profiles

Chronoamperometry profiles for an iron electrode in a blank electrolyte and in electrolyte solutions containing 2 mM CT or 4 mM TCE are shown in Figure 2. In the CT solution, there is an increase in current compared to that in the blank electrolyte. This increase in current is equal to the CT reduction current determined from a chloride balance on the solution. This indicates that CT dechlorination occurred via direct electron transfer. In contrast to CT, the presence of TCE in the solution resulted in a decrease in current compared to the blank electrolyte. This indicates that TCE on the electrode surface interfered with water reduction, and that TCE was reduced more slowly than water at an electrode potential of -780 mV.

The apparent independence of CT and water reduction can be attributed to a low adsorbed surface coverage of CT on the electrode. The amount of CT adsorbed on the electrode surface was determined from an Anson analysis of early time CA profiles for blank and CT containing solutions (*22*). The charge transferred during the first 0.1 seconds of CA profiles for blank and CT solutions is shown in Figure 3. For the time span shown in Figure 3, virtually all the current was due to electrical double-layer charging and to reduction of compounds adsorbed on the electrode surface before the start of the experiment.

The similar slopes of the two profiles in Figure 3 indicates that the double-layer charging currents were the same in the blank and CT containing solutions. The 15.5 µC difference in intercepts between the two profiles was due to reduction of CT adsorbed on the electrode surface under open circuit conditions (22). Assuming a surface area of 0.32 nm^2 for each adsorbed CT molecule, only 15% of the 1 cm^2 geometric electrode surface area was covered with adsorbed CT under open circuit conditions. However, since the geometric surface area ignores molecular scale roughness, and thus underestimates the surface area available for adsorption, the actual portion of the iron surface covered with adsorbed CT may be significantly less than 15%.

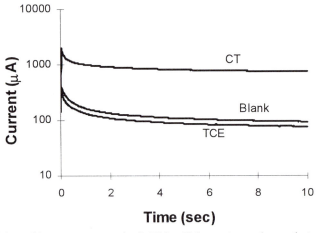

Figure 2. CA profiles at a potential of –780 mV for an iron electrode in a blank 10 mM CaSO$_4$ electrolyte solution and in electrolyte solutions with 2 mM CT or 4 mM TCE.

Although the adsorbed surface coverage of CT was much less than one monolayer at an aqueous concentration of 2 mM, reaction rates for CT deviated from pseudo-first order kinetics at concentrations below 0.2 mM. Figure 4 shows that the rate of CT reduction (R$_{CT}$) did not increase in a linear fashion with the CT concentration. This indicates that CT reduction was not first order in CT concentration. For surface mediated reactions, this type of behavior is normally indicative of reactive site saturation effects. Previous investigators have observed that CT reduction by corroding zerovalent iron also deviates from pseudo-first order kinetics at CT concentrations as low as 100 µM (6). As suggested by the data in Figure 3, CT concentrations in this range would be associated with adsorbed surface coverages less than 1% of a monolayer. This behavior suggests that there are varying reactivities for different areas of the iron surface.

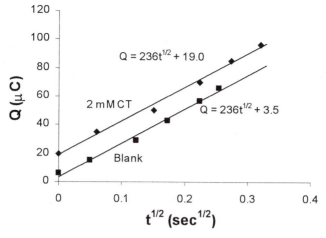

Figure 3. Anson analysis of CA profiles generated by an iron electrode at –985 mV in 10 mM CaSO₄ solutions, with and without 2 mM CT. At time=0, the electrode potential was stepped from its open circuit potential of -565 mV (blank) or –482 mV (CT) to -985 mV.

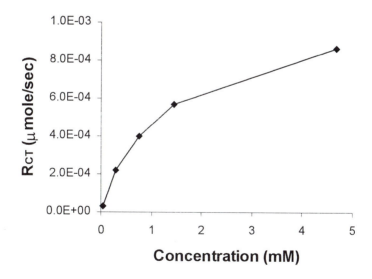

Figure 4. Reaction rate for dechlorination of CT to chloroform by a rotating disk electrode at a potential of –635 mV as a function of the bulk solution concentration.

This variation in surface reactivity can likely be attributed to molecular scale heterogeneity of the electrode surface. It is well known that rates of electron transfer on electrode surfaces vary depending on the crystal face (*23*). For corroding iron systems, corrosion rates are known to vary with location on the iron surface due to crystal defects, grain boundaries, stress differences and surface energy effects (*24*).

Transfer Coefficient Analysis

In multistep electron transfer reactions, the overall reaction rate may be limited by rates of reactant chemisorption, rates of bond breaking, or rates of molecular rearrangement. These are chemical, rather than potential, dependent factors. If the observed reaction rates are limited by chemical dependent factors, the measured transfer coefficients may be only fitting parameters, or apparent transfer coefficients. In this case, the apparent transfer coefficient will be temperature dependent (*25-27*). Conversely, for reactions that are limited by potential dependent factors, *i.e.*, the rate of outer-sphere electron transfer, the transfer coefficient should be independent of temperature (*25-27*).

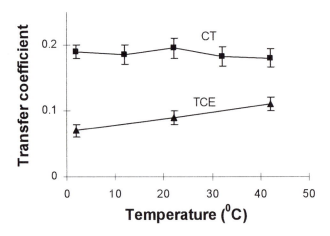

Figure 5. Electron transfer coefficients for reductive dechlorination of CT and TCE by an iron disk electrode. Transfer coefficients were determined using equation 1 and measured rates of dechlorination over the potential range from –600 to –1200 mV.

To determine whether the rate limiting step for CT reduction involved electron transfer, the transfer coefficient for CT reduction was determined over the temperature range from 2 to 42 °C. Figure 5 shows that the transfer coefficient for CT reduction was independent of temperature, at the 95% confidence level. This suggests that the rate of electron transfer controlled the rate of CT dechlorination. Further supporting this conclusion is the temperature dependence of the apparent activation energy (E_a) for CT reduction. Figure 6 shows that the apparent activation energy for CT reduction decreased with decreasing electrode potential. This is consistent with electrochemical theory which predicts that a fraction of the overpotential (E-E_{eq}) goes towards overcoming the chemical activation energy (23). Thus, for an electron transfer reaction, the apparent E_a should depend on the electrode potential according to (23):

$$E_a = E_a^{eq} + \vec{\alpha}F(E - E_{eq}) \tag{2}$$

where E_a^{eq} is the activation energy at the equilibrium potential. In this study, all overpotentials for CT reduction were negative, and thus E_a values decreased with increasingly negative electrode potentials.

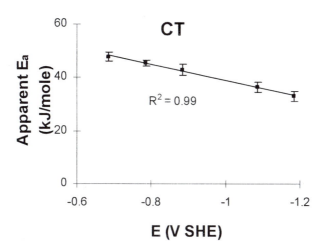

Figure 6. Apparent activation energies for reductive dechlorination of CT at an iron electrode. Activation energies were determined from Arrhenius analysis of dechlorination rates at temperatures of 2, 12, 22, 32 and 42 °C.

In contrast to CT, the transfer coefficient for TCE reduction was temperature dependent at the 95% confidence level, as shown in Figure 5. This suggests that the rate of TCE reduction was not limited by an electron transfer step. The temperature dependence of the E_a for TCE reduction shown in Figure 7 also supports this hypothesis. If the rate of TCE dechlorination was limited by the rate of electron transfer, the E_a should have declined with increasingly negative potential, as indicated by equation 2.

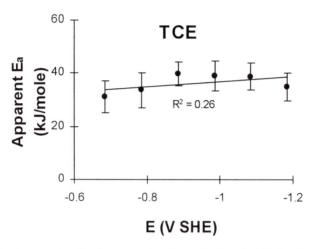

Figure 7. Apparent activation energies for reductive dechlorination of TCE at an iron electrode. Activation energies were determined from Arrhenius analysis of dechlorination rates at temperatures of 2, 22, and 42 °C.

Direct and Indirect Reduction

Comparison of amperometrically with analytically measured reaction rates for TCE and PCE suggest that there was an indirect mechanism involved in chloroethene reduction. Figure 8 shows the current going towards TCE and PCE dechlorination via direct electron transfer. Assuming a 6 electron transfer for TCE dechlorination and an 8 electron transfer for PCE, the currents for direct reduction at a concentration of 0.9 mM indicate that the direct reduction rate for PCE should have been 5 times faster that that for TCE. However, analytical determination of the rates of TCE and PCE dechlorination indicate that the rate constant for PCE reduction was only 2.8 times that for TCE. This suggests that in addition to direct electron transfer, there is an indirect reaction mechanism that is faster for TCE than for PCE. This mechanism likely involves reaction with atomic hydrogen produced from water reduction.

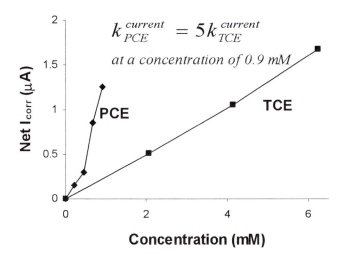

Figure 8. Corrosion current (I_{corr}) associated with TCE and PCE reduction via direct electron transfer from an iron wire suspended in a 10 mM CaSO$_4$ electrolyte solution with either TCE or PCE present at the indicated concentrations.

The indirect reduction mechanism becomes increasingly more important with decreasing solution concentration. For example, at aqueous concentrations of 0.01 mM, the overall rate for TCE dechlorination was 20% faster than that for PCE. This indicates that the relative rates of TCE and PCE dechlorination depend on the concentration range, and further supports the hypothesis that there are two reaction mechanisms for dechlorination.

The concentration dependence of the indirect reduction mechanism for TCE can be understood from the data in Figure 9. At low TCE concentrations, there is sufficient atomic hydrogen produced from water reduction to measurably contribute to TCE dechlorination. For example, Figure 9 shows that at concentrations less than 1 mM, the current for water reduction was greater than the direct current for TCE reduction, whereas, at higher TCE concentrations, the water reduction current was small compared to the direct reduction current for TCE.

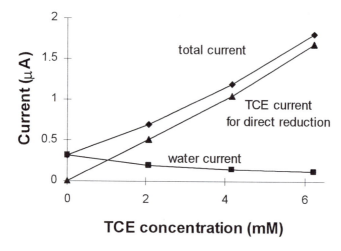

Figure 9. Total corrosion current, water reduction current and direct TCE reduction current for an iron wire electrode suspended in a 10 mM $CaSO_4$ background electrolyte solution as a function of the TCE concentration.

The results presented in Figures 8 and 9 may help explain some apparently contradictory data in the literature. Although most investigators have reported faster rates of PCE versus TCE dechlorination (2), one investigation reported significantly faster rates for TCE dechlorination (16). Arnold and Roberts reported that reaction rates for a series of chlorinated ethenes decreased with increasing degree of halogenation, a trend contrary to virtually all previously reported data (2). These apparently contradictory results may be explained in terms of direct and indirect reaction mechanisms. The Arnold and Roberts study was conducted at very low concentrations (in the 1-100 μM range), whereas other investigators have conducted their studies in the 0.1 to 10 millimolar range. At low concentrations, indirect reduction by atomic hydrogen is the dominant reaction pathway, while the direct reduction mechanism dominates at high chlorocarbon concentrations. This hypothesis is supported by data from the Arnold and Roberts study which showed that TCE and PCE reaction rates decreased with increasing concentration, and the ratio of the TCE rate constant to that for PCE decreased with increasing concentration (16). For example, at a concentration of 4 μM, the rate constant for TCE was 12 times greater than that for PCE, while at 80 μM, the difference was only a factor of 7.

The fact that less chlorinated compounds react faster via the indirect mechanism is consistent with studies of chlorocarbon dechlorination by atomic hydrogen in palladium catalyzed systems (18, 19). This may be due to the fact that less chlorinated compounds are more easily chemisorbed than are more chlorinated homologues. Decreasingly favorable chemisorption with increasing halogenation may be attributed to repulsions between the negatively charged chlorine atoms and electrons in the metal surface. This hypothesis is supported by studies of ethene hydrogenation on metal surfaces. *Ab initio* modeling indicates that the energy barrier for chemisorption is dominated by repulsions between electrons in the hydrocarbon and those in the metal surface (*28*).

Results from this study indicate that reduction of CT involves direct electron transfer, and that the rate of electron transfer controls the CT dechlorination rate. In contrast, reaction of TCE involves both direct and indirect reduction mechanisms, and the rate determining step does not involve electron transfer. The indirect reduction mechanism for TCE is faster than the direct mechanism, but its contribution to the overall rate of dechlorination decreases with increasing concentration.

Acknowledgment

This research has been supported by the United States Environmental Protection Agency (USEPA) office of Research and Development. Although the research described in this article has been funded by the USEPA through grant R-825223-01-0 to J.F., it has not been subjected to the Agency's required peer and policy review, and therefore does not necessarily reflect the views of the Agency, and no official endorsement should be inferred.

References

1. Gillham, R. W.; O'Hannesin S. F. International Association of Hydrologists Conference, "Modern Trends in Hydrology," Hamilton, Ontario, Canada, **1992**; pp 10-13.
2. Tratnyek, P. G.; Timothy, L. J.; Michelle, M. S.; and Gerald, R. E. *Ground Water Monitoring and Remediation*, Fall, **1997**, 108-114.
3. Matheson, L.J.; Tratnyek, P. G. *Environ. Sci. Technol.* **1994**, 28, 2045-2053.
4. Roberts, A. L.; Totten, L. A.; Arnold, W. A.; Burris, D. R.; Campbell, T.J. *Environ. Sci. Technol.* **1996**, 30, 2654-2659.
5. Campbell, T. J.; Burris, D. R.; Roberts, A. L.; Well, J. R. *Environ. Toxicol. Chem.* **1997**, 16, 625-630.

6. Johnson, T. L.; Scherer, M. M.; Tratnyek, P. G. *Environ. Sci. Technol.* **1996**, 30, 2634-2640.

7. U.S. Environmental Protection Agency, *"Permeable reactive subsurface barriers for the interception and remediation of chlorinated hydrocarbon and chromium (VI) plumes in ground water,"* Rep. No. EPA-600-f97-008, Washington, D. C., **1997**.

8. Wust, W. F.; Kober, R.; Schlicker, O.; Dahmke, A. *Environ. Sci. Technol.* **1999**, 33, 4304-4309.

9. Farrell, J.; Kason, M.; Melitas, N.; Li, T. *Environ. Sci. Technol.* **2000**, 34, 514-521.

10. Burris, D. R.; Campbell, T. J.; Manoranjan, V. S. *Environ. Sci. Technol.* **1995**, 29, 2850-2855.

11. Orth, W. S.; Gillham, R. W. *Environ Sci. Technol.* **1996**, 30, 66-71.

12. O'Hannesin, S.F.; Gillham, R.W. *Ground Water*, **1998**, 36, 164-170.

13. Cheng, I. F.; Muftikian, R.; Fernando, Q.; Korte, N. *Chemosphere*, **1995**, 35, 2689-2695.

14. Hardy, L. I.; Gillham, R. W. *Environ. Sci. Technol.* **1996**, 30, 57-65.

15. Brewster, J. H. *J. Amer. Chem. Soc.* **1954**, 76, 6361-6363.

16. Arnold, W. A.; Roberts, A. L. *Environ. Sci. Technol.* **2000**, 34, 1794-1805.

17. Li, T.; Farrell, J. M. *Environ. Sci. Technol.* **2000**, 34, 173-179.

18. Lowry, G. V.; Reinhard, M. *Environ. Sci. Technol.* **1999**, 33, 1905-1910.

19. Muftikian, R.; Fernando, Q.; Korte, N. *Wat. Res.* **1995**, 29, 2434-2439.

20. Schuth, C.; Reinhard, M. *Applied Catalysis B: Environmental,* **1998**, 18, 215-221.

21. Schreier, C. G.; Reinhard, M. *Chemosphere.* **1995**, 31, 3475-3487.

22. Bard, A. J.; Faulkner, L. R. *Electrochemical Methods.* John Wiley and Sons: New York, **1980**.

23. Bockris, J. O'M.; Reddy, A. K. *Modern Electrochemistry*, Vol. 2; Plenum Press: New York, **1970**.

24. Uhlig, H.H.; Revie, R.W. *Corrosion and Corrosion Control, third ed.*; John Wiley: New York, **1985**.

25. Savéant, J. M. *Adv. Phys. Org. Chem.* **1990**, 26, 1-130.

26. Savéant, J. M. *Acc. Chem. Res.* **1993**, 26, 455-461.

27. Schmickler, W. *Interfacial Electrochemistry*, Oxford University Press: New York, **1996**.

28. Blowers, P.; Ford, L.; Masel, R. *J. Phys. Chem. A.* **1998**, 102, 9267-9277.

Chapter 24

Enhanced Bioaccumulation of Heavy Metals by Bacterial Cells with Surface-Displayed Synthetic Phytochelatins

Wilfred Chen, Weon Bae, Rajesh Mehra, and Ashok Mulchandani

Department of Chemical and Environmental Engineering, University of California, Riverside, CA 92521

A novel strategy using synthetic phytochelatins was described for the purpose of developing microbial adsorbents for enhanced bioaccumulation of toxic metals. Synthetic genes encoding for several metal-chelating phytochelatin analogs $(Glu-Cys)_n Gly$ (ECs) were synthesized and displayed on the surface of *E. coli* using different anchor systems. Cells displaying ECs exhibited chain-length dependent increase in cadmium accumulation. A similar increase in Hg^{2+} accumulation was observed with cells expressing EC20 on the surface. The ability to genetically engineer ECs with precisely defined chain length could provide an attractive strategy for developing high-affinity bioadsorbents suitable for heavy metal removal.

Introduction

Because of their intrinsically persistent nature, heavy metal ions (As, Cd, Cr, Cu, Hg, Ni, Pb, Se, V and Zn) are major contributors to pollution of the biosphere (1). These metals when discharged or transported into the environment may undergo transformations and can have a large environmental, public health, and economic impact (2). The increasingly restrictive Federal regulation of allowable levels of heavy metal discharge and accelerated requirements for the remediation of contaminated sites make necessary the

412

development of new approaches and technologies for heavy metal removal. Conventional technologies such as precipitation-filtration, ion exchange, reverse osmosis, oxidation-reduction, and membrane separation are often inadequate to reduce metal concentration to acceptable regulatory standards. Recent research has focused on the development of novel bioadsorbents with increased affinity, capacity, and selectivity for target metals.

Immobilization is the major mechanism employed by nature (animals and plants) for counteracting heavy metal toxicity (3). Metallothioneins (MTs) and phytochelatins (PCs) are two groups of naturally occurring, cysteine-rich peptides that are synthesized for bind a wide range of heavy metals (4). Expression of MTs in *E. coli* to improve the bioadsorption of heavy metals is a promising technology for the development of microbial-based biosorbents (5). However, metal removal by intracellular MTs has been problematic because of the limited metal uptake (6). One approach to circumvent this problem is to directly anchor the MTs onto the cell surface (7). Suggestions have been made to use PCs in a similar manner as MTs since PCs offer many advantages, particularly the higher metal-binding capacity on a per cysteine basis (8). Because of the presence of a γ bond between amino acids, the exact biochemical and genetic mechanisms for their synthesis and chain elongation have not been elucidated.

The *de novo* design of metal-binding peptides is a promising alternative to MTs or PCs as they offer the potential of better affinity and selectivity for heavy metals. One particular strategy is to develop organisms harboring protein analogs of PC with the general structure (Glu-Cys)nGly (ECs) that can be synthesized using the ribosomal machinery. Detailed experiments with EC2 and EC4 have shown that these peptides bind a variety of metals in a manner similar to that exhibited by PC2 and PC4 (9). Although the metal-binding stoichiometries for ECs with higher cysteine content are still to be established, it is easy to envision that they might work in a similar fashion as EC2 and EC4. More importantly, it is possible to produce large quantities of ECs with any defined chain length of interest.

In this paper, we describe the construction and characterization of recombinant *E. coli* strains that anchor and display functional synthetic phytochelatins ranging from 7 to 20 cysteines (EC7, EC8, EC11, and EC20) onto the cell surface of *E. coli*. We demonstrate that these synthetic phytochelatins confer cadmium and mercury binding capability on the host cells and the resulting novel bioadsorbents accumulate a substantially higher amount of cadmium than the wild-type cells.

Experimental

Strains, plasmids, media, and general procedures. *E. coli* strain JM105 (*endA*1, *thi, rps*L, *sbcB*15, *hsdR*4, Δ(*lac-proAB*), [F', *traD*36, *proAB, lacI*q ZΔM15]) was used as the recipient of all plasmids. Plasmids pLO7, pLO8, pLO11, and pLO20 are derivatives of the vector pUC18, which bear the sequence coding for Lpp-OmpA-EC7, Lpp-OmpA-EC8, Lpp-OmpA-EC11, and Lpp-OmpA-EC20, respectively. Plasmid pINP20 was used to express the INPNC-EC20 fusion. Cultures were grown in low-phosphate MJS medium supplemented with 50 μg/ml ampicillin at 30°C to an OD_{600} of 0.3 when 1 mM IPTG was added to induce the expression of the fusion proteins. 100 μM $CdSO_4$ or $HgCl_2$ was subsequently added for metal binding experiments. General molecular biology procedures followed standard protocols unless specified otherwise (10).

Construction of Lpp-Omp- and INPNC-EC20 fusions. The synthetic gene encoding for (Glu-Cys)$_{20}$Gly (EC20) was prepared using two oligonucleotides (Research genetics, Huntsville, AL): ec-a) 5'TTT*GGATCC*ATGGAATGTGAATGTGAATGTGAATGTGAATGTGAAT GTGAATGTGAGTGTGAATGT<u>GAGTGCGAATGCGAA</u>3' and ec-b) 5'TTT*AAGCTT*TTAACCACATTCACATTCACATTCACATTCACATTCACA TTCGCATTCACATTCGCA<u>TTCGCATTCGCACTC</u>3'. The two oligonucleotides were mixed, boiled, and cooled to hybridize at the underlined sequence. The preferred codons for glutamic acid (GAA) and cysteine (TGT) have been changed at some locations to less frequently used codons, GAG and TGC, respectively, to prevent unwanted hybridization. Double strand synthesis was accomplished using the Klenow fragment (Promega). The synthetic gene was digested with *Bam*HI (italicized) and *Hind*III (italicized) and the resulting fragment was cloned into the yeast-*E. coli* shuttle vector pVT102-U, digested with the same restriction enzymes to generate pVT20. The resulting clone was sequenced to confirm the presence of the correct *ec20* fragment. Construction of synthetic genes for EC7, EC8 and EC11 followed similar procedures.

To construct the *lpp-ompA-ec20* fusion, the *lpp-ompA* fragment (481bp) was PCR amplified as described previously (11), digested with *Eco*RI and *Kpn*I and cloned into pUC18 to generate plasmid pLO. The *ec20* fragments were PCR amplified from plasmid pVT20 using the primers: ec-c) 5' GCT*GGATCC*TATGGAATGTG 3' and ec-d) 5' GCAAGGTAGACAAGCCG 3'. The primer ec-c contains an extra base (bold) just behind the *Bam*HI site in order to generate an in-frame fusion with *lpp-ompA*. The amplified fragment was digested with *Bam*HI and *Hind*III, gel-purified and subcloned into pLO to generate pLO20. The cloning of the *ec20* fragment was again confirmed by DNA sequencing. The recombinant plasmid pLO20, coding for Lpp-OmpA-EC20, was used for all subsequent experiments. Plasmids pLO7, pLO8 and pLO11 containing the genes encoding Lpp-OmpA-EC8 and Lpp-OmpA-EC11 were generated similarly. Plasmid pINP20, coding for the INPNC-EC20 fusion, was constructed by inserting the ec20 fragment into *Bam*HI/*Hind*III digested pINCOP (12).

Radiolabeling the target proteins and SDS-PAGE analysis. Radiolabeled cysteine (^{35}S, 1075 Ci/mmol, ICN) was added at the time of induction (final concentration of 5 μCi/ml). After desired time of induction, 1.5 ml aliquot of each culture was centrifuged. The extracted total proteins were boiled in sample buffer (Sambook et al., 1989) for 5 min and separated by SDS-PAGE (12.5% (w/v) polyacrylamide) (Laemmli, 1970). The gel was dried and exposed to X-ray film.

Bioaccumulation of Cd^{2+} and $Hg2^+$. Cells were grown in MJS medium and induced with 1 mM IPTG for the expression of fusion proteins. $CdSO_4$ (100μM) or $HgCl_2$ (5μM) was added to the culture in order to allow expression of ECs in the presence of Cd^{2+} or $Hg^{2+.}$ Cells did not show any significant reduction in growth at this concentration of the metal. Cells were harvested after desired time of induction, washed twice with double distilled water, and treated overnight with concentrated nitric acid. Disrupted cells were then diluted with double distilled water and centrifuged for 10 min at 4°C. The concentration of metals in the soluble fraction was directly measured through atomic absorption spectrophotometer (Perkin Elmer AAS3100) or by a mercury analyzer (Bacharach).

Results and Discussion

Expression of synthetic phytochelatins on the cell surface. A gene fusion system consisting of the signal sequence and the first nine amino acids of lipoprotein joined to a transmembrane domain from Outer Membrane Protein A (Lpp-OmpA), which has been used successfully to express a variety of proteins and enzymes onto the cell surface, was employed to anchor the different synthetic phytochelatins (11, 13). Synthetic genes coding for several synthetic phytochelatins were synthesized, linked to *lpp-ompA* fusion gene and displayed on the surface of *E. coli*. The ability to genetically engineer ECs with precisely defined chain lengths enables us to demonstrate for the first time the metal-binding capability of any PC or EC containing up to 20 cysteines. The high cysteine content of the synthetic phytochelatins, when labelled with ^{35}S cysteine, enables their ready detection by autoradiography. In the presence of 1 mM IPTG, synthesis of full-size Lpp-OmpA-EC8 (18.5 kDa), Lpp-OmpA-EC11 (19 kDa), and Lpp-OmpA-EC20 (21 kDa) was detected (Fig. 1A).

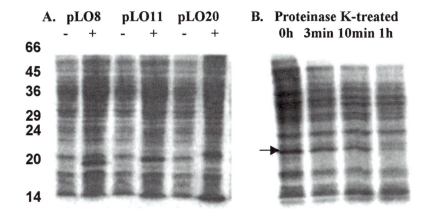

Figure 1. (A) Expression of Lpp-OmpA-EC fusions. [³⁵S]cysteine was added to the cultures at OD₆₀₀ = 0.3. The cultures were further grown for 24 hours. Total cell proteins were separated on SDS-PAGE (12.5% (w/v) polyacrylamide). The gel was dried and autoradiographed. Expression from uninduced (-) and induced (+) cultures were shown. The molecular weight markers are shown in the far left lane. (B) Protease accessibility experiments. Autoradiogram of radiolabeled proteins from E. coli cells harboring pLO20 after proteinase K treatment.

Protease accessibility experiments were carried out to ascertain the presence of ECs on the surface. Cultures grown on ^{35}S labeled cysteine were incubated with and without proteinase K and the total protein was analyzed by SDS-PAGE. For the cells incubated with proteinase K, the intensity of Lpp-OmpA-EC20 fusions continued to decrease and was no longer detectable after 1 hr (Fig. 1B). In contrast, no observable decline in the intensity was detected from cells overexpressing the MBP-EC20 fusions even after 21 hrs of incubation (data not shown).

Bioaccumulation of Cd^{2+} and Hg^{2+}. The metal-binding ability of whole cells expressing EC20 was tested by monitoring the binding of Cd^{2+} to *E. coli* expressing ECs on the cell surface through atomic absorption spectrometry. As shown in Fig. 2, strains expressing ECs on the cell surface accumulated a significantly higher amount of Cd^{2+} than cells carrying pUC18. This result confirmed that all ECs retain metal-binding capability independent of chain length. However, the chain length of ECs did influence the overall Cd^{2+} accumulation. The amount of Cd^{2+} accumulated increased with increasing cysteine residues in the ECs (Fig. 2). Cells with EC20 expressed on the surface (ca. 60 nmol Cd^{2+}/mg dry weight of cell) accumulated almost twice the amount of Cd^{2+} compared to cells expressing EC11. This result is consistent with the increasing number of metal-binding centers present.

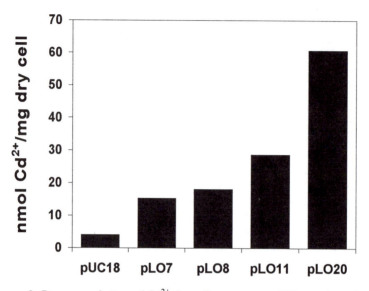

Figure 2. Bioaccumulation of Cd^{2+} by cells expressing ECs on the cell surface. Plasmids pUC18 were used as negative controls. The data were obtained from 5 independent experiments.

To demonstrate the binding capability of ECs to other heavy metals, Hg^{2+} accumulation of various *E. coli* strains were investigated. As shown in Figure 3, *E. coli* strain carrying pUC18 accumulated a very low level of Hg^{2+} and whole cell accumulation of Hg^{2+} was again increased with EC20 expressed on the surface.

Bioaccumulation of Cd^{2+} by cells expressing INPNC-EC20. One of the problems associated with surface expression using the Lpp-OmpA fusion system is the severe growth inhibition. In the case of EC20, cell growth was virtually stopped after induction. To explore the possibility of improving cell growth, a different anchor system based on the truncated ice nucleation protein (INPNC) was used. Cells carrying pINP20 showed no sign of growth inhibition even after induction. The final cell density was more than 2-fold higher than cells carrying pLO20. However, whole cell accumulation of Cd^{2+} was only 50% that of cells carrying pLO20. This difference may reflect a lower level of EC20 expressed on the cell surface (Figure 4). Recently, we demonstrated that expression of proteins on the cell surface was 60-fold more efficient in *Pseudomonas* than in

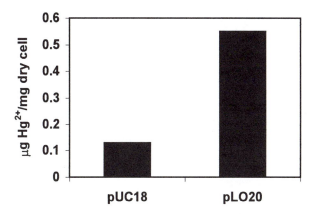

Figure 3. Bioaccumulation of Hg²⁺ by cells expressing EC20 on the cell surface. Plasmids pUC18 were used as negative controls. The data were obtained from 2 independent experiments.

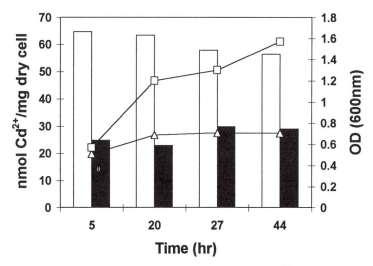

Figure 4. Cell growth (△, pLO20 and □, pINP20) and Cd²⁺ accumulation (white bar, pLO20 and black bar, pINP20) for cells expressing EC20 on the surface using either the Lpp-OmpA or INPNC anchor.

418

E. coli using the ice nucleation protein anchor. An alternative strategy to further improve the whole cell accumulation of heavy metal is to express EC20 on the surface of *Pseudomonas* or related species (14). The resulting recombinants could be immobilized onto solid supports for continuous removal of heavy metals.

Acknowledgments

This work was supported by the UC Biotechnology Research and Education Program and the US EPA (R827227).

References

1. Nriagu, J.O. and Pacyna, J.M. *Nature* **1989**, 333, 34.
2. Gadd, G.M. and White, C. *Trends Biotechnol.* **1993**, 11, 353.
3. Mehra, R.K. and Winge, D.R. *J. Cell. Biochem.* **1991**, 45, 30.
4. Stillman, MJ, Shaw III, FC, Suzuki. KT. **1992**. Metallothioneins. VCH Publishers.
5. Pazirandeh, M.; Chrisey, L.A.; Mauro, J.M.; Campbell, J.R. and Gaber, B.P. *Appl. Microbiol. Biotechnol.* **1995**, 43, 1112.
6. Chen, S. and Wilson, D.W. *Appl. Environ. Microbiol.* **1997**, 63, 2442.
7. Sousa, C.; Kotrba, P.; Ruml, T.; Cebolla, A. and de Lorenzo, V. *J. Bacteriol.* **1998**, 180, 2280.
8. Zenk, MH. *Gene* **1996**, 179, 21.
9. Bae, W. and Mehra, R.K. *J. Inorg. Biochem.* **1997**, 68, 201.
10. Sambrook, J, Fritsch, EF, Maniatis, T. **1989**. Molecular Cloning - A Laboratory Manual, 2nd ed. Cold Spring Harbor Laboratory Press, Cold Spring Harbor, New York.
11. Richins, R.; Kaneva, I.; Mulchandani, A. and Chen, W. *Nat. Biotechnol.* **1997**, 15, 984.
12. Shimazu, M.; Mulchandani, A. and Chen, W. *Biotechnol. Prog.* **2001**, In Press.
13. Francisco, J.A., Earhart, C.F. and Georgiou, G. *Proc. Natl. Acad. Sci. USA* **1992**, 89, 2713.
14. Shimazu, M.; Mulchandani, A. and Chen, W. Unpublished Results.

ome cases contained up to 20% by weight water! To understand these soil econtamination results, many studies were carried out on chlorinated model ompounds dissolved in liquid NH₃, free of soil. For example, adding ndividual PCBs or PCB mixtures to Na/NH$_{3(l)}$ gives biphenyl immediately equation 1) and subsequently further reduction products are formed depending n conditions.

$$(1)$$

Dechlorination with deficient amounts of sodium

CCl_4 was converted quantitatively to CH_4 when treated with excess Na in NH₃. All of the chlorine was mineralized to NaCl. What would happen if insufficient Na was present to completely dechlorinate CCl_4? Sodium-deficient reactions were carried out to find the answer to this question. Remarkably, even under conditions of rapid stirring, the only products observed were CH_4 and unreacted CCl_4.[9] No partially dehalogenated products (e. g. CCl_3H, CH_2Cl_2 or CH_3Cl) were detected. This suggests that all reduction steps are occurring under diffusion control. CCl_4 is reduced by dissociative electron transfer where a solvated electron is transferred to CCl_4 as the carbon-chlorine bond simultaneously breaks[14] (Scheme 1). The resulting $^\bullet CCl_3$ radical adds another electron and the anion formed, $^-CCl_3$, may eject chloride or capture a proton (from dissolved water). Subsequent dechlorination steps all occur before these intermediate species can diffuse away from the vicinity of the dissolving Na particle.

Diffusion control was not limited to CCl_4. Other CAHs also produced no partially dechlorinated products in Na/NH₃ reductions. The same was true for multiply chlorinated aromatic compounds. Example sodium-deficient dechlorinations are summarized in Table 3. Partially dechlorinated intermediates were not observed when 1,2-dichlorobenzene, CCl_4 and CH_3CCl_3 were reduced in NH$_{3(l)}$ with a stoichiometric deficiency of Na (Table 3), either with or without H_2O present. For example, 3,4-dichlorotoluene gave only toluene and recovered starting material when treated with Na/NH₃ at Na/substrate ratios of 1, 2 and 3 in the presence of 5 equivalents of water. Dechlorinations of CCl_4 with 1, 2 and 3.4 equivalents of Na in anydrous NH$_{4(l)}$ gave only CH_4, recovered CCl_4 and traces of unknown products. The material balances were almost complete. No $CHCl_3$, CH_2Cl_2 or CH_3Cl were detected by GC analysis. Similarly, dechlorination of CH_3CCl_3 with 2 equivalents of Na produced ethane (62%) and CH_3CCl_3 (38%) was recovered unchanged. The material balances confirm that NH₃ or $^-NH_2$ is not reacting with partially dechlorinated intermediates to give aminated products or elimination products.

Chapter 25

Solvated Electron (Na/NH₃) Dechlorination of Model Compounds and Remediation of PCB- and CAH-Contaminated Wet Soils

Charles U. Pittman, Jr.[*], Jinbao He, and Guang-Ri Sun

Department of Chemistry, Mississippi State University,
Mississippi State, MS 39762

Soils, sludges and aggregates contaminated with polychlorobiphenyls (PCBs) and chlorinated aliphatic hydrocarbons (CAHs) have been remediated by simply slurrying in liquid NH₃, at ambient or low temperature, followed by addition of Na. Na dissolves generating solvated electrons which dechlorinate PCBs and CAHs at diffusion controlled rates. Na/NH$_{3(l)}$ remediation treatments work well on wet soils and sludges because the lifetime of a solvated electron in 20% H_2O/80% NH₃ is ~100 sec. and longer when less water is present. This lifetime sufficient so that the reactions of NH₃ , H_2O and impurities with the solvated electron do not compete efficiently with dechlorination until the concentrations of the chloroorganic compounds are very low. Reductions of the model compounds, CCl_4, dichlorotoluenes, dichlorobenzenes and CH_3CCl_3, produced only completely dechlorinated products and unreacted starting materials when a deficiency of sodium was used. The consumption of Na (per chlorine removed) was studied as a function of water concentration and the chlorocarbon contaminant's concentration in model compound reductions and in soil remediations. Sample remediations of soils and sludges from superfund sites are presented. It seems likely that Na/NH$_{3(l)}$ treatment technology can simultaneously react and destroy PCBs, CAHs, polynuclear aromatic hydrocarbons, pentachlorophenol, dioxins/furans and possibly organic nitro/nitrates (munitions wastes) in soils or sludges while also lowering the concentrations of several metals in these samples.

Polychlorinated biphenyls (PCBs) and other chlorinated compounds are distributed in soils, sludges, estuaries, etc. at over 400 sites in the United States. Chlorinated aliphatic hydrocarbons (CAHs) occur as serious contaminants at 358 major hazardous waste sites. CAHs migrate vertically through soils to form dense nonaqueous phase liquids (DNAPLs) on aquifer bottoms. Both *ex-situ* methods to remediate PCBs, CAHs are critically needed. Therefore, our goal has been to develop a generalized technology to decontaminate soils (*ex-situ* and *in-situ*) contaminated with these compounds. Ideally, such methodology should operate very rapidly at ambient temperatures. Chlorinated organics are also found frequently in mixed wastes (those containing radioactive contaminants). Usually such hazardous organic wastes must be removed before dealing, finally, with the radionucleotides. Thus, several treatment options are needed for all the reasons listed above. These considerations highlight the need to provide many economical methods, both *ex-situ* and *in-situ*, to decontaminate soils, sludges and aggregates containing chlororganics.[1-6]

Such remediations have been a high priority research area at the US Environmental Protection Agency and also at the Department of Energy and Department of Defence. In addition to chlorinated organic pollutants, many soils are contaminated with polynuclear aromatic hydrocarbons (PAHs) from coal tar operations, wood treatment processes (e. g. creasote treatments of telephone poles, railroad ties etc.). Other locations are contaminated with nitrated organic compounds associated with munitions and explosives manufacture. Metal ion contamination constitutes another troublesome problem. For example, lead, chromium, vanadium, cadmium, mercury as well as the radionucleotides are serious problems at many contaminated sites. Our recent studies[7-10] and those at Commodore Solution Technologies[11-13] have now demonstrated a generalized solvated electron technology which can decontaminate both neat samples and soils or sludges which are contaminated with PCBs, CAHs, CFCs, explosive's wastes, cyanides, chemical warfare agents and even reduce the contamination levels of some metals.

Results

PCBs, PCB-contaminated soils, CAHs and CAH-contaminated soils (as received clay loam and sandy soils) were readily decontaminated in our laboratory with Na/NH$_3$ at ambient temperature. Example decontaminations of PCBs from soils are illustrated in Table 1. Typically, the soil is slurred in liquid NH$_3$ and then metallic Na is added. No special effort is made to dry the soil. Soil slurry reactions are completed within about 30 sec. and the NH$_3$ is flashed off and recovered for recycle. PCB-destruction efficiencies of >99.9 have been demonstrated both in our laboratory and at Commodore Solutions Technologies. In these reactions all chlorine is converted to NaCl. Similar results have been achieved with CAH-contaminated soils. This is illustrated in Table 2 for soils contaminated with carbon tetrachloride, CCl$_4$. These soils were not dried and in

Table 1. Destruction of PCBs in various soils with Na/NH$_3$

Soil Type	Pre-Treatment PCB Level (ppm)	Post-Treatm Level
Sand, clay[a]	0	
Sand, silt, clay[b]	77	
Sand, silt[c]	1250	
Sand, silt, clay[d]	8.8	

Treatment temperature in liquid NH$_3$: (a) 32 °C; (b) –33 °C. The range wt. Percents in NH$_3$ used was 1.27 to 3.3% in these examples.

Table 2. Decontamination of CCl$_4$-Contaminated soils with Na/NH$_3$[a]

Soil Type	Pre-Treatment CCl$_4$ Level (ppm)	Post-Treatment Level (ppm)
Sand, silt	1200	<1.0
Sand, clay	226	<1.0
Clay	109	<1.0

[a]Reactions carried out at 24-30 °C.

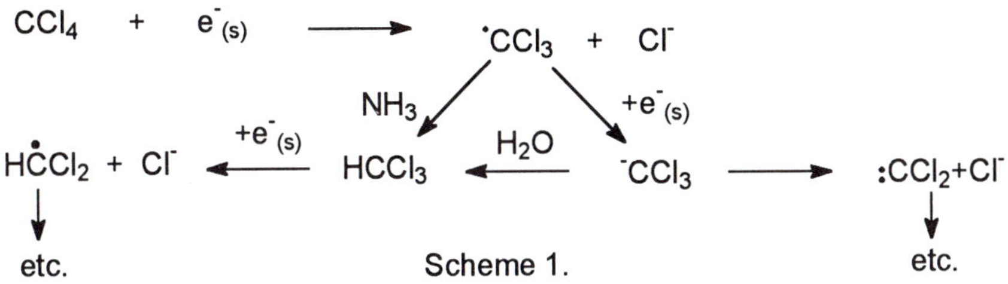

$$CCl_4 \ + \ e^-_{(s)} \ \longrightarrow \ ^\bullet CCl_3 \ + \ Cl^-$$

Scheme 1.

Reactions competing with dechlorination

Excavated contaminated soils could be added to preformed solvated electron solutions or, alternatively, they could be preslurried in $NH_{3(l)}$ and then Na could be added to form the solvated electrons. The latter method is often more effective because it might take some significant time for solvated electrons and NH_3 to diffuse into soil particles to bring the electrons into contact with the intercalated chlorinated organics. During this time the solvated electrons could react with water, with dissolved oxygen, with NH_3 or with other compounds in the soil.[15] The reaction with NH_3 is catalyzed by dissolved Fe^{3+} ions which are plentiful in most soils (see equation 2) and by oxygen.[16,17] Thus, experience has taught that adding Na to preslurried soils leads to lower Na consumption at equivalent decontamination levels, particularly if weathered clay soils are being treated.

$$Na \ + \ NH_3 \ \xrightarrow{Fe^{3+}} \ 1/2\,H_2 \ + \ Na^+ \ +\ ^-NH_2 \qquad (2)$$

As sodium dissolves into liquid NH_3, solvated electrons are formed which react to dechlorinate chloroorganic toxicants at enormous rates. However, to decontaminate a soil containing 2000 ppm of PCBs or CAHs to a level of 1 ppm or lower when 20% by weight of water is present, requires the solvated electron's rate ratio, k_{RCl}/k_{H2O} to be huge. If not, reactions of solvated electrons with water will consume large amounts of Na because, at 10 ppm and 0.5 ppm of PCB, the mole ratio of H2O/PCB is ~240,000 and ~5x10^6, respectively. Thus, for the technology to be successful, the relative rate of reaction of solvated electron with R-Cl versus that with water must be enormous. Therefore, soil-free experiments have been run to determine the amount of Li, Na, K or Ca which is consumed in order to give 99.9% dechlorination of model chlorinated compounds. Some sample results for Na are shown in Table 4. These experiments were run both in anhydrous NH_3 and in NH_3/H_2O solutions. As can be seen, the consumption of Na does not become excessive in the presence of water. This proves that the k_{RCl}/k_{H2O} ratio must be huge and we estimate this

Table 3. Products obtained in sodium deficient reactions in liquid NH_3 at room temperature

Substrate	Na/Substrate (mole ratio)	H_2O/Substrate (mole ratio)	Products
3,4-Dichlorotoluene	2	5	60% SM,[a] 40% toluene
	3	5	25% SM, 75% toluene
1,2-Dichlorobenzene	3.5	5	7% SM, 93% benzene
CCl_4	1	0 or 4	79% CCl_4 20% CH_4
	2	0 or 4	54% CCl_4 45% CCl_4
	3.4	0 or 4	33% CH_4 64% CH_4
CH_3CCl_3	2	0 or 4	38% CH_3CCl_3, 62% CH_3CH_3

[a]SM = starting material

Table 4. Minimum amount of Na required to completely dechlorinated

Substrate	Na/Substrate Mole Rations and (Na/Per Cl Mineralized) Required For Complete Dechlorination at Different H_2O/Substrate Mole Ratios		
H_2O/Substrate	0/1[a]	5/1	50/1
4-Chlorotoluene	1.5 (1.5)	2.3	2.5 (2.5)
3,4-Dichlorotuene	2.5 (1.25)	4.2	6.0 (3)
1,2,3-Trichlorobenzene	4.0 (1.33)	5.5	7.0 (2.33)
2,4,6-Trichlorophenol	4.6 (1.53)	7.4	10 (3.33)
1,1,1-Trichloroethane	3.6 (1.2)	4.2	5.1 (1.7)
Carbon tetrachloride	4.6 (1.15)	5.1	6.4 (1.6)
Lindane	6.7 (1.11)	6.7	7.4 (1.23)

rate at ~10^7 in 15% H_2O/85% NH_3. The metal consumption required per chlorine removed was found to be lower for Na than for Li, K or Ca.[9] This was especially true as the amount of water present in the reactions increased. Na consumption was far lower than Ca consumption, clearly indicating that Na is the metal of choice for environmental remediations. This is treated in more detail in reference nine to which the reader is referred.

The amount of Na per chlorine mineralized required for complete dechlorination followed the order: tetrachloro- <trichloro- <dichloro- <monochloro-. Similarly, the most efficiently reduced CAHs were those which contained the most chlorine. The efficiency of Na utilization also depends on the Na particle size. Preformed solvated electron solutions required more Na consumption than high surface area, thin Na mirrors. Chunks of Na required the least Na consumption for complete dechlorination of several model compounds although the differences were not large.

Remediating Soils Containing 1,1,1-Tricethane and Tetrachloroethylene

Standard soils obtained from the Agronomy Department at MSU were purposely contaminated with from 3000 to 5000 ppm of the chloro compounds and then treated with Na/NH_3. 1,1,1-Trichloroethane was one of the compounds selected since it is a widely disperse pollutant in soils and DNAPLs. The quantity of Na required to lower the chlorocarbon level in the soil to a series of specific lower concentrations was determined and this amount was expressed as the moles of Na consumed per mole of chlorine that was mineralized (e. g. RCl + nNa → NaCl + RH). The Na consumption was then compared for dry versus wet soil samples to see the effect of water. As shown in Table 5, high clay content B1 soil and organic surface soil, B2-type, contaminated with 3000 ppm CH_3CCl_3 or $Cl_2C=CCl_2$, were remediated with Na/NH_3 treatments. The effect of adding water to these soils was then examined. While CH_3CCl_3 studies were not taken to sub ppm levels of decontamination, lowering CH_3CCl_3 from 3000 to 1-8 ppm required only 5 equiv. of Na in dry B2 soil and only 22 equiv. of Na in the presence of 15% water. Similar results were obtained for tetrachloroethylene where 9 equivalents of Na were needed per equivalent of Cl to lower contamination levels from 3000 to 0.9 ppm in dry B1 soil versus a ratio of 26 Na/Cl in wet (7% H_2O) B1 soil to remediate from 3000 to 0.8 ppm.

These studies show that the amount of Na required overall is rather modest. As the concentration of chlorocarbon gets smaller and smaller, the side reactions with water, NH_3 and soil components become increasingly competitive. Thus, many moles of Na are consumed at low chloro compound concentration in order to dechlorinate each additional chloroorganic molecule. However the Na consumption is modest since so few moles of chloroorganic are present at these low concentrations. This point is illustrated below by the use of 1-chloroctane as a model soil contaminant.

Table 5. Remediation of CH_3CCl_3 or $Cl_2C=CCl_2$ in B1 (clay) and B2 (organic) soils. Sodium Consumption per Chlorine Mineralized at Different Levels of Water.[a]

CAH Contaminant	Soil Type and Water Content	NH₃ Vol. per g. soil (ml)	CH_3CCl_3 Level In Soil (ppm)		Na Consumed per Cl Removed (mole ratio)
			Start	Finish	
CH_3CCl_3	B1 Dry	10	3000	7.5	7.4
	B1 7% H_2O	10	3000	131	15
		10	3000	2.2	22
	B1 15% H_2O	10	3000	361	22
	B2 Dry	5	3000	1.7	5
	B2 15% H_2O	5	3000	4.1	22
$Cl_2C=CCl_2$	B1 Dry	5	3000	0.9	9
	B1 7% H_2O	5	3000	0.8	26
	B1 15% H_2O	5	3000	2.0	35
	B2 Dry	5	3000	4.0	9
	B2 15% H_2O	5	3000	2.6	27

[a]B1 is a clayey, mixed thermic typic tapludults soil containing shale; 2.1 type (hydroxy interlayered vermiculite/semectite) with minor amounts of mica, kaolinite and quartz. B2 is a high organic soil with 10.2% clay, 39.1% silt, 50.7% sand, 3.4% organic matter, pH+4.67, base saturation 2.8%, caution exchange capacity 8.79 C mol_c/kg. Citratedithionate extractable Fe=6.44g/kg.

1-Chlorooctane is not highly soluble in $NH_{3(l)}$ and its dechlorination is slower than that of the other models used. Thus competiting reactions become more serious. Dry B1 soil and B1 soil with 5% and 10% water were contaminated with varying amounts of 1-chlorooctane and then remediated with deficient amounts of Na so all of the Na was consumed. The final concentrations of 1-chlorooctane were determined and the Na consumed/chlorine removed ratios were determined (see Table 6). Decontaminating dry soil from 5000 to 0.6 ppm of 1-chlorooctane required 30 moles of Na per mole of chlorine mineralized overall. But the portion of the reaction between 50 ppm and 0.6 ppm of 1-chlorooctane required 1200 moles Na per mole of chlorine mineralized. As water (5% and 10%) is added the Na consumption further increases but the trends remain the same. Decontaminating from 5000 to 0.4 ppm in the presence of 5% wt. water required 60 moles Na overall per mole of chlorine removed and going from 5000 to 0.6 ppm with 10% water required 80 moles Na overall per mole of chlorine removed. As can be seen in Table 6, the portions of these decontamination reactions between 200-100 ppm down to 0.6-0.4 ppm required in the neighborhood of 1300 to 1500 moles Na per mole chlorine mineralized. Nevertheless, the overall sodium consumption is not excessive and a major cost consideration is the need to excavate the soil.

Decontaminating PCBs from Soils, Sludges and Surfaces

Na/NH_3 treatments of PCB-contaminated soils from several locations in the U. S. are summarized in Table 7. These reactions are typical of commercial treatment and were performed by Commodore Solutions Technologies, Inc. Of particular interest is the PCB- and dioxin-contaminated sludge from the New Bedford Harbor Sawyer Street site in Massachusetts. It initially contained 32,800 ppm of PCBs and 47 ppm of dioxin/furan compounds. These decontaminations seem to occur equally well at low or ambient temperatures and good results are obtained over all variety of soil types investigated. The technology works on acidic or basic soils, high ion exchange capacity soils, soils/sludges with high water contents and high iron contents. Some soils can be treated as-received. Some, however, required some pre- or post processing such as size reduction or washing to improve remediation effectiveness. After treatment, the soils pass all TCLP criteria for putting it back into the excavation site or for non-hazardous waste landfill disposal.

The Bedford Harbor sample discussed in Table 7 shows another advantage of $Na/NH_{3(l)}$ treatment. The levels of toxic metals present in the sample were greatly reduced in this remediation. Methods using Na/NH_3 to decontaminate metals from soils are discussed in US Patents 5,516,968 (1996) and 5,613,238 (1997).

The destruction of PCBs and dioxins in waste oils is readily accomplished by the Na/NH_3 technique. This is illustrated in Table 8 for motor, mineral and transformer oils, hexane and a waste oil from the McCormick and Baxter site,

Stockton, CA. Typically, 2 to 4% weight Na in NH_3 was used in these remediations.

Why Do These Dechlorinations Work?

Solvated electrons are destroyed at very fast rates in water so why should wet soil decontaminations be successful in $Na/NH_{3(l)}$? In pure water the half-life of the solvated electron is very short ($t_{1/2} = \sim100$ μsec.).[18] However, in 20% H_2O/80% NH_3 the solvated electron's half life increases dramatically ($t_{1/2} = \sim100$ sec.).[19,20] In pure liquid NH_3 the $t_{1/2}=300$ h.[20] In fact, the reaction of solvated electron with water (equation 3) has a much higher kinetic barrier than electron transfer to chlorinated molecules. Therefore, the kinetics of dechlorination are much faster than reaction with water (equation 3). All that is required is that the solvated electrons survive long enough for mixing, penetration into the soil or extraction of the chloroorganics from the soils. If solvated electrons have $t_{1/2}=100$ sec. when 20% water is present (and longer when only 5 or 10% water is present) then the very fast dechlorination processes dominate. If electron transfer to RCl occurs in ~1 μsec and the solvated electron's half life in the medium is 10 to 100 sec, a huge rate ratio (k_{RCl}/k_{H2O}) exists favoring dechlorination.

$$e^-{}_{(s)} + H_2O \longrightarrow 1/2\ H_2 + {}^-OH \qquad (3)$$

When dissolved oxygen and ions such as Fe^{3+} are present, solvated electrons are also consumed in reactions with NH_3 (see equation 2). Other reactions undoubtably occur with components of the soil. These reactions, together with reaction with water, compete for the solvated electron. However, this competition only becomes serious at low concentrations of PCBs or CAHs as was illustrated in Tables 5 and 6. At low pollutant concentrations, not many moles of chloroorganic remain to be remediated. Therefore, even if the number of moles of Na required to mineralize each additional mole of chlorine is high at low RCl concentrations, the actual consumption of Na is modest. Thus, these kinetic factors lead to the success of this technology.

Other Applications

Electron transfer to polynuclear aromatic hydrocarbons (PAHs) and nitro compounds takes place very rapidly. Thus, remediation of soils contaminated with these species might be an attractive target. Work at Commodore Solutions Technologies and Mississippi State University have shown that a wide range of PAHs are rapidly destroyed (products unknown) when treated with Na/NH_3. Similarly, soils contaminated with the explosives, HMX and RDX, and 1,2-dinitrobenzene have been effectively remediated with destruction efficiencies of

Table 6. Remediation of 1-Chlorooctane from B1 Soil. The requisite Na consumption per chlorine mineralized versus the 1-chlorooctane concentration range.

Water in Soil	1-Chlorooctane Concentration in Soil		Na Consumed per Cl Mineralized
(% wt)	Initial (ppm)	Final (ppm)	(Na/Cl mole ratio)
Dry	5000	870	7
	5000	560	13
	1000	22	65
	100	7.4	300
	100	1	650
	50	0.6	1200
	5000	0.6	30
5%	5000	381	20
	5000	21	30
	5000	14	45
	200	4	375
	100	5	750
	100	1	1125
	100	0.4	1500
	5000	0.4	60
10%	5000	333	30
	5000	54	35
	5000	8.8	45
	200	3.8	1100
	200	0.6	1300
	5000	0.6	80

Table 7. Destruction of PCBs and Dioxins in Soils and Sludges with Na/NH$_3$ Treatments

Source	Soil or Sludge	Pre-Treatment PCB Level (ppm)	Post-Treatment PCB Level (ppm)
Harrisburg, PA[a]	Sand, clay	777	<1.0
Los Alamos, NM[b]	Sand, silt, clay	77	<2.0
New York[c]	Sand, silt	1250	<2.0
New Bedford Harbor, MA	Sludge	32,800	1.3
		47[d]	0.012[d]
		Hg 0.93	0.02
		Pb 73	0.2
		Se 2.5	0.2
		As 2.8	0.1

Treatment temperature in NH$_3$ (a) °C, (b) 20 °C, (c) –33 °C, (d) 47 ppm of dioxin/furan present before treat and 0.12 ppm after treatment; metals present pre- and post-treatment are also shown.

Table 8. Decontamination of PCB- and Dioxin-Contaminated Oils

Material	Pre-Treatment ppm		Post-Treatment ppm
Motor oil	23,339	PCB	<1.0
Transformer oil	509,000	PCB	20[a]
Mineral oil	5,000	PCB	<0.5
Hexane	100,000	PCB	0.5
(McCormick-Baxter	418	PCB	2.3
Site Waste Oil)	14.1	Dioxin	1.3
Transformer oil spill at New York Utility Site	1200	PCB	1.4

[a]Sodium feed was deficient. Addition of more sodium reduced PCB level to <1.0 ppm.

99.99% or more using 2.8% wt. Na in one liter of $NH_{3(l)}$ per 50g of soil at 39 °C. Thus, it appears this chemistry offers a single treatment method to simultaneously remove many different classes of pollutants.

Safe Handling of Sodium and Ammonia

Standard laboratory methods of handling sodium were used in the small laboratory reactions reported in this chapter. However, applications of larger scale remediation reactions requires further comment. Commodor Solution Technologies, Inc., Marengo, OH has patented many Na/NH_3 remediation applications under the trade name SET.^{TM} They have carried out commercial 1200 liter scale reactions in a system consisting of a sodium transfer station which warms sodium, cast in shipping drums, to the liquid state and then pumps liquid sodium into a reactor tank. Anhydrous NH_3 from an ammonia storage tank is then pumped in, dissolving the molten sodium. Liquid contaminants to be remediated can be pumped into the Na/NH_3 solutions at the rate of about 1,600 pounds per day. Solids have been treated in a 10 ton per day unit. The contaminated solids and liquid ammonia are preslurried and then either liquid or solid sodium is added. Thus, this technology has now been safely employed on a commercial scale.

The major supplier of sodium in the USA is DuPont, Inc. DuPont requires its sodium customers to undergo rigorous instruction on the safe handling of sodium as part of its product stewardship. Details of the safe handling of sodium can be found in DuPont's product literature. DuPont document number E-92775-2 entitled "Sodium. Properties, Uses, Storage and Handling" is especially useful. Special assistance can be provided by V. Marcant of DuPont (phone: 716-278-5378) who can supply an Adobe Acrobat: PDF version of the document mentioned above. The company also supplies a video on handling sodium.

Finally, a useful chapter entitled "Sodium and Sodium Alloys" is available in the Kirk Othmer "Encyclopedia of Chemical Technology", Fourth Edition, Volume 22, pages 327-354, 1997 published by John Wiley and Sons.

The Ammonia Safety Training Institute, Manteca, CA, provides gratis information on the handling and use of ammonia. Their web site is **www.ammonia-safety.com/**.

Acknowledgements

Support of this research was provided by the US Environmental Protection Agency, grant No. GAD# R826180 and by the Department of Interior, US Geological Survey, Grant no. HQ 96GR02679-12.

References

1. Subsurface Contaminants Focus Area: Technology Summary, Rainbow Series, U. S. Department of Energy, Office of Science and Technology, August 1996. Report No. DOE/EM-0296 (NTIS Order No. DOE/EM-0296; available through EM Helpline, Telephone 1-800-736-3282).

2. Dense Non-Aqueous Phase Liquids: A Workshop Summary, Dallas, TX, April 16-18, 1991, Ada, OK: Environmental Protection Agency, February 1992. EPA/600/R-92/030 (NTIS Order No. PB 92-178938).

3. FY 1995 Technology Development Needs Summary, pp. 204, U. S. Department of Energy, Office of Environmental Management, March 1994. Report No. DOE/EM-0147P (NTIS Order No. DE 94012580).

4. LSFA: Landfills Stabilization Focus Area. "National Technology Needs Assessment." Working Draft, U. S. Department of Energy, January 31, 1996.

5. PFA: Plumes Focus Area, "National Technology Needs Assessment." Review Draft, U. S. Department of Energy, May 17, 1996.

6. Plumes and Landfills Stabilization Focus Areas. Progress Report, U. S. Department of Energy, Winter, 1996.

7. C. U. Pittman, Jr. and S. M. H. Tabaei, Preprint Extended Abstracts of the Special ACS Symposium: Emerging Technologies in Hazardous Waste Management V, Vol II, 557-560, 1993, Sept. 27-29, 1993, Atlanta, GA, D. W. Tedder, Ed.

8. C. U. Pittman, Jr. and M. K. Mohammed, "Solvated Electron Reductions: Destruction of PCBs, CFCs, CAHs and Nitro Compounds by Ca/NH$_3$ and Related Systems," Preprint Extended Abstracts of the IC & E Special Symposium, Emerging Technologies in Hazardous Waste Management VIII, Sept. 9-16, 1996, Birmingham, AL, pp. 720-723.

9. G.-R. Sun, J.-B. He and C. U. Pittman, Jr. *Chemosphere* 41, 907-916 (2000).

10. C. U. Pittman, Jr., Abstract of the Spring AIChE Meeting, Atlanta, GA, March 5-9, 2000.

11. N. Weinberg, D. J. Mazer and A. E. Abel, *U. S. Patent.* 4,853,040 Aug 1, 1989 and *U. S. Patent* 5,110,364, May 5, 1992.

12. A. E. Heyduk, A. E. Able and R. W. Mouk, U. S. Patent 5,678,231 1997 and U. S. Patent 5,613,238, 1997.

13. A. E. Abel, *U. S. Patent* 5,495,062, 1996 and *U. S. Patent* 5,516,968, 1996.

14. T. Holm, *J. Am. Chem. Soc.*, 121, 515, 1999.

15. M. Smith, Dissolving Metal Reductions, in Techniques and Application in Organic Synthesis, R. L. Augustine, Ed., Marcel Dekker, New York, 1968, pp. 99-170.

16. J. E. Eastam and D. R. Larkin, *J. Am. Chem. Soc.*, 81, 3652-3655, 1959.

17. P. W. Rabideau, D. M. Wetzeland and D. M. Young, *J. Org. Chem.*, 49, 1544-1549, 1984.

18. R. F. Gould, Advances in Chemistry, American Chemical Society, Washington, D. C., 50, 1965.

19. R. M. Crooks and A. J. Bard, *J. Phys. Chem.*, 91 (5), 1274 1987.

20. U. Schindewolf in Metal-Ammonia Solutions, J. J. Lagowski and M. J. Sienko, Eds., Butterworths, London, 1970, pp. 199-218.

Chapter 26

Groundwater and Soil Remediation Using Electrical Fields

Jiann-Long Chen[1], Souhail R. Al-Abed[2], James A. Ryan[2], and Zhenbin Li[2]

[1]Department of Civil and Environmental Engineering, North Carolina Agriculture and Technical State University, Greensboro, NC 27411
[2]National Risk Management Research Laboratory, U.S. Environmental Protection Agency, Cincinnati, OH 45268

Enhancement of contaminant removal and degradation in low permeability soils by electrical fields are achieved by the processes of electrical heating, electrokinetics, and electrochemical reactions. Electrical heating increases soil temperature resulting in the increase of contaminant volatility and diffusivity and decrease of partitioning coefficients. The main electrokinetic phenomena include electroosmosis and electromigration. Electroosmosis facilitates groundwater flow and transport of nonionic contaminants in low permeability regions while electromigration increases the removal efficiency of ionic contaminants from the regions. Electrochemical reactions utilize high and low redox potentials at the anode and the cathode to oxidize or reduce contaminants, such as trichloroethylene (TCE), to less toxic forms. This article demonstrates the feasibility of reducing TCE to chloromethane at a granular graphite cathode. The results warrant further research to identify the degradation pathways and mechanisms.

Introduction

Remediation of contaminated groundwater and soil has been a major challenge for engineers and scientists since the passages of the Comprehensive Environmental Response, Compensation, and Liability Act (CERCLA) in 1980 and revised Resource Conservation and Recovery Act (RCRA) in 1984. The predominant method for remediating groundwater and soil is pump-and-treat (P&T). P&T operates by pumping contaminated groundwater from an array of wells, removing the contaminants from the water with on-site treatment systems, and re-injecting the treated water back into the aquifer or disposing of it off-site. Despite the popular use of P&T as the treatment method, the available data shows that P&T is inefficient and sometimes unable to restore the groundwater quality to acceptable levels (*1-3*). The ineffectiveness of P&T on groundwater and soil remediation can be attributed to the presence of low permeability soil, non-uniform flow patterns, and sorption of contaminants.

The presence of low permeability soil causes low water recovery rate and requires more pumping power. This results in indefinite pumping periods from wells and excessive operational cost. Non-uniform flow patterns are due to the inter-bedding of soils with high and low permeability, the presence of fracture, or other heterogeneities. When water flows through heterogeneous media, the flow rate in the high permeability is much greater than in the low permeability region. As a result, advective transport, which is the major mechanism for contaminant transport in a P&T system, in the low permeability soil is insignificant. Instead, the dominant transport mechanism of contaminants in these regions is diffusion (*1*).

Sorption processes occur when a solute is bonded to the surface of soil particles either by adsorption, chemisorption, or ion exchange (*4*). The mass of contaminants sorbed is proportional to the surface area of the soil particles. Since the surface area to mass ratio of low permeability soils, such as clay, is orders of magnitude greater than high permeability soils (*5*), greater amounts of contaminants may accumulate in the low permeability regions. Thus, during P&T operations, the low permeability regions become reservoirs of contaminants, which are slowly released to the flushing flow by diffusion and desorption mechanisms. The combined effects of non-uniform flow patterns and sorption greatly reduce the amount of contaminants removed and prolongs the remediation period (*3,6*). Typically, contaminant concentrations rapidly decrease at the beginning of P&T operations as dissolved contaminants are easily removed. After the initial removal period, the contaminant concentration reaches a level in which further pumping is unable to remove the contaminant because of diffusion.

The drawbacks of P&T has prompted the development of new treatment technologies that are capable of addressing the shortcomings of P&T. The application of electrical fields is one of many methods applied to facilitate the removal of contaminants (3). When an electrical field is applied to a clay-rich aquifer, the mobilization and removal of contaminants can be accomplished by electrical heating, electrokinetics, and electrochemical degradation. These processes are described in details in the following sections.

Electrical Heating

When current flows through a medium, heat energy is generated by resistive heating and causes soil temperature to rise. The magnitude of heat energy generated is a function of the soil electrical conductivity and the gradient of the electrical field (7,8) and can be written as Eq.(1).

$$Q = \sigma \|\nabla V\|^2 \tag{1}$$

where Q = heat energy generated, σ = electrical conductivity (S/m), and V = electrical field (V).

Electrical heating usually is combined with other remediation methods, such as soil vapor extraction, air sparging, or P&T, to mobilize and facilitate the transport of contaminants in the low permeability regions. By heating and thus, increasing the temperature of soils with electrical fields, the removal efficiency of contaminants may be increased. First, the volatility of volatile organic compounds increases with temperature (7,9). This would greatly increase the efficiency of techniques, such as soil vapor extraction, that extract contaminants from the gaseous phase. Secondly, the diffusion coefficient of organic contaminants increase with temperature (9,10). Therefore, heating soil, especially the low permeability zones, can increase the mass transfer rate of contaminants from the diffusion-controlled regions to the advection-controlled regions. Finally, the sorption of neutral organic compounds such as trichloroethylene (TCE) and pyrene, on natural particles decrease with temperature (9,10). As a result of electrical heating, more of the sorbed contaminants partition into the aqueous phase and can be removed by the flushing flow. Thus, the retardation factors of contaminants are reduced and the removal efficiency is increased.

It should be emphasized that although other thermal-based remediation methods, such as steam injection, offer some advantages as electrical heating, they suffer from the same difficulties as P&T when low permeability soil is present as the non-uniform flow paths limit the pathways of the steam (11,12). Electrical heating, however, is not limited by the presence of low permeability soil. In fact, the efficiency of electrical heating benefits from the presence of clay minerals that are major constituents in low permeability regions (5). The reason for this is clay minerals increase the electrical conductivity of soil by providing additional paths for current flow through the mineral surface and the double layer, which contains excess cations (13). The low permeability regions, thus, are preferential paths for current flow and the heat generated (Eq. 1) is greater than in high permeability regions.

The ability to increase the temperature of the low permeability regions by electrical current has been recognized and applied to enhance contaminant removal efficiency. Buettner and Daily (7) combined electrical heating with air stripping to remove TCE from an aquifer that contained a clay layer. The target volume (89 m^3) was heated with six cylindrical electrodes (3.05 m in length and 10.2 cm in diameter). The electrodes were equally spaced and powered by a three-phase source. After 9,600 kW-h of energy input, the TCE concentration in the exit air flow was decreased from 70 ppm$_v$ to below 1.5 ppm$_v$. Due to difficulties in sampling and insufficient monitoring points, the effect of electrical heating on the TCE removal efficiency and mechanisms were not characterized in this study. They estimated that approximately 50% of the energy was applied to the target volume due to the setup of the electrodes. Other limiting factors include the potential hazard of electrical shock to the operator, which resulted in periodical rather than continuous operation of the process.

In a more controlled laboratory setup, Heron et al. (9) used electrical heating to enhance the TCE removal efficiency by soil vapor extraction. The authors showed that the steady-state TCE flux in the exit gas was increased from 0.13 to 0.35 g/d when the soil was heated from 23 to 85 °C. The removal rate was further increased to 2.5 g/d when the soil was heated to approximately 100 °C. At this temperature, however, steam was generated and the column started to dry out, which would cause significant decrease in electrical conductivity if the soil moisture was not replenished. At the end of the test, 99.8% of TCE was removed. The enhancement of TCE removal efficiency by electrical heating was attributed to the changes in Henry's constant (H), partitioning coefficient (K_d), and gaseous diffusion coefficient (D). These changes are summarized in Table 1.

Although electrical heating has been proven to be a feasible method in mobilizing and transporting volatile and semi-volatile contaminants in low permeability soils, sorption of contaminants by the soil particles can still cause the tailing of effluent concentrations to proceed for a long period of time. Even by heating soil to around 100 °C, more TCE was sorbed than dissolved (9). Therefore, sorption/desorption processes, which are least affected by temperature (Table 1), could be the limiting mechanisms for mobilizing contaminants by electrical heating.

Table 1 Physical Property changes of TCE with Soil Temperature

Soil Temperature °C	Observed TCE flux ratio (flux/flux$_{23}$)	Henry's constant ratio (H/H$_{23}$)	Partitioning coeff. ratio (K$_{d,23}$/K$_d$)	Gaseous diffusion coeff. ratio (D/D$_{23}$)
23	1	1	1	1
85	2.65	6.8	1.15	1.33
95	18.8	9.0	1.18	1.41

SOURCE: Adapted with permission from reference 9. Copyright 1998 American Chemical Society.

Electrokinetics

Electrokinetic remediation of groundwater and soil usually involves the use of electroosmosis and electromigration. Electroosmosis describes the phenomenon of water flow induced by application of an electrical field on porous media (*14*). Electromigration describes the transport of ions under the influence of an electrical field (*15*). The electroosmotic flux, q_{eo}, can be written as

$$q_{eo} = v_{eo}\tau n_e \tag{2}$$

where v_{eo} = electroosmotic velocity, τ = tortuosity, and n_e = effective porosity. The electroosmotic velocity can be described by the Helmholtz-Smoluchowski equation (*14*).

$$v_{eo} = -\frac{\varepsilon\zeta}{\mu}\nabla V \tag{3}$$

where ε = permittivity of the pore fluid, ζ = zeta potential, and μ = viscosity of the pore fluid.

According to Eq.(3), the electroosmotic velocity is directly proportional to the zeta potential and the electrical gradient, and inversely proportional to the pore fluid viscosity. It also indicates that unlike hydraulic flow, electroosmotic flow is independent of pore size (*16*). Thus, theoretically the electroosmotic flow is unaffected by the presence of low permeability soil with small pore sizes. As clay minerals that possess high zeta potential and electrical conductivity are preferentially found in low permeability regions, the electroosmotic flow tends to be greater than in the high permeability regions. This gives electroosmosis the advantage over conventional methods, such as P&T, in mobilizing and transporting contaminants in the low permeability regions.

Cations move from the anode toward the cathode, whereas anions move from the cathode to the anode under the influence of an electrical field. The migration velocity of an ion, v_i, in a dilute solution is (*17*)

$$v_i = u_i \nabla V \tag{4}$$

where u_i = ionic mobility of ion i.

The ionic mobility ,defined as the average velocity per unit electrical field strength, is determined from (*17*)

$$u_i = \frac{\Lambda_i}{|z_i| F} \tag{5}$$

where Λ_i = molar ionic conductivity (S m^2/mol), z_i = the valence of the ion, and F = the Faraday constant (96,490 C/mol). Eq.(4) and (5) suggest that the migrating velocity of ions is independent of pore size. As mentioned above, the electrical gradient in low permeability regions is greater than in high permeability regions due to the high clay contents. As a result, ions migrate faster in low permeability than in high permeability regions.

Electrokinetic technique has been applied to remove inorganic and organic contaminants from low permeability soils. It should be pointed out that the major mechanisms in removing inorganic contaminants (ionic) are electroosmosis and electromigration, whereas electroosmosis is the main mechanism for removing organic contaminants (non-ionic). Laboratory-scale experiments showed that electrokinetics can removed 75 to 92% of metals, such as Cd^{2+}, Pb^{2+}, and Zn^{2+} (*18-20*). Acar et al. (*21*) showed 85 to 95% of phenol was removed using electrokinetics in a test column and the retardation was minimal. In another study, Bruell et al. (*22*) observed 7 to 25% removal of petroleum hydrocarbons and TCE by electroosmosis after 2 to 25 days of processing. Al-Abed and Chen (*23*) showed more than 90% of TCE mobilization by electroosmosis after 28 days of processing.

In larger pilot- and field- scale applications, electrokinetics is combined with other in situ treatment methods to form a treatment system called Lasagna® process (*24*). The setup in a Lasagna® process includes placing a sheet-like electrodes on each end of the target volume and creating in situ treatment zones between the electrodes. The designing concept is to mobilize and transport contaminants by electrokinetics toward the treatment zones, where they can be degraded. The configuration of the electrodes and treatment zones can be either vertical or horizontal (Fig. 1). Pilot-scale tests conducted by Ho et al. (*24*) using

440

Horizontal Configuration

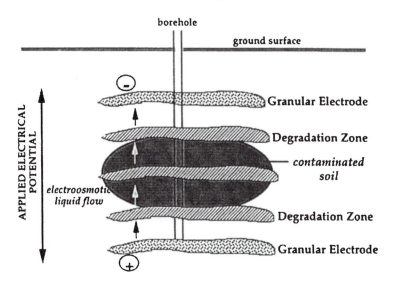

Vertical Configuration

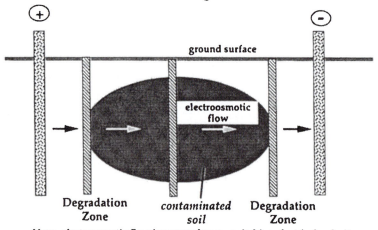

Note: electroosmotic flow is reversed upon switching electrical polarity.

Figure 1. Horizontal and Vertical Configurations of the Lasagna® Process. Reproduced with permission from reference 24. Copyright 1997 Elsevier Science B.V.

a vertical Lasagna® process showed 98% of paranitrophenol (PNP) was mobilized after 0.94 pore volume of electroosmotic flow. After additional 1.6 pore volume of electroosmotic flow, PNP was non-detect (less than 10 ppb) in the soil. Additional field-scale tests (vertical Lasagna® process) by Ho et al. (25,26) achieved TCE removal of 95 to 99% by electroosmosis. They also reported an estimated cost of $45 to $80/yd³ for remediating soil between depths of 15 to 45 ft by electrokinetics.

The horizontal Lasagna® process offers advantages that 1) can be applied at greater depths than the vertical Lasagna® process; and 2) utilizes both electroosmotic and hydraulic flows to remove contaminants. However, the application of such setup has been limited to uncontaminated aquifers (27). Reports that document the removal of contaminants by using horizontal Lasagna® process is unavailable at this time. Researchers at the US EPA are currently conducting a field-scale test using the horizontal Lasagna® process to remove TCE at Offutt AFB near Omaha, NE (28). The results from this study should provide significant performance data for the removal of contaminants with the horizontal Lasagna® process.

The first reported field tests on removing heavy metals from groundwater and soil using electrokinetics were conducted by Lageman (29,30). The contaminants included Pb, Cu, Zn, Cd, and As. The tests showed average removal efficiency of 74% for Cu and Pb, 50% for As and 20% for Zn. It should be noted that out the removal percentages were variable across the test site due to non-uniform electrical fields. Pilot-scale tests conducted by Acar and Alshawabkeh (31) showed that 90% of Zn was mobilized by electrokinetics and 75% of Zn precipitated within 2 cm of the cathode compartment. Precipitation of metals close to the cathode was also observed by Hicks and Tondorf (20) and Eykholt and Daniel (32). This was attributed to the low solubility of metals at alkaline conditions close to cathode during electrokinetic remediation. The precipitation of metals at cathode can be prevented by rinsing away hydroxyl ions generated at the cathode (20).

Electrochemical degradation

Electrochemical degradation utilizes the high redox potential at an anode and the low redox potential at a cathode to oxidize or reduce contaminants. Most applications of electrochemical degradation involved cathodic reduction or anodic oxidation of chlorinated organic contaminants. Schmal et al. (33) showed that pentachlorophenol was sequentially dechlorinated (i.e. reduced) at a carbon fiber cathode. Experiments conducted by Cheng et al. (34) indicated that the composition of the cathode was critical for cathodic reduction of 4-

chlorophenol. Reduction of 4-chlorophenol was not observed when a carbon cloth or graphite rod were used as cathodes. Palladized carbon cloth cathode was shown to dechlorinate 4-chlorophenol to phenol. They hypothesized three possible mechanisms for reduction of 4-chlorophenol to phenol: 1) direct reduction at the cathode, 2) hydrogenation (i.e. reduced by hydrogen gas) at the palladium catalyst surface, or 3) adsorption of 4-chlorophenol at the cathode surface followed by hydrogenation at the palladium/electrode surface.

Anodic oxidation was also shown to be a feasible technique for chlorophenol degradation (35). The mechanisms for this process is not well understood. It is hypothesized that chlorophenol radical cations are first formed and subsequently deprotonated. The product undergoes further oxidation to a benzoquinone derivative and ring-opening to acids, such as muconic, naleic, and oxalic and finally to CO_2 (36-38). Just like cathodic reduction, the efficiency of anodic oxidation of chlorophenol depends on the material of the electrodes. Generally, oxide-based anodes perform better than metals as they are less prone to the formation of oligomers, which cause the inactivation of the electrodes (35,38,39).

Electrochemical Degradation of TCE

In an experiment of removing TCE from a natural soil with electrokinetics, Al-Abed and Chen (23) observed dichloroethylene (DCE) isomers at the vicinity of the cathode. The mechanism of TCE reduction is unclear, however, the low redox potential and ample electrons available at the cathode suggest that TCE might be reduced at the surface of the graphite cathode. To our best knowledge, cathodic reduction of TCE, although theoretically possible, has not been experimentally proven. The authors have conducted experiments to prove the occurrence of cathodic reduction of TCE using graphite cathodes. The rest of this article describes their approaches and results.

Experimental Setup and Approach

The setup included two major chambers: a reactor where cathodic reduction occurred and a solution reservoir. The reactor was a glass cylinder 15 cm in length and 5 cm in diameter. The reactor was filled with 10 cm of granular graphite (30-50 mesh) as the anode and 4 cm of granular graphite as the cathode. A 1-cm layer of silicate sand separated the anode and cathode. The solution reservoir was a 500 mL gas-tight glass cylinder. The reactor and reservoir were hydraulically connected with a peristaltic pump and Teflon tubing. The whole system was kept gas-tight during operation to prevent loss of

TCE and its degradation products. The solution matrix was 400 mL of 0.1 M ammonium acetate with 250 mg/L of sodium azide. The addition of sodium azide was to prevent the TCE from being degraded by microbial activities. The solution was put inside the reservoir, spiked with 2.8 µl of pure TCE, and shaken at 100 rpm for at least 12 h. The solution was pumped at the designated flow rate (3.8 mL/min) from the reservoir to anode, through cathode, and back to the reservoir. The TCE concentration was monitored by taking samples from the reservoir. Electrical power was not applied until the TCE concentration reached steady-state, which usually took 12-24 h. Three different voltages (5, 10 and 20 V) were applied between the anode and cathode to evaluate the effects of voltage on the reduction rate. The durations of the experiments were between 14 and 60 h. The test under 10 V lasted for only 14 h because of an unrepairable leak in the system.

Analytical Methods

At each sampling period, 5 mL of solution and 50 µl of head space gas from the solution reservoir was extracted and analyzed. The solution was spiked with 50 µL of 1000 ppm benzene-D6 as a internal standard. Afterwards, the sample was shaken in a head space sampler (HP 7694E) at 70 °C for 20 minutes. The concentrations of vinyl chloride, DCE isomers, TCE, and other chlorinated products that could be daughter products of TCE were analyzed with a gas chromatograph (HP 6890) equipped with a mass selective detector (HP 5973). After the analysis of organic compounds, pH was measured and the chloride concentration in the sample was analyzed with a Capillary Ion Analyzer (Waters CIA). The 50 µl of head space gas was directly injected into a GC (HP 6890) with a HP Plot/Al_2O_3 column (50m x 0.53 mm x 15 µm) and the concentrations of methane, ethane, ethylene, and acetylene were quantified with a FID detector.

Results and Discussion

TCE monitoring before the initiation of electrical power indicated TCE was sorbed on the granular graphite. Assuming a linear isotherm for the sorption of TCE, we estimated the average partitioning coefficient, K_d, to be 2.2±0.52 mg/g (n=3). The concentration of TCE decreased with time under all the voltage applied (Fig. 2). The degradation rate of TCE increased with applied voltage. Assuming the reduction of TCE is pseudo-first-order with respect to the aqueous concentration of TCE, the reaction rate constants determined from Fig. 2 at applied voltages of 5, 10, and 20 V are 0.01, 0.04, and 0.06 h^{-1}, respectively. The half-life, t_{50}, of TCE at applied voltages of 5, 10, and 20 V are 69, 17, and 12 h, respectively. Thus, increasing applied voltage from 5 to 20 V decreases the half-life of TCE by more than 80%.

444

The major product of TCE reduction was chloromethane (CM) and no other chlorinated daughter products were observed. This is unexpected as reduction of TCE usually results in the formation of DCE isomers, which are further reduced to vinyl chloride and to ethane or ethylene (40). The pathway of CM formation from TCE reduction is unclear. Judging from the fact that CM was the only chlorinated product, we hypothesize that TCE was sorbed on the graphite surface at the cathode and both the carbon-carbon and carbon-chlorine bonds were cleaved. Additional studies are needed to determine whether carbon-carbon bond is cleaved before carbon-chlorine bond, or vise versa.

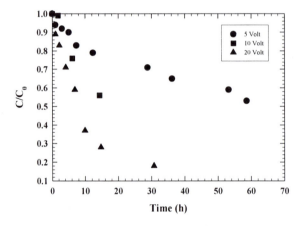

Figure 2. Disappearance of TCE with Time at Applied Voltages of 5 (circle), 10 (square), and 20 V (triangle).

The rate of CM generation is proportional to the applied voltages (Fig. 3). The mole of CM generated in Fig. 3 includes CM that is in the aqueous phase and in the head space. To estimate the amount of CM in the head space, we used Henry's Law constant of 9.55 L·atm/mole for CM (10). The amount of CM sorbed on the granular graphite was not included since the K_d for CM was unavailable. Therefore, the amount of CM generated shown in Fig. 3 is an underestimation of the total amount of CM in the system.

At 5V, the moles CM during the test were less than the moles TCE added. At applied voltages of 10 and 20 V, however, the moles CM generated were larger than the moles TCE added after 10 h of operation (Fig. 3). This suggested that for each mole of TCE degraded, more than one mole of CM was generated at 10 and 20 V of applied voltages. Thus, this supports our hypothesis that the carbon-carbon bond of TCE is cleaved. The products underwent further reactions to form CM. The ratio of moles CM generated to moles of TCE degraded was estimated to between 2 and 3. A ratio of 2 would suggest that the carbon in CM comes only from TCE, whereas a ratio of more than 2 suggests the carbon in CM might come from other sources, such as graphite or acetate. Graphite or acetate as the sources of carbon were confirmed in several control experiments. The solutions in controls contained no TCE, however, methane and ethane formations were observed when an electrical field was applied. The maximum ratio is 3 as each mole of TCE contains 3 moles of chlorine. Further studies that quantify the amount of CM sorbed on granular graphite are needed to determine the moles of CM generated per mole of TCE degraded.

In addition to the formation of CM, chloride concentration increased with operation time. With the measurements of TCE, CM, and chloride concentrations in the solution and head space, it is possible to conduct a mass balance of the test based on the total mass of chloride. The procedures are as the following: 1) Calculate the mass of chloride in initial TCE; 2) Calculate the mass of chloride in TCE; 3) Calculate the mass of chloride in CM; and 4) Subtract the results of 2) and 3) from the result of 1). The resulting mass of chloride was converted to concentration and taken as the estimated chloride produced. It should be noted that the estimation assumed no sorption of CM on the graphite. The estimated chloride production was compared to the measured chloride concentration (Fig. 4). The results indicate the estimated and measured chloride concentrations agree with each other. This agreement also suggests that sorption of CM on graphite was minimum and hence, can be neglected.

pH is one of the important solution parameters that changes as a result of an applied electrical field. The change is due to electrolysis at the anode and cathode. Protons and oxygen are produced at the anode, whereas hydroxyl ions and hydrogen are produced at the cathode. The pH with time at all applied voltages had similar patterns. It increased from pH 6.7 to 7.7 shortly after the initiation of power (Fig. 5). The increase of pH was due to hydroxyl ions that were pumped from the cathode compartment into the solution chamber. The magnitude of pH jump increased with the applied voltage. This is because the rate of hydroxyl ion production increased with applied voltage. After 6 to 7 h of power application, the pH started decreasing with time suggesting protons generated at the anode finally reached the solution chamber (Fig. 5).

One of the objectives of this study was to degrade TCE to non-toxic compounds, such as ethane and chloride, using cathodic reduction. This was not achieved because most of the product was CM. However, long term application of electrical field might change the compositions of the degradation products as CM could undergo further reduction once TCE is completely depleted from the system. Experiments are being conducted at US EPA, Cincinnati, OH to investigate the long-term effects of electrical field on the compositions of TCE degradation products.

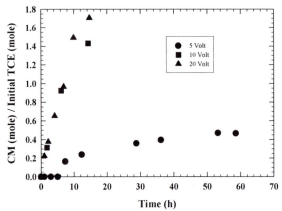

Figure 3. Ratio of mole CM generated to the initial mole of TCE as a function of time at applied voltages of 5 (circle), 10 (square), and 20 V (triangle).

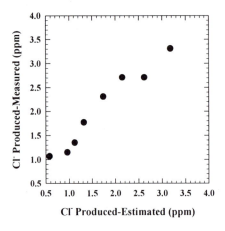

Figure 4. Comparison of Estimated Chloride with Measured Chloride Production at an Applied Voltage of 5 V.

References

1. Mackay, D.M; Cherry, J.A. *Environ. Sci. Technol.* **1989**, 23, 630-636.
2 MacDonald, J.A.; Kavanaugh, M.C. *Environ. Sci. Technol.* **1994**, 28, 362-368.
3. NRC, *Groundwater and Soil Clean Up: Improving Management of Persistent Contaminants;* National Academy Press: Washington, DC, 1999, 285pp.

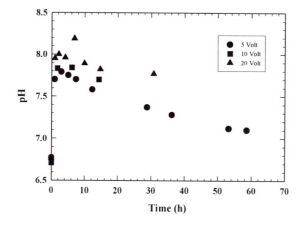

Figure 5. pH of Solution with Time at Applied Voltages of 5 (circle), 10 (square), and 20 V (triangle).

4. Fetter, C.W. *Contaminant Hydrogeology;* Macmillan Pub.: New York, NY, 1993, 458pp.

5. Sparks, D.L. *Environmental Soil Chemistry;* Academic Press: San Diego, CA, 1995, 267pp.

6. Keely, J.F. *Performance Evaluation of Pump-and-Treat Remediations;* EPA Ground Water Issue, EPA/540/4-89/005, 1989.

7. Buettner, H.M.; Daily, W.D.; *J. Environ. Engrg. ASCE.* **1995**, 121, 580-589.

8. Ould El Moctar, A.; Peerhossaini, H.; Bardon, J.P. *Int. J. Heat Mass Transfer.* **1996**, 39, 975-993.

9. Heron, G.; van Zutphen, M.; Christensen, T.H.; Enfield, C.G. *Environ. Sci. Technol.* **1998**, 32, 1474-1481.

10. Schwarzenbach, R.P.; Gschwend, P.M.; Imboden, D.M. *Environmental Organic Chemistry;* John Wiley and Sons, Inc., New York, NY, 1993, 681pp.

11. Udell, K.S. In *Subsurface Restoration;* Ward, C.H., Cherry, J.A., Scalf, M.R., Eds.; Ann Arbor Press: Chelsea, MI, 1997, pp 251-271.

12. Carrigan, C.R.; Nitao, J.J. *Environ. Sci. Technol.* **2000**, 34, 4835-4841.

13. Rhodes, J.D.; Manteghi, N.A.; Shouse, P.J.; Alves, W.J. *Soil Sci. Soc. Am. J.* **1989**, 53, 433-439.

14. Hunter, R.J. *Zeta Potential in Colloid Science;* Academic Press, New York, NY, 1981, 386pp.

15. Acar, Y.; Alshawabkeh, *Environ. Sci. Technol.* **1993**, 13, 2639-2647.

16. Probstein, R.F.; Hicks, R.E. *Science,* **1993**, 260, 498-503.

17. Rieger, P.H. *Electrochemistry;* Chapman and Hall, New York, NY, 1994, 483pp.

448

18. Acar, Y.B.; Hamed, J.T.; Alshawabkeh, A.N.; Gale, R.J. *Geotechnique,* **1994**, 44, 239-254.
19. Hamed, J.; Acar, Y.B.; Gale, R.J. *J. Geotech. Engrg. ASCE.* **1991**, 117, 241-271.
20. Hicks, R.E.; Tondorf, S. *Environ. Sci. Technol.* **1994**, 28, 2203-2210.
21. Acar, Y.B.; Li, H.; Gale, R.J. *J. Geotech. Engrg. ASCE.* **1992**, 1837-1852.
22. Bruell, C.J.; Segall, B.A.; Walsh, M.T. *J. Environ. Engrg. ASCE.* **1992**, 118, 68-83.
23. Al-Abed, S.; Chen, J.-L. In *Physico-Chemical Remediation of Contaminated Subsurface Environment;* Smith, J.; Burn, S., Eds.; Kluwer Academic Publishers: Hingham, MA, 2001, in press.
24. Ho, S.V.; Athmer, C.J.; Sheridan, P.W.; Shapiro, A.P. *J. Hazard. Mater.* **1997**, 55, 39-60.
25. Ho, S.V.; Athmer, C.; Sheridan, P.W.; Hughes, B.M.; Orth, R.; McKenzie, D.; Brodsky, P.H.; Shapiro, A.; Thornton, R.; Salvo, J.; Schultz, D.; Landis, R.; Griffith, R.; Shoemaker, S. *Environ. Sci. Technol.* **1999**, 33, 1086-1091.
26. Ho, S.V.; Athmer, C.; Sheridan, P.W.; Hughes, B.M.; Orth, R.; McKenzie, D.; Brodsky, P.H.; Shapiro, A.; Sivavec, T.M.; Salvo, J.; Schultz, D.; Landis, R.; Griffith, R.; Shoemaker, S. *Environ. Sci. Technol.* **1999**, 33, 1092-1099.
27. Chen, J.-L.; Murdoch, L. *J. Geotech. Engrg. ASCE.* **1999**, 125, 1090-1098.
28. Roulier, M.; Kemper, M.; Al-Abed, S.; Murdoch, L.; Cluxton, P.; Chen, J.-L.; Davis-Hoover, W. *J. Hazard. Mater.* **2000**, B77, 161-176.
29. Lageman, R.; Pool, W.; Seffinga, G. *Chem.& Industry.* September 18, 1989, p585.
30. Lageman, R. *Environ. Sci. Technol.* **1993**, 27, 2648-2650.
31. Acar, Y.B.; Alshawabkeh, A.N. *J. Geotech. Engrg. ASCE.* **1996**, 122, 173-185.
32. Eykholt, G.R.; Daniel, D.E. *J. Geotech. Engrg. ASCE.* **1994**, 120, 797-815.
33. Schmal, D.; van Erkel, J.; van Duin, P.J. *The Institution of Chemical Engineers Symposium Series, no. 98 (Electrochemical Engineering);* 1986, pp 281-291
34. Cheng, I. F.; Fernando, Q.; Korte, N. *Environ. Sci. Technol.* **1997**, 31, 1074-1078.
35. Rodgers, J.D.; Jedral, W.; Bunce, N.J. *Environ. Sci. Technol.* **1999**, 33 1453-1457.
36. Comninellis, Ch.; Vercesi, G.P. *J. Appl. Electrochem.* **1991**, 21, 335-345.
37. Comninellis, Ch.; Pulgarin, C. *J. Appl. Electrochem.* **1993**, 23, 108-112.
38. Trabelsi, F.; Ait-Lyzaid, H; Ratsimba, B; Wilhelm, A.M.; Delmas, H.; Fabre, P.L.; Berlan, *J. Chem. Eng. Sci.* **1996**, 51, 1857.
39. Gattrell, M.; Kirk, D.W. *J. Electrochem. Soc.* **1993**, 140, 1534-1540.
40. Criddle, C.S; McCarty, P.L. *Environ. Sci. Technol.* **1991**, 25, 973-978.

Chapter 27

Influence of Nonionic Surfactants on the Bioavailability of Hexachlorobenzene for Microbial Reductive Dechlorination

Kurt D. Pennell, Spyros G. Pavlostathis, Ahmet Karagunduz, and Daniel H. Yeh

School of Civil and Environmental Engineering, Georgia Institute of Technology, Atlanta GA 30332

Experimental and mathematical modeling studies were performed to evaluate the potential benefits and limitations associated with the use of nonionic surfactants to enhance the microbial transformation of hexachlorobenzene (HCB) by a dechlorinating mixed culture enriched from a contaminated sediment. In general, Tween series surfactants were shown to have little impact on methanogenesis, whereas, polyoxyethylene (POE) alcohols, Triton X-100 and SDS were found to strongly inhibit methanogenesis and HCB dechlorination. Subsequent experiments conducted with Tween 80 illustrated the ability of this surfactant to enhance the solubility of HCB and to reduce the HCB-soil distribution coefficient. Model simulations demonstrated, however, that the aqueous phase mass fraction of HCB was substantially reduced in micellar solutions, which corresponded with observed reductions in HCB dechlorination. These results indicate that the impacts of surfactants on both biological activity and contaminant phase distributions should be evaluated in order to accurately assess the potential for biotransformation of hydrophobic contaminants in the presence of surfactants.

Hydrophobic organic compounds (HOCs) are commonly found in contaminated soil and sediment systems. A two-phase pattern of HOC desorption, an initial fast stage (in hours) followed by longer, slow phase (in days), has been observed (e.g., *1-4*). The resulting low aqueous-phase concentrations limit the availability of HOCs to indigenous microorganisms for biodegradation. Chlorinated organic compounds, many of which are toxic and carcinogenic, are especially resistant to biodegradation due to stability resulting from chlorine substituents. Prytula and Pavlostathis (*4*) reported that sediment-bound chlorobenzenes exhibited very low bioavailability, with less than 10% of the contaminant mass was reductively dechlorinated over an incubation period of 183 days. Conclusions from this and other studies indicate that efforts to successfully bioremediate contaminated soils and sediments should include means to increase the bioavailability of the sorbed-phase contaminants.

Surfactants have been utilized to enhance the removal of both sorbed-phase HOCs and non-aqueous phase liquids (NAPLs) from subsurface systems by increasing the aqueous-phase concentration of HOCs. For example, Pennell *et al.* (*5*) demonstrated that surfactant flushing dramatically improved the recovery of residual dodecane from aquifer materials. The ability of surfactants to increase the aqueous solubility of HOCs results from the tendency of surfactants to aggregate in solution to form micelles (*6*). Surfactants are amphiphilic compounds, possessing both hydrophobic and hydrophilic moieties. At low concentrations, surfactants exist as monomers in solution and usually exhibit minimal effects on the solubility of HOCs (*7*). As the concentration of surfactant in solution is increased, however, the hydrophobic moieties of the monomers tend to associate with one another, eventually forming micelles in solutions. The concentration at which monomer aggregation occurs is referred to as the critical micelle concentration (CMC). It should be noted that, as the surfactant concentration is increased above the CMC the number of monomers in solution remains constant, while the number of micelles continues to increase. In aqueous solutions, surfactant micelles consist of a hydrophobic core surrounded by a hydrophilic mantle. The observed capacity of surfactant solutions to increase the apparent aqueous solubility of HOCs is attributed to the incorporation or partitioning of HOCs within the hydrophobic core of micelles (e.g., *7, 8*). This general behavior is depicted in Figure 1, which illustrates a linear increase in HOC aqueous solubility above the CMC. This relationship can be expressed as the weight solubilization ratio (WSR);

$$WSR = \frac{C_{org} - C_{org,cmc}}{C_{surf} - C_{surf,cmc}} \qquad (1)$$

where C_{org} is the concentration of organic species in the aqueous phase (M/L^3), $C_{org,cmc}$ is the concentration of the organic species at the CMC (M/L^3), C_{surf} is the

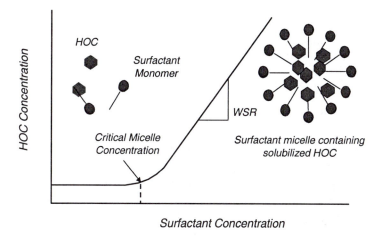

Figure 1. Relationship between surfactant concentration and HOC solubility.

concentration of surfactant in solution (M/L^3), and $C_{surf,cmc}$ is the concentration of surfactant at the CMC (M/L^3).

A number of researchers have evaluated the use of surfactants to facilitate bioremediation for both *in situ* and *ex situ* applications. However, these studies have yielded mixed results, showing either enhanced, reduced or no effect on biodegradation following surfactant addition (9). Such conflicting data can be attributed, in large part, to the numerous interactions that can occur between microorganisms, surfactant monomers and micelles, the solid phase, and the HOC, as depicted in Figure 2. Surfactant monomers can accumulate at the solid-liquid interface and partition into the solid phase (step 1). The latter process may cause swelling of the organic and clay fractions of the solid phase, thereby increasing the rate of contaminant diffusion within the solid matrix (*1*). In addition to sorption by the solid-phase, surfactant monomers can also sorb to biomass (step 2). It has been hypothesized that the association of surfactants with cell membranes may facilitate mass transfer of HOCs across the membrane, thus enhancing biotransformation (*10*). The HOC may also participate in sorption processes associated with the solid phase and sorbed surfactant (step 3). As noted above, partitioning or incorporation of the HOC into surfactant micelles leads to enhanced contaminant solubility (step 4). In most cases, the exchange of HOCs between the aqueous phase and surfactant micelles is considered to be rapid relative to HOC desorption from the solid phase (*1,11*). Finally, HOCs will tend to accumulate on biomass due to the lipophilic nature of cell membranes. Under proper conditions, the HOC will then be transported into the cell and transformed (step 5).

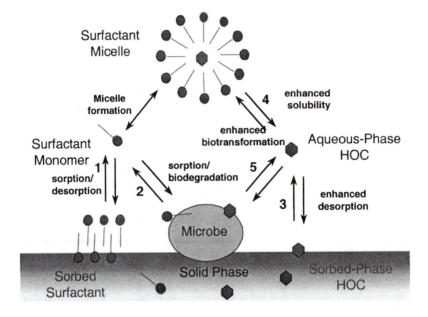

Figure 2. Conceptual model of coupled interactions among HOCs and the solid phase, biomass and surfactant.

To date, the majority of surfactant-enhanced bioremediation studies have focused on aerobic, rather than anaerobic, processes. Thus, the research described herein was specifically designed to evaluate the influence of surfactants on the availability of hexachlorobenzene (HCB) for microbial reductive dechlorination. Hexachlorobenzene was selected for study due to its widespread occurrence in sediments (*12, 13*), low aqueous solubility, and tendency to be strongly sorbed by soils and sediments. In addition, HCB is classified as a probable human carcinogen by the U.S. Environmental Protection Agency (EPA), and is known to accumulate in the food chain (*14*). Several classes of nonionic surfactants were evaluated, as well as a representative anionic surfactant. Although this research involved a large matrix of batch, column and modeling studies, the intent of this chapter is to present representative results and to highlight important findings. Specific topics that will be addressed include: a) the impact of surfactants on the methanogenesis and reductive dechlorination of HCB; b) the influence of surfactants on the aqueous solubility and phase distribution of HCB; and c) the development and evaluation of mathematical models to describe coupled processes governing HCB bioavailability.

Materials and Methods

Materials

A total of sixteen surfactants were evaluated. Fourteen of the surfactants represented two main classes of nonionic surfactants; linear polyoxyethylene (POE) alcohols (Brij 30, and 35; Witconol SN-70, 90, and 120), and food grade POE sorbitan esters (Tween 20, 21, 40, 60, 61, 65, 80, 81, and 85). Within each class, the selected surfactants represent variations in carbon chain length and type, number of ethylene oxide (EO) groups and hydrophile-lipophile balance (HLB). In addition, a representative octylphenol ethoxylate (Triton X-100) and an anionic surfactant, sodium dodecyl sulfate (SDS), were also included in surfactant screening tests. The surfactants were obtained from Aldrich Chemical Co. (Milwaukee, WI) with the exception of Tween 21, 61, 65, and 81, which were obtained from ICI Americas, Inc. (Wilmington, DE) and the Witconol SN series surfactants, which were obtained from Witco Corporation (Houston, TX). The surfactants were used as received from the suppliers without purification. Selected surfactant properties are given in Table I. Surfactant solutions were prepared in deionized, distilled water that had been treated with a Barnstead Nanopure II filter system. Hexachlorobenzene (99+% purity) was obtained from Aldrich Chemical Co. (Milwaukee, WI). Hexachlorobenzene (C_6Cl_6) has a molecular weight of 284.8 g/mole, an aqueous solubility of $5\mu g/L$ at 25°C (15) and a logarithmic octanol-water partition coefficient (Log K_{ow}) of 5.73 (16).

An HCB-contaminated sediment was collected from the Bayou d'Inde, a tributary of the Calcasieu River near Lake Charles, Louisiana. The sediment was used as the inoculum for the HCB-dechlorinating enrichment cultures, as well as one of the media for the examination of the effect of surfactants on HCB desorption and bioavailability. A natural soil (Appling) was also used in batch sorption experiments. Appling soil is classified as a loamy coarse sand of the Appling series (clayey, kaolinitic thermic Typic Hapludult). The organic carbon content of the Appling soil is 7.54 g/kg, and the specific surface area is 3.5 m²/g based on nitrogen adsorption.

The HCB-dechlorinating culture was developed using anaerobic media and the contaminated sediment as inoculum. The resulting mixed, methanogenic culture was maintained in 9-L sealed glass reactors (6-L liquid volume) with an average hydraulic retention time of 84 days. At each feeding cycle (7-9 days), glucose (333 mg/L), yeast extract (17 mg/L) and HCB (0.025 mg/L) were added to the culture medium. The culture was maintained at an oxidation-reduction potential (ORP) of -330 mV and a pH range of 6.9-7.1. Details of the culture development and maintenance procedures are provided by Yeh et al. (18, 19).

Table I. Selected properties of surfactants evaluated in this study

Surfactant	Structure[a,b]	MW (g/mole)	CMC[c] (mg/L)	HLB[d]	N_a^e	ThOD[f] (g O_2/g)
Polyoxyethylene (POE) Sorbitan Fatty Acid Esters						
Tween 20	$C_{12}S_6EO_{20}$	1226	44-58	16.7		1.92
Tween 21	$C_{12}S_6EO_4$	522	13	13.3		2.05
Tween 40	$C_{16}S_6EO_{20}$	1282	30-51	15.6		1.98
Tween 60	$C_{18}S_6EO_{20}$	1310	26-55	14.9	67	2.02
Tween 61	$C_{18}S_6EO_4$	606	32	9.6		2.24
Tween 65	$3(C_{18})S_6EO_{20}$	1842	46	10.5		2.34
Tween 80	$C_{18}S_6EO_{20}$	1308	33-45	15.0	110	2.01
Tween 81	$C_{18}S_6EO_5$	648	72	10.0		2.20
Tween 85	$(3C_{18})S_6EO_{20}$	1836	40	11.0		2.32
Polyoxyethylene (POE) Alcohols						
Brij 30	$C_{12}EO_4$	362	7-14	9.7	120	2.48
Brij 35	$C_{12}EO_{23}$	1198	70-110	16.9	53	2.02
Witconol SN-70	$C_{10-12}EO_5$	392	25	11.2	129	2.36
Witconol SN-90	$C_{10-12}EO_6$	436	36	12.1	118	2.31
Witconol SN-120	$C_{10-12}EO_9$	568	54	13.9	107	2.20
Alkylphenol Ethoxylate and Anionic Surfactant						
Triton X-100	$C_8\varphi EO_{9.5}$	624	110-150	13.5	140	2.19
SDS	$C_{12}OSO_3Na$	288		40.0		1.81

[a]Sorbitan ring (S_6). [b]Phenolic ring (φ). [c]Critical micelle concentration, obtained from the supplier and Rosen (6). [d]Hydrophile-lipophile balance, calculated as HLB=%wt EO/5 (6). [e]Aggregation number (8). [f]Theoretical oxygen demand, calculated using method of Metcalf and Eddy, Inc. (17).

Experimental Methods

The first phase of biotic screening was designed to assess the impact of surfactant additions on methanogenic activity of the mixed culture. A serum bottle assay was conducted for all sixteen surfactants at an initial surfactant concentration of 200 mg/L. Two experimental systems were prepared: 1) surfactant plus glucose, to assess the effect of surfactants on methanogenesis; and 2) surfactant only, to assess the anaerobic biodegradability of the surfactants when present as the only carbon source. Over the 82-day incubation period, total gas and methane production from the surfactant-amended series were measured from the bottle headspace and compared to those from a reference series (containing glucose and no surfactant) and a seed blank series (no carbon source added).

The second phase of biotic screening evaluated the effect of Tween surfactants on the reductive dechlorination of HCB by the mixed culture. Experiments were conducted in 28-mL serum tubes sealed with Teflon-lined rubber septa and aluminum crimps. Each tube contained HCB (~140 μg/L) dissolved in methanol, glucose, and Tween surfactant at concentrations of either 10, 50, 200, or 1,000 mg/L. At each sampling time, the contents of the tubes were extracted with iso-octane followed by an analysis for HCB, dechlorination products and excess gas production. The following parameters were measured: total gas and methane production, pH, ORP, particulate organic carbon, volatile fatty acids (VFAs), and total chemical oxygen demand (COD).

The solubility of HCB in aqueous solutions of Tween 60, Tween 80 and Triton X-100 was measured in completely mixed batch reactors over a surfactant concentration range of 0 to 2,000 mg/L. Each 35-mL glass reactor contained excess HCB which was deposited as a thin film on the bottom of the reactor. Analysis of HCB in aqueous surfactant solutions was achieved using a direct injection gas chromatography technique (8). This analytical method yielded more accurate and reproducible results than conventional solvent extraction methods, which resulted in the formation of persistent macroemulsions between the organic solvent and aqueous phases.

Batch sorption experiments were conducted using Tween 80 and Appling soil. The batch reactors consisted of either 25-mL glass centrifuge tubes or 30-mL polypropylene centrifuge tubes. The initial concentration of surfactant solutions added to each reactor ranged from 100 to 2,500 mg/L, and contained 0.005 M CaCl$_2$ as a background electrolyte and 500 mg/L NaN$_3$ to minimize biological activity. The contents of the reactors were mixed for periods ranging from 1 day to 4 weeks in order to assess adsorption rates and to establish equilibrium sorption capacities. Following mixing, the solid phase was separated by centrifugation (3,000 rpm X 45 minutes) and the resulting supernatant was analyzed for surfactant using a high pressure liquid chromatography (HPLC) system equipped with a diode array detector (DAD) and an evaporative light scattering detector (ELSD).

Sorption of HCB by Appling soil was measured in the absence and presence of surfactant. The HCB sorption experiments were conducted in 25 mL Corex glass centrifuge tubes sealed with aluminum-lined caps. The aqueous phase contained 0.005 M CaCl$_2$ as a background electrolyte and 500 mg/L NaN$_3$ to minimize biological activity. The effect of Tween 80 on HCB sorption by Appling soil was evaluated as a function of surfactant concentration and mixing time.

Tween 80 and HCB were added simultaneously, and for each surfactant concentration, the initial concentration of HCB was also varied to allow for the determination of individual HCB sorption isotherms as a function of surfactant loading. In these experiments, the final aqueous phase concentrations of both HCB and Tween 80 were measured independently by GC and HPLC analysis, respectively. Individual and coupled HCB and Tween 80 sorption experiments were conducted in triplicate, with duplicate reference blanks (no solid phase).

Results and Discussion

Effect of Surfactants on Methanogenesis

The effects of surfactant additions on the rate and extent of electron donor utilization were evaluated based on methane production in the presence and absence of surfactant. Methane production profiles for the mixed culture augmented with 200 mg/L of surfactant and 1,800 mg/L glucose are shown in Figure 3. Although none of the Tween surfactants negatively affected the ultimate extent of methanogenesis compared to the reference series (no surfactant), the presence of Tween 21 did result in a slightly lower initial rate of methanogenesis. In addition, methane production in the presence of Tween surfactants alone was greater than that of the seed blank control, which did not contain electron donor. The linear POE alcohols, Triton X-100, and SDS greatly inhibited the utilization of glucose for methanogenesis by the mixed-culture (Figure 3). The observed inhibition of methanogenesis in the presence POE alcohols, which have been used extensively in surfactant enhanced biodegradation studies (9), was not anticipated. Previous studies conducted by Federle and Schwab (20) showed that POE alcohols were degraded by mixed culture derived from anaerobic sediment. These contrasting results suggest that surfactant-microbial interactions and compatibility are system specific.

Effect of Surfactants on Reductive Dechlorination

Hexachlorobenzene concentration profiles and total gas production for the glucose-fed culture exposed to three initial surfactant (Tween series) loadings are shown in Figure 4. At surfactant concentrations of 10 and 200 mg/L, total gas production was identical to that of the reference (without surfactant amendment). As the initial surfactant concentration was increased to 1,000 mg/L (Figure 4C), total gas production was equal to or exceeded the reference case. These data suggest that many of the Tween surfactants were subject to biodegradation. Both the rate and extent of HCB disappearance decreased, however, as the initial surfactant concentration was increased from 10 to 1,000 mg/L. The HCB concentration in azide-amended controls did not vary significantly over the

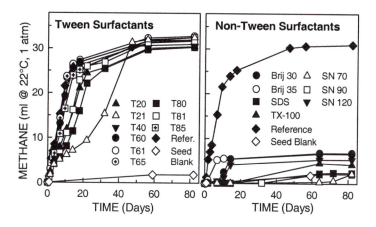

Figure 3. Methane production by glucose-fed cultures in presence of Tween and non-Tween surfactants at an initial concentration of 200 mg/L; from Yeh et al. (*18*).

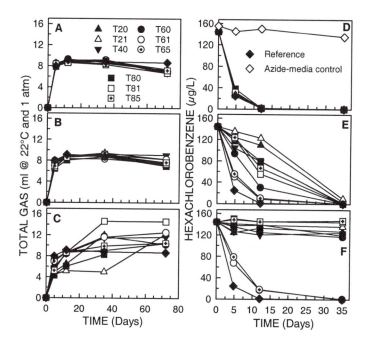

Figure 4. Effects of Tween surfactants on total gas production and HCB disappearance at initial concentrations of 10 (A, D), 200 (B, E) and 1,000 mg/L (C, F); from Yeh et al. (*19*).

incubation period, indicating that the reduction in HCB concentration was due to microbial activity. At the lowest Tween surfactant loading (10 mg/L), no change in HCB dechlorination behavior was observed, while at the intermediate loading (200 mg/L) reductions in HCB dechlorination rates were apparent. At the highest surfactant loading (1,000 mg/L) HCB dechlorination was minimal for all cases except Tween 61 and Tween 65. In summary, HCB dechlorination in the presence of Tween surfactants never exceeded that of the reference (Figure 4D-F), and was, in most cases, greatly reduced at the highest surfactant loading. However, the fact that methane production (Figure 3) and total gas production (Figure 4A-C) were maintained, and in some cases exceeded the reference at the highest surfactant loading, indicate that the presence of Tween surfactants did not directly inhibit methanogenesis, as was the case for the non-Tween surfactants. Nevertheless, the observed reductions in HCB dechlorination suggest that the Tween surfactants altered the biological activity of the mixed culture, either through (a) inhibition or toxicity toward HCB degraders or (b) preferential or competitive utilization of surfactant as a substrate. In addition, the possibility exists that the fraction of HCB incorporated within surfactant micelles was not directly available to the microbial population. The latter hypothesis will be explored in more detail in subsequent sections of this chapter.

HCB Solubilization

Detailed HCB solubility studies were performed with three surfactants (Tween 60, Tween 80 and Triton X-100) to provide for a range in surfactant properties and biological compatibility. In the absence of surfactant, the solubility of HCB at 25°C was found to be approximately 7 μg/L, which is similar to values reported in the literature (15). Relationships between surfactant concentration and the apparent solubility of HCB are shown in Figure 5. Interestingly, the aqueous-phase concentration of HCB increased by approximately one-order-of-magnitude as the CMC of each surfactant was approached. Similar trends were reported by Kile and Chiou (7), who observed an increase in the apparent solubility of DDT from 5.5 μg/L in pure water to approximately 70-80 μg/L at the CMC of three Triton series surfactants. This behavior is attributed to interactions between the dissolved HOC and the hydrophobic moiety of surfactant monomers.

Above the CMC of each surfactant, linear enhancements in HCB solubility were observed, similar to trends reported for HOCs in micellar solutions (7,8,15). The corresponding WSR values for Tween 60, Tween 80 and Triton X-100, calculated using Equation 1, were 0.59 g/kg, 0.63 g/kg and 0.35 g/kg, respectively. The lower HCB solubilization capacity of Triton X-100 is consistent with solubility correlations developed by Pennell et al. (8) for a range of surfactants and HOCs. This behavior is attributed to the greater alkyl chain length, and hence larger micelle size of Tween 60 and Tween 80 relative to that of Triton X-100.

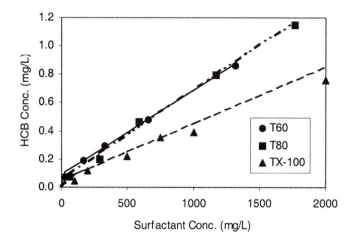

Figure 5. Aqueous solubility of HCB as a function of Tween 60, Tween 80 and Triton X-100 concentration.

Surfactant Sorption

A matrix of surfactant sorption experiments was conducted as a function of mixing time, initial concentration, and soil type for Tween 60, Tween 80 and Triton X-100. Although only data for Tween 80 and Appling soil are presented here, all of the soil-surfactant systems investigated exhibited Langmuir-type sorption behavior. The Langmuir equation may be written as:

$$S_s = \frac{S_m b C_s}{1 + b C_s} \qquad (2)$$

where S_s is the amount of surfactant sorbed (M/M), S_m is the maximum or limiting sorption capacity (M/M), b is the Langmuir sorption parameter representing the ratio of the adsorption and desorption rates (L^3/M), and C_s is the equilibrium surfactant concentration in solution (M/L^3). Sorption isotherms for Tween 80 and Appling soil at mixing times of 1 day, 3 days and 7 days are shown in Figure 6. To obtain values of S_m and b, sorption data were fit to the Langmuir equation using a nonlinear, least-squares regression procedure (SYSTAT 5.0). Although the observed values of b were essentially the same after 1, 3 and 7 days of mixing (0.008 L/mg), the maximum or limiting sorption capacity (S_m) increased substantially, from 1.44 mg/g to 5.99 mg/g. These data indicate that sorption of

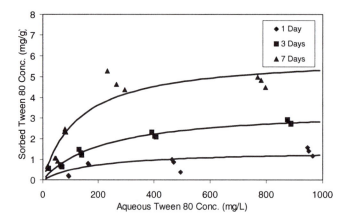

Figure 6. Sorption of Tween 80 by Appling soil as a function of mixing time.

Tween 80 by Appling soil was strongly rate limited. Based on a theoretical cross-sectional molecular area of 300 A^2/molecule for Tween 80, calculated following the approach of Adeel and Luthy (21), and the specific surface area of Appling soil (3.47 m^2/g), the limiting sorption capacities were expressed in terms of surface coverage. This analysis yielded Tween 80 surface coverages of 0.6, 1.4 and 2.9 monolayer equivalents after 1, 3 and 7 days of mixing, respectively. These results suggest that surfactant loadings corresponding to monolayer surface coverage were achieved within 48 hours, while the formation of surfactant bilayers or hemicelles on the surface, and interactions between Tween 80 and soil organic matter, occurred over longer time frames.

Coupled Surfactant and HCB Sorption

The sorption of HCB by Appling soil was evaluated as a function of surfactant (Tween 80) concentration and mixing time. Tween 80 was selected for these experiments because it readily dissolves in water and remains in solution. Although several of the Tween surfactants (i.e., Tween 60, 61 and 65) were less inhibitory toward HCB dechlorination (Figure 4D-F), these surfactants were more difficult to dissolve in water and were susceptible to the formation of separate phases (e.g., precipitate). The formation of surfactant-rich phases by Tween 60 has also been reported by Shaiu et al. (22).

In the absence of surfactant, HCB sorption by Appling soil yielded linear sorption isotherms, with the observed distribution coefficient (K_D) increasing from 385 L/kg to 527 L/kg for mixing periods of 1, 3, 15, and 22 days. In the presence of surfactant, the initial concentration of HCB was varied to allow for the determination of individual HCB sorption isotherms. The addition of Tween 80 at aqueous phase concentrations above the CMC dramatically reduced the sorption of HCB by Appling soil (Figure 7). As the surfactant concentration was increased, the observed HCB distribution coefficient (K_D) decreased substantially. For a given surfactant loading, the HCB distribution coefficient increased with time, indicating that HCB sorption was rate-limited in the presence of surfactant. Since the aqueous phase concentration of Tween 80 also decreased with mixing time due to sorption, these results illustrate the importance of considering the effects of multiple rate limitations on the distribution of hydrophobic organic compounds between the aqueous, micellar and solid phases.

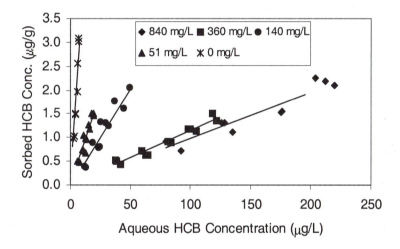

Figure 7. Hexachlorobenzene sorption isotherms as a function of Tween 80 concentration after 7 days of mixing.

Mathematical Modeling of HCB and Surfactant Phase Distributions

The overall or apparent solubility of an HOC in the presence of surfactant can be represented as the amount of HOC associated with surfactant monomers plus the amount associated with surfactant micelles. This relationship can be expressed in the following form (23):

$$\frac{C^*}{C} = 1 + C_{mn} K_{mn} + C_{mc} K_{mc} \tag{3}$$

where, C^* is the apparent solubility of the solute at the total surfactant concentration (M/L^3), C is the intrinsic solubility of the solute in pure water (M/L^3), C_{mn} is the concentration of surfactant monomers (M/L^3), C_{mc} is the concentration of surfactant micelles (M/L^3), K_{mn} is the solute distribution coefficient between surfactant monomers and water (L^3/M), and K_{mc} is the solute distribution coefficient between the surfactant micelles and water (L^3/M). This approach can be extended to include the effect of sorbed-phase surfactant on the distribution of the solute between the solid and aqueous phases:

$$K^* = \frac{K_D(1 + C_{s/oc}K_{s/oc})}{(1 + C_{mn}K_{mn} + C_{mc}K_{mc})} \tag{4}$$

where, K^* is the apparent solute soil-water distribution coefficient (L^3/M), K_D is the intrinsic solute soil-water distribution coefficient (L^3/M), $C_{s/oc}$ is the concentration of sorbed surfactant per unit mass of native soil organic carbon (M/M), $K_{s/oc}$ is the solute distribution coefficient between sorbed surfactant and organic carbon (K_s/K_{oc}), and K_s is the solute distribution coefficient between the sorbed surfactant and water (L^3/M).

A comparison of measured and predicted values of the HCB soil-water distribution coefficient, K^*, for the Tween 80 and Appling soil system is presented in Figure 8. The measured values of K^* (data points) were obtained from the HCB sorption isotherms shown in Figure 7. All of the model input parameters, except for $K_{s/oc}$, were derived from independent measurements. The fitted value of $K_{s/oc}$ was not statistically different from zero, indicating that interactions between HCB and the sorbed-phase surfactant were minimal, and thus, solid-phase sorption of HCB was dominated by natural soil organic matter. It should be noted, however, that the effects of sorbed-phase surfactant on HOC distributions are most apparent below the CMC, and for solids containing no organic carbon (23,24). In the absence of HCB-monomer interactions (K_{mn} = 0 L/kg), the value of K^* decreased exponentially only after the CMC of Tween 80 (35 mg/L) was exceeded. When the value of K_{mn} increased, however, the reduction in K^* occurred immediately upon addition of Tween 80, as the monomers acted to increase the tendency for HCB to remain in solution. The best-fit value of K_{mn} (0.09 L/kg) was smaller than the measured value (K_{mn} =0.257 L/kg), which was attributed to a nonlinear increase in HCB solubility as the CMC was approached (7). Nevertheless, these data clearly demonstrate that interactions between HCB and surfactant monomers influenced the distribution of HCB below the CMC, and that the presence of Tween 80 micelles greatly reduced the tendency for HCB to is associate with the solid phase.

The simulated distribution of HCB mass among surfactant monomers, micelles, solid phase, and aqueous (water) phase as a function of Tween 80 concentration is shown in Figure 9. Here, model input parameters were identical

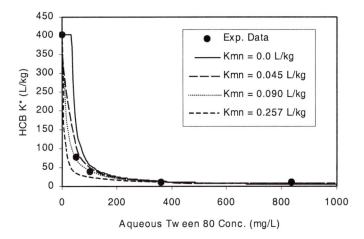

Figure 8. Simulated and measured values of the apparent HCB-soil distribution coefficient (K^*) as a function of Tween 80 concentration and the HCB-monomer partition coefficient (K_{mn}). Input parameters: CMC = 35 mg/L; b = 8.0 L/g; S_m = 5.99 g/kg; K_D = 403.4 L/kg; $K_{s/oc}$ = 0.0 L/kg; K_{mc} = 0.09 L/kg.

to those given in Figure 8, with a solid-liquid ratio equivalent to that used in the batch reactors (1 g of soil:20 mL of solution). For this scenario, the initial (no surfactant) aqueous phase concentration of HCB was assumed to be the solubility limit (7 μg/L), and the corresponding solid-phase concentration was 2.82 mg/kg. Although the HCB mass fraction associated with the solid phase decreased from 95 to 20% as the aqueous phase concentration of Tween 80 increased from 0 to 1,000 mg/L, the corresponding fraction of HCB incorporated within surfactant micelles increased from 0 to 75%. As a result, the mass fraction of HCB in the aqueous (water) phase actually decreased from 5% to 0.9% (7 μg/L to 1.3 μg/L) in the presence of Tween 80. To allow for direct comparisons with the HCB dechlorination data discussed previously, the HCB distribution in the absence of a solid phase was also simulated (Figure 10). In this scenario, the mass fraction of HCB in the aqueous (free water) phase decreased from 100% in the absence of Tween 80 to 1.1% at a Tween 80 concentration of 1,000 mg/L. These results suggest that the sequential reduction in HCB dechlorination observed in the presence of Tween 80 (Figure 4D-F) was due primarily to a decrease in the aqueous phase concentration of HCB available to the microbial population.

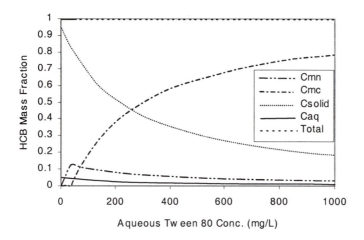

Figure 9. Simulated distribution of HCB among surfactant monomers, micelles, solid phase, and aqueous (free water) phase.

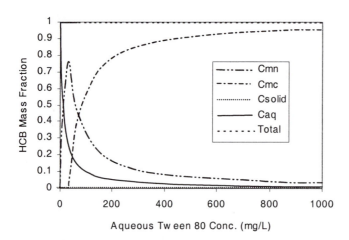

Figure 10. Simulation distribution of HCB among surfactant monomers, micelles and aqueous (free water) phase in the absence of soil.

Conclusions

Results of surfactant biological compatibility studies indicate that methane and total gas production in the presence of Tween series surfactants was similar to, and in some cases exceeded that of, the reference standard (no surfactant). In contrast, POE alcohols, Triton X-100 and SDS were shown to markedly reduce methane production and to inhibit HCB dechlorination. Within the Tween series, substantial variations in HCB dechlorinating activity were also observed despite similar molecular structures. These results suggest that surfactant compatibility with the HCB-degrading mixed culture was highly surfactant specific. Although methanogenesis was not adversely impacted by the Tween surfactants, rates of HCB disappearance were sequentially reduced as the Tween surfactant concentration was increased from 0 to 1,000 mg/L. These results initially suggested that a) HCB dechlorinating activity of the mixed culture was altered by the Tween surfactants or b) the fraction of HCB available to the microbial population was reduced in micellar solutions. To investigate the latter hypothesis, detailed solubilization, sorption, and HCB phase distribution experiments were performed. Experimental data obtained from these studies showed that Tween 80 enhanced the apparent solubility of HCB both above and below the CMC, Tween 80 was strongly sorbed by a natural soil, and that the HCB-soil distribution coefficient was reduced substantially as the concentration of Tween 80 was increased above the CMC. A mathematical model, developed to simulate HCB phase distributions in the presence of surfactants, clearly demonstrated that the amount of HCB present in the aqueous (water) phase decreased dramatically in the presence of Tween 80. These findings suggest that although Tween 80 was not toxic or inhibitory toward the mixed culture with respect to methanogenesis, micellar solubilization of HCB greatly reduced the mass of HCB available for reductive dechlorination.

Acknowledgements

This research was supported by the U.S. Environmental Protection Agency, National Center for Environmental Research and Quality Assurance (Contract No. R-825404-01-0). The contents of this publication has not been subject to agency review, and does not necessarily represent the view of the agency.

References

1. Deitsch, J.J.; Smith, J. A. *Environ. Sci. Technol.* **1995**, 29, 1069-1080
2. Pavlostathis, S.G.; Jaglal, K. *Environ. Sci. Technol.* **1991**, 25, 274-279.
3. Pignatello, J. J.; Xing, B. *Environ. Sci. Technol.* **1996**, 30, 1-11.

4. Prytula, M.; Pavlostathis, S.G. *Water Res.* **1996**, 30, 2669-2680.

5. Pennell, K.D.; Abriola, L.M.; Weber, W.J. Jr. *Environ. Sci. Technol.* **1993**, 27, 2332-2340.

6. Rosen, M.J. *Surfactants and Interfacial Phenomena,* 2nd ed.; Wiley: New York, NY, 1989; p 431.

7. Kile, D.E.; Chiou, C.T. *Environ. Sci. Technol.* **1989**, 23, 832-838.

8. Pennell, K.D.; Adinolfi, A.M.; Abriola, L.M., Diallo, M.S. *Environ. Sci. Technol.* **1997**, 31, 1382-1389.

9. Rouse, J.D.; Sabatini, D.A.; Suflita, J.M.; Harwell, J.H. *CRC Crit. Rev. Environ. Sci. Technol.* **1994**, 24, 325-370.

10. Van Hoof, P.L.; Jafvert, C.T. *Environ. Toxicol. Chem.* **1996**, 15, 1914-1924.

11. Pennell, K.D.; Abriola, L.M. In *Bioremediation: Principles and Practice*; Sikdar, S.K.; Irvine, R.L. Ed.; Technomic Publ.: Lancaster, PA, 1997; Vol. 1, pp. 693- 750.

12. Baker, J.E.; Eisenreich, S.J.; Johnson, T.C.; Halfman, B.M. *Environ. Sci. Technol.* **1985**, 19, 854-861.

13. Keller, M. *Wat. Sci. Technol.* **1994**, 29, 129-131.

14. Burkhard, L.P.; Sheedy, B.R.; McCauley, D.J.; DeGraeve, G.M. *Environ. Toxicol. Chem.* **1997**, 16, 1677-1686.

15. Jafvert, C.T.; Van Hoof, P.L.; Heath, J.K. *Wat. Res.* **1994**, 28, 1009-1017.

16. deBruijn, J.; Busser, F.; Seinen, W.; Hermens, J. *Environ. Toxicol. Chem.* **1989**, 8, 499-512.

17. Metcalf and Eddy, Inc. *Wastewater Engineering: Treatment, Disposal, and Reuse,* 3rd ed.; McGraw-Hill: New York, NY, 1991; p. 1334.

18. Yeh, D.H.; Pennell, K.D.; Pavlostathis, S.G. *Wat. Sci. Technol.* **1998**, 38, 55-62.

19. Yeh, D.H.; Pennell, K.D.; Pavlostathis, S.G. *Environ. Toxicol. Chem.* **1999**, 18, 1408-1416.

20. Federle, T.W.; Schwab, B.S. *Wat. Res.* **1992**, 26, 123-127.

21. Adeel, Z.; Luthy, R.G. *Environ. Sci. Technol.* **1995**, 29, 1032-1042.

22. Shiau, B.J.; Sabatini, D.A.; Harwell, J.W. *Environ. Sci. Technol.* **1995**, 29, 2929-2935.

23. Sun, S.; Inskeep, W.P.; Boyd, S.A. *Environ. Sci. Technol.* **1995**, 29, 903-913.

24. Ko, S.-O.; Schlautman, M.A.; Carraway, E.R. *Environ. Sci. Technol.* **1998**, 32, 2769-2775.

Indexes

Author Index

Subject Index

471